Springer Proceedings in Materials 96

Series Editors

Springer Proceedings in Materials publishes the latest research in Materials Science and Engineering presented at high standard academic conferences and scientific meetings. It provides a platform for researchers, professionals and students to present their scientific findings and stay up-to-date with the development in Materials Science and Engineering. The scope is multidisciplinary and ranges from fundamental to applied research, including, but not limited to:

- Structural Materials
- Metallic Materials
- Magnetic, Optical and Electronic Materials
- Ceramics, Glass, Composites, Natural Materials
- Biomaterials
- Nanotechnology
- Characterization and Evaluation of Materials
- Energy Materials
- Materials Processing

To submit a proposal or request further information, please contact one of the following Springer Publishing Editors according to your affiliation location:

European countries: **Mayra Castro** (mayra.castro@springer.com)
India, South Asia and Middle East: **Swati Meherishi** (swati.meherishi@springer.com)
South Korea: **Smith Chae** (smith.chae@springer.com)
Southeast Asia, Australia and New Zealand: **Ramesh Nath Premnath** (ramesh.premnath@springer.com)
The Americas: **Michael Luby** (michael.luby@springer.com)
China and all the other countries or regions, as well as topics in materials chemistry: **Maggie Guo** (maggie.guo@cn.springernature.com)

This book series is indexed in **SCOPUS** and **EI Compendex** database.

Yuyuan Zhao
Editor

Progress in Advanced Manufacturing and Materials

Selected Contributions of ICAMM 2025

Springer

Editor
Yuyuan Zhao
School of Mechanical and Automotive Engineering
Ningbo University of Technology
Ningbo, China

ISSN 2662-3161 ISSN 2662-317X (electronic)
Springer Proceedings in Materials
ISBN 978-3-032-09861-0 ISBN 978-3-032-09862-7 (eBook)
https://doi.org/10.1007/978-3-032-09862-7

This Springer imprint is published by the registered company Springer Nature Switzerland AG
The registered company address is: Gewerbestrasse 11, 6330 Cham, Switzerland

Preface

This book presents the final versions of papers accepted and presented at the 9th International Conference on Advanced Manufacturing and Materials (ICAMM 2025), held in the historic and intellectually vibrant city of Oxford, United Kingdom, from July 8 to 11, 2025. The conference brought together cutting-edge research and innovative advancements from academia and industry, fostering collaboration among experts in the fields of advanced manufacturing and materials engineering. ICAMM 2025 continued its tradition of excellence, with all submitted papers undergoing rigorous peer review by an international committee of renowned scholars and practitioners.

The longstanding reputation of ICAMM underscores its pivotal role in disseminating groundbreaking research and facilitating global knowledge exchange. Delegates from over a dozen countries, including the United Kingdom, United States, Italy, France, China, Japan, and Korea, participated in three days of vibrant discussions, technical sessions, and networking opportunities. Beyond the stimulating academic program, attendees enjoyed the rich cultural heritage of Oxford and the inspiring atmosphere of its world-class university, further enhancing the collaborative spirit of the conference.

The manuscripts compiled in this ICAMM 2025 Proceedings represent the final revised versions, incorporating valuable feedback from reviewers. The proceedings are organized into six thematic chapters.

Chapter 1 features pioneering research on aerospace systems, addressing critical challenges in design, safety, and performance under demanding operational conditions. Chapter 2 explores advancements in metal processing and additive manufacturing, highlighting innovative techniques that enhance material properties and production efficiency. Chapter 3 delves into the multiscale behavior of composite structures, offering insights into their performance under extreme environments. Chapter 4 focuses on the in-service behavior and failure mechanisms of engineering materials, providing essential knowledge for durability and reliability improvements. Chapter 5 showcases breakthroughs in the synthesis and characterization of specialty materials, unlocking new possibilities for high-performance applications. Finally, Chapter 6 presents cutting-edge research on fluid dynamics and thermodynamic systems, optimizing performance across a range of industrial applications.

Together, these 30 papers reflect the breadth and depth of research in advanced manufacturing and materials, offering valuable contributions to both theoretical understanding and practical innovation. The diversity of topics and the high quality of the work underscore the dynamic and interdisciplinary nature of the field.

I extend my deepest gratitude to the authors, reviewers, and organizing committee members of ICAMM 2025 for their dedication and expertise, which made this conference a resounding success. As the ninth installment in this esteemed conference series, ICAMM continues to serve as a vital platform for sharing knowledge, fostering international collaborations, and driving progress in advanced manufacturing and materials engineering.

We look forward to the continued growth of this conference and its enduring impact on the global research community.

Yuyuan Zhao
Conference Chair, ICAMM 2025

Organization

Conference Chairs

Yuyuan Zhao	Ningbo University of Technology, China
Alexander M. Korsunsky (Vice President)	Trinity College, University of Oxford, UK

Program Chairs

Bingang Xu	The Hong Kong Polytechnic University, HKSAR, China
Teik-Cheng Lim	Singapore University of Social Sciences, Singapore
Raul Campilho	Instituto Superior de Engenharia do Porto, Portugal
Edwar M. Sabolsky	West Virginia University, USA
Anja Pfennig	HTW-Berlin University of Applied Sciences, Germany

Program Co-chairs

Hasan Gokce	Istanbul Technical University, Türkiye
Zuruzi Abu Samah	Alfaisal University, Kingdom of Saudi Arabia

Publicity Committee Chairs

Richard (Yu) Jiang	Cambridge University, UK
Emmanuelle Biglete	Colegio de Muntinlupa, Philippines
Steven Frederick Durrant	Instituto de Ciência e Tecnologia de Sorocaba, UNESP, Brazil
Changhong Ke	State University of New York at Binghamton, USA
Maximiliano Delany Martins	Centro de Desenvolvimento da Tecnologia Nuclear, Brazil
Vu Ngoc Pi	Thai Nguyen University of Technology, Vietnam

Technical Committee

Technical Committees

Richard (Yu) Jiang	Cambridge University, UK
Siti Norasmah Surip	Universiti Teknologi MARA, Malaysia
Viranjay M. Srivastava	Birmingham City University, UK
Xuefeng Zhou	Southeast University, China
Yan Huang	Brunel University London, UK
Yanming Zhang	The University of Tokyo, Japan

Contents

Metal Processing and Additive Manufacturing

Multiscale Characterization and Extreme Environment Impact Response of Composite Structures

Aerospace System Design and Structural Safety Analysis

Dynamic Characteristic Analysis of Pump Inlet Pressure of Aircraft Hydraulic Power System Based on AMESim

Yuange Kong and Liang Zhou(✉)

Shanghai Aircraft Design and Research Institute State Key Laboratory of Airliner, Integration Technology and Flight Simulation Shanghai, Shanghai, China
{kongyuange,zhouliang}@comac.cc

Abstract. The Aircraft hydraulic power system provides hydraulic energy for aircraft. When the aircraft hydraulic system provides large flow to the hydraulic users, the suction pressure of the engine driven pump will be sharply reduced, and the pump inlet pipeline is prone to cavitation. Base on the problem of low pump suction pressure in aircraft hydraulic system, the main factors affecting the suction pressure are studied through theoretical analysis and simulation analysis. The AMESim simulation analysis and calculation can explore the factors that affect the suction pressure of the pump, such as the accumulator, pipe diameter and reservoir pressure, and then the cavitation phenomenon can be avoided as much as possible through reasonable parameter setting.

Keywords: pump · cavitation · simulation · aircraft

1 Introduction

Aviation hydraulic engine driving pump (EDP) is generally installed in the engine short cabin fan hood, and the hydraulic reservoir is generally installed in the aircraft drum (wing fairing). When the hydraulic user needs to achieve large flow actuation, the swash plate of the EDP will be suddenly adjusted, and the oil suction demand will be rapidly enhanced. A local low pressure area will be formed in the pipeline near the oil suction port, making the pressure lower than the separation pressure of the bubbles [1]. After the pressure is restored, the bubbles will collapse, resulting in cavitation [2, 3]. Cavitation mostly occurs when the piston just enters or leaves the suction area [4]. The cavitation phenomenon seriously damages the service life and reliability of the EDP, and at the same time will produce the problem of insufficient pressure supply of the EDP pressure port, affecting the function and performance of the downstream hydraulic users. Compared with other models, the distance between the hydraulic bootstrap reservoir and the oil suction port of the EDP of the long-range wide-body aircraft is longer, and the user flow demand is greater. Therefore, a detailed analysis of the cavitation problem of the hydraulic system of the long-range wide-body aircraft is required. The large flow users in the hydraulic system include the main flight control, the back push, the landing gear

Y. Zhao (Ed.): ICAMM 2025, SPM 96, pp. 3–12, 2025.
https://doi.org/10.1007/978-3-032-09862-7_1

and the high lift, involving many scenarios [5]. The largest flow demand user is the landing gear, and the largest flow rate change is the go-around stage, involving the high lift and the landing gear retraction. According to the above analysis of the scenario, due to the large flow of user actions in too many scenarios, cavitation phenomenon involves a wide range of scenarios, and mostly concentrated in the main landing gear retractable time period, so the decision to the main landing gear hydraulic subsystem analysis.

2 Theoretical Analysis of Cavitation in Hydraulic Power System

2.1 Cavitation Scenarios Introduction

Large flow users of the hydraulic system include main flight control, reverse thrust, landing gear and high lift, which involve many scenarios, among which the largest flow demand user is the landing gear, and the scenario with the largest flow change rate is the go-around stage, involving high lift and landing great retraction. According to the above scenario analysis, because there are too many scenarios for high-traffic users to operate, the cavitation phenomenon involves a wide range of scenarios, and most of them are concentrated in the main landing gear retracting period, so it is decided to analyze the hydraulic subsystem where the main landing gear is located.

2.2 Configuration of Pump Inlet Pressure

The design of the pump inlet line, reservoir location, and reservoir pressurization are major system elements affecting hydraulic pump life. Once the basic needs of pump inlet pressure as in (1) and the inevitable consequences of air and cavitation are understood, attention to the components in the pump inlet line becomes more meaningful.

It follows then that pump inlet pressure should be determined by the following elements:

$$P_{in} = P_{resv} - \Delta P_{acc} - \Delta P_{lin} - \Delta P_{com} - \Delta P_{sta} \tag{1}$$

Where Pin is the pump inlet pressure, P_{resv} is the reservoir pressure, ΔP_{acc} is the pressure to accelerate fluid mass in pump response time, ΔP_{lin} is the frictional pressure loss in suction line, ΔP_{com} is the pressure loss of inlet line components, and ΔP_{sta} is the static pressure head.

According to the low pressure chamber reservoir pressure of other similar models, as well as the low pressure chamber

reservoir pressure reply from the supplier, the following calculation assumes that the low pressure chamber reservoir pressure is 71 psi, because the pressure difference caused by the height difference is within 1.5 psid, the following calculation does not consider the static pressure.

The transient or acceleration pressure (ΔP_{acc}) is that pressure necessary to accelerate the fluid mass in the inlet line to the maximum required flow within the response time of the pump, which is shown as (2). The pressure can be determined from the equation as follows:

$$\Delta P_{acc} = 21.2\frac{\rho L}{D^2}\frac{dQ}{dt} \tag{2}$$

Where ρ is the mass density of the fluid, L is the length of tubing, D is the tube inside diameter, Q is the maximum flow change and t is the pump response time.

The next element in the sum of inlet pressure considerations is the fluid friction loss (ΔP_{lin}) in the line (including hoses) itself, which is shown in (3). The pressure drop in a line for maximum steady state flow can be calculated by the standard formulas used for turbulent or laminar flow conditions. For most airborne hydraulic power systems, laminar flow may be present at large flow rates, and should be considered in the system analysis. The line friction loss can be determined from the equation as follows:

$$\Delta P_{lin} = 2.16 \times 10^{-4} f \rho L \frac{Q^2}{D^5} \tag{3}$$

Where f is the friction factor.

From the hydraulic reservoir to the EDP inlet port, the only device that will produce flow resistance is the fire wall shut off valve. Referring to the flow resistance of the fire wall shut off valve of others, the value of ΔP_{com} is only 1 psid, which can be ignored in the formula calculation.

2.3 Conclusion of Analysis and Calculation

According to the above analysis and calculation, the main loss of pump inlet pressure is acceleration pressure and the fluid friction loss in the suction line. The acceleration pressure is determined by the rate of change of user flow, and the loss in the suction line is determined by the steady state value of user flow. The numerical superposition of the two at the corresponding time can obtain the approximate oil suction pressure loss.

3 Simulation Model Building

3.1 Modeling Environment

The software version used for EDP inlet pressure analysis modeling is LMS Imagine Lab AMESim Rev. 16. The components used in modeling are all from the standard library of the software, mainly including the standard hydraulic library Hydraulic, the hydraulic component design library Hydraulic Component Design, the mechanical library Mechanical and the signal control library Signal Control, etc.

3.2 EDP Inlet Pressure Analysis Modeling and Parameter Setting

Parameter setting is shown as Table 1. The total simulation time is set to 140 s and the simulation time of each scenario to simulate user action is 5 s. In addition, the influence of oil viscosity should be considered when analyzing the pressure of oil suction pipe. The oil viscosity at −29 °C is obtained by interpolating the V-type oil viscosity parameters provided by relevant information.

Based on the above modeling of hydraulic components, pipelines and user flow requirements, the model of the entire hydraulic system can be obtained. The pressure simulation model of the hydraulic system is shown in Fig. 1.

Table 1. Parameter Setting

Parameter	Unit	Value
Length of suction line	m	23
Mass density of the fluid	kg/m^3	1020
Tube inside diameter	mm	37
Fluid temperature	°C	−29
Fluid viscosity (kinematic)	Cst	232
EDP rated speed	rpm	4000
EDP response time	ms	50

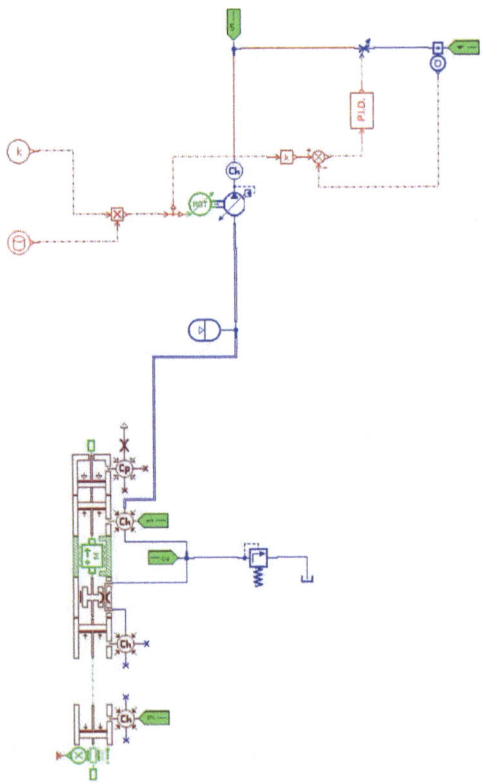

Fig. 1 Pump inlet pressure simulation model

The upper part of the figure is divided into the simulation model of the pump, including the speed signal module and the oil return flow control module of the case drain. The second half is the simulation model of the reservoir, which mainly defines the area ratio and pressure ratio of two chamber.

The simulation result of EDP inlet pressure of hydraulic system is shown in Fig. 2. It can be found that when the time is 35 s, 45 s, 100 s and 105 s, that is, when the aircraft is in the lift-off, climb and go-around flight scenario, the EDP inlet pressure is lower, which corresponds to the time period when the main landing gear is retracted.

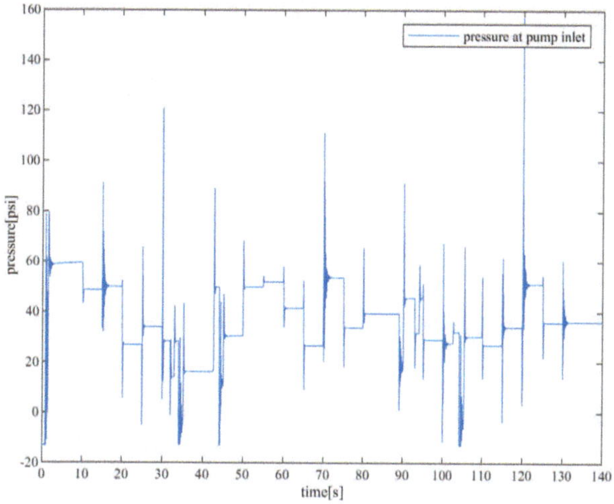

Fig. 2 Simulation result of EDP inlet pressure

3.3 Conclusion of EDP Inlet Pressure Simulation

According to the dissolution characteristics of hydraulic oil and the suction characteristics of EDP, when the pump inlet pressure is lower than cavitation pressure (−4.5 psi or 10 psia), that is, the pressure is lower than the gas-liquid separation pressure of the hydraulic oil itself, the gas will precipitate from the hydraulic oil and produce cavitation phenomenon on EDP. Therefore, there is a risk of cavitation in the aircraft hydraulic system under this design state, and a hydraulic cavitation control scheme needs to be proposed.

4 Cavitation Control Scheme of Hydraulic System

4.1 Set Up the Accumulator

The EDP inlet pressure with accumulator is shown in Fig. 3. According to relevant information, an oil accumulator with a volume of 40in3, a gas pre-compression pressure of 29 psi and an oil temperature of −29 °C is added to the oil suction line.

Through simulation calculation, the addition of the accumulator in the oil suction line can effectively alleviate the sudden change of oil suction pressure caused by the sudden actuation of large flow users. However, for the case that large flow users need to operate for a long time, the improvement of the minimum pressure value is not obvious.

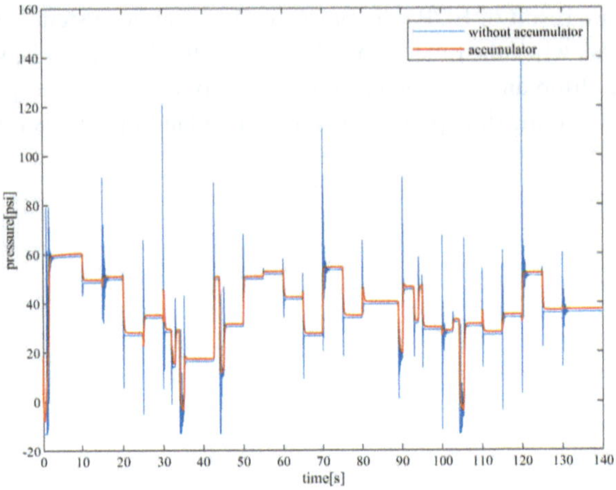

Fig. 3 EDP inlet pressure under different accumulator settings

On the one hand, the oil viscosity is too high at the limit low temperature, resulting in reduced fluidity, and thus reducing the oil suction pressure of EDP; on the other hand, the oil accumulator has a small volume, which cannot meet the pressure demand of the oil suction line in the case of long-term large flow.

4.2 Increase the Low Pressure Chamber Reservoir Pressure

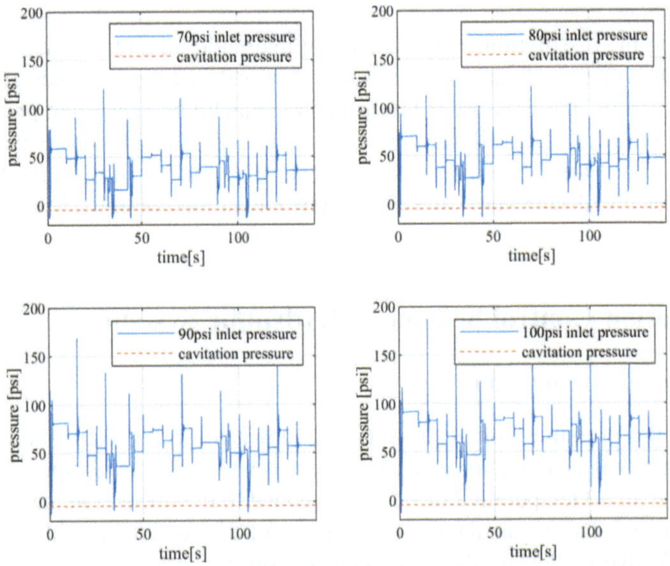

Fig. 4 EDP inlet pressure under different low pressure chamber pressure

The low chamber pressure of the self-supercharged reservoir is increased from 70 psi to 80 psi, 90 psi and 100 psi respectively, and the simulation results of EDP inlet pressure are shown in Fig. 4. With the continuous increase of reservoir pressure, the minimum value of EDP inlet pressure is constantly increasing. According to the dissolution characteristics of hydraulic oil and the EDP suction characteristics of other aircraft, when the EDP inlet pressure is lower than cavitation pressure (−4.5 psi or 10 psia), gas will be precipitated from the hydraulic oil and produce cavitation on the EDP.

Quantitative comparative analysis of the minimum EDP inlet pressure under different low pressure chamber pressure of the reservoir as shown in Fig. 5.

According to the polynomial fitting curve, when the minimum suction pressure is −4.5 psi, the corresponding low pressure chamber pressure of the reservoir is approximately 101.1 psi. In addition, due to the large demand for hydraulic user flow, when the low pressure chamber pressure of the reservoir is lower than 85 psi, the improvement of the EDP suction pressure is not obvious. Therefore, at a temperature of −29 °C, if no other measures are taken, when the low pressure chamber pressure of the reservoir is controlled above 101.1 psi, it can be assumed that the EDP suction port has a highly probability of not occurring cavitation phenomenon.

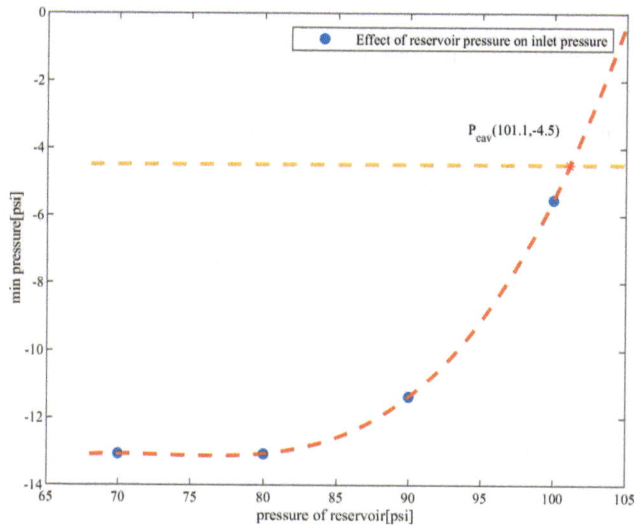

Fig. 5 Minimum EDP inlet pressure trend under different reservoir pressure

4.3 Increase the Diameter of Suction Line

The simulation results of the EDP suction pressure are obtained by increasing the inner diameter of the suction pipeline from 36 mm to 40 mm, 45 mm and 50 mm, as shown in Fig. 6. With the increase of the diameter of the suction pipeline, the minimum value of the EDP suction pressure is constantly increased. According to the dissolution characteristics of hydraulic oil and the EDP suction characteristics of other aircraft, when the oil hydraulic pressure is lower than cavitation pressure (−4.5 psi or 10 psia), gas will be precipitated from the hydraulic oil and produce cavitation on the EDP.

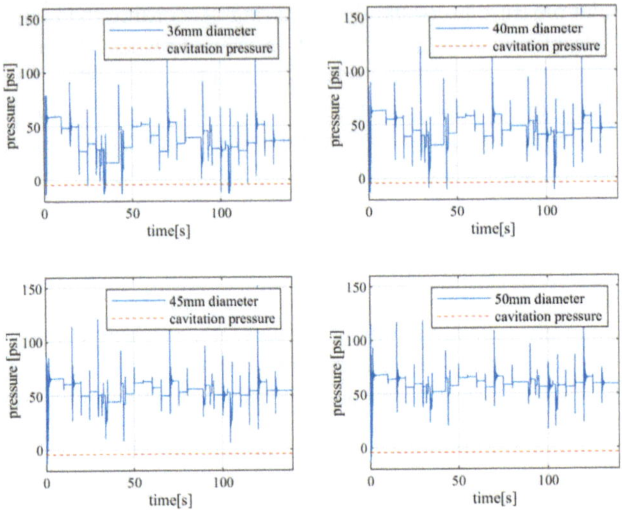

Fig. 6 EDP inlet pressure under different suction line diameter

Through quantitative comparative analysis of different suction pipe diameters under EDP suction pressure, as shown in Fig. 7, according to the polynomial fitting curve, the minimum suction pressure is −4.5 psi, and the corresponding suction pipe diameter is approximately 41.8 mm. Therefore, at −29 °C, if no other measures are taken, when the suction pipe diameter is set above 42 mm, it can be considered that the EDP suction port has a highly probability of not occurring cavitation phenomenon.

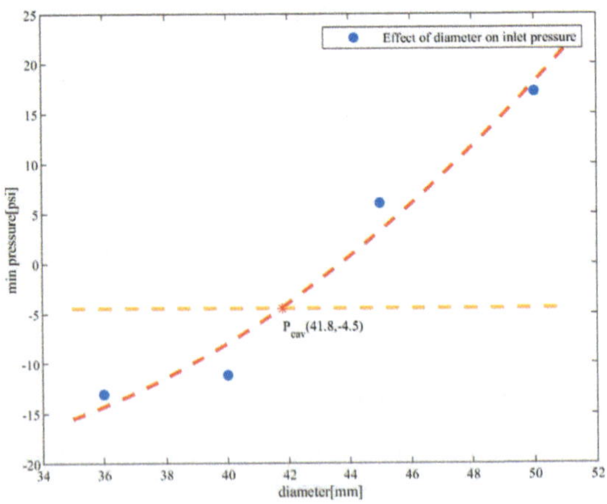

Fig. 7 Minimum EDP inlet pressure trend under different diameter

4.4 Conclusion of Cavitation Control Analysis

After the above simulation analysis, the proposed improvement schemes are as follows: first, by setting up a pressure accumulator in the suction pipe to supplement the instantaneous flow; second, by increasing the pressurized pressure of the reservoir to increase the pressure value of the outlet of the pipe; third, by adjusting the pipe diameter to improve the transient minimum pressure at the outlet of the suction pipe. The three schemes have their own advantages and disadvantages: although the suction accumulator can reduce the pressure impact of the pipe, the flow provided by it is limited, and can only improve the minimum suction pressure of the instantaneous large flow of the oil suction moment; increasing the low pressure chamber pressure of the reservoir can improve the minimum suction pressure of the whole period, but the high pressure of the reservoir will lead to the reduction of the pressure drop at the user end, and then lead to the decrease of the power at the user end; increasing the suction pipe diameter can improve the minimum suction pressure of the whole period, but the excessive diameter will increase the overall weight of the hydraulic system, which is not economical.

After comprehensive analysis, three schemes can be adopted at the same time. For example, when the low pressure chamber pressure of the reservoir is set to 90 psi, the diameter of the suction pipe is set to 40 mm, and the suction accumulator is added, the minimum suction pressure is also higher than cavitation pressure (−4.5 psi or 10 psia).

5 Conclusion

Based on the current status, this paper conducts a simulation analysis of the cavitation control of the hydraulic system of the long-range wide-body aircraft, analyzes the calculation method of acceleration pressure and pressure loss along the way, and establishes the dynamic simulation model of the aircraft hydraulic energy system based on AME Sim. By comparing the presence or absence of the accumulator, the pressure of the low pressure chamber of the reservoir and the diameter of the suction pipe, the suction pressure of the pump is analyzed, so as to evaluate whether there is a risk of cavitation. It was found that the suction accumulator can reduce the pressure impact of the pipe, the flow provided by it is limited, and can only improve the minimum suction pressure of the instantaneous large flow of the oil suction moment; increasing the low pressure chamber pressure of the reservoir can improve the minimum suction pressure of the whole period, but the high pressure of the reservoir will lead to the reduction of the pressure drop at the user end, and then lead to the decrease of the power at the user end; increasing the suction pipe diameter can improve the minimum suction pressure of the whole period, but the excessive diameter will increase the overall weight of the hydraulic system. The parameter settings in the analysis process are estimated according to the current design status. Subsequently, the finite element analysis will be used to determine the value of the oil-gas separation pressure in the actual pipeline configuration and conduct in-depth research as required.

References

1. Liang, D., Yuqi, Z., Cui, D., Yong, W.: Research on cavitation acoustic characteristics of centrifugal pump based on fluid-acoustic field coupling method. Adv. Mech. Eng. **10**(5), 1687814018773665 (2018)
2. Tian, Q., Lin, C., Hu, R., Ya, C., Xu, S.: Numerical and experimental investigations on cavitation characteristics of inducer and impeller combination fuel pumps. In: CSAA/IET International Conference on Aircraft Utility Systems (AUS 2022), pp. 1369–1376, Nanchang, China (2022)
3. Yan, H., Liu, M., Zhao, W., Wu, Y., Lin, P.: Study on the hydraulic performance of large axial-flow pumping stations. In: 2014 ISFMFE-6th International Symposium on Fluid Machinery and Fluid Engineering, pp. 1–5, Wuhan (2014)
4. Jia, C.: Structural performance analysis of piston pump valve plate for pump-controlled cylinder system. In: CSAA/IET International Conference on Aircraft Utility Systems (AUS 2022), pp. 693–698, Nanchang, China (2022)
5. Yang, B.: Research of hydraulic pump cavitation on aircraft. In: CSAA/IET International Conference on Aircraft Utility Systems (AUS 2020), Online Conference, pp. 1179–1183 (2020)

Design and Analysis Method of Downlock Spring for Landing Gear of a Certain Aircraft

Ziyang Song[1(✉)] and Kuan Wang[2]

[1] Shanghai Aircraft Design and Research Institute, Shanghai, China
songziyang1@comac.cc

[2] State key Laboratory of Airliner Integration Technology and Flight Simulation, Shanghai, China

Abstract. The downlock spring of the aircraft is a key component that can achieve the function of lowered locked position of the landing gear. In the design process, it is necessary to consider its position relationship with the lock link mechanism and the influence of the retraction load. This article discusses the scenarios that need to be considered in the design of downlock spring, introduces a novel spring design process and method, and provides an example to complete the design and calculation process of the downlock spring for a certain aircraft.

Keywords: Downlock Spring · Landing Gear · Lock Link

1 Introduction

When the aircraft is taxiing, landing and parking on the ground, the landing gear must rely on the downlock to realize the lowered locked position, and the downlock spring is the key component of the downlock to achieve the intended function [1], due to its special location, making its design process is quite complex, and the performance of the downlock spring will have a direct impact on the flight safety. Therefore, the design of the spring of the landing gear downlock is particularly important.

In order to solve the problem that the boundary conditions of the downlock spring in the linkage mechanism are difficult to set, a reasonable and effective design and calculation method of the landing gear downlock spring of a certain aircraft is proposed based on its typical structure and working principle, performance requirements and selected working conditions as design inputs.

2 Downlock Spring Design Method

2.1 Principle of Operation

When the landing gear is normal extension or alternative extension, the spring pulls the lock link to the over-center position and keeps it locked so as to realize the extended and downlocked position of the landing gear [2]; during landing or ground taxiing state, the

Y. Zhao (Ed.): ICAMM 2025, SPM 96, pp. 13–22, 2025.
https://doi.org/10.1007/978-3-032-09862-7_2

spring overcomes the impact load of the ground on the landing gear and ensures that the lock link is not opened; during ground maintenance state, it overcomes the mis-operation load of the ground personnel and prevents the opening of the lock. Common downlock spring arrangement is shown in Fig. 1.It mainly includes an upper support assembly, a lower support assembly, a number of spring tie rods and springs, wherein the upper support assembly is connected with the lock link, and the lower support assembly is connected with the side stay, and the corresponding mounting holes need to be processed on the lock link and the side stay.

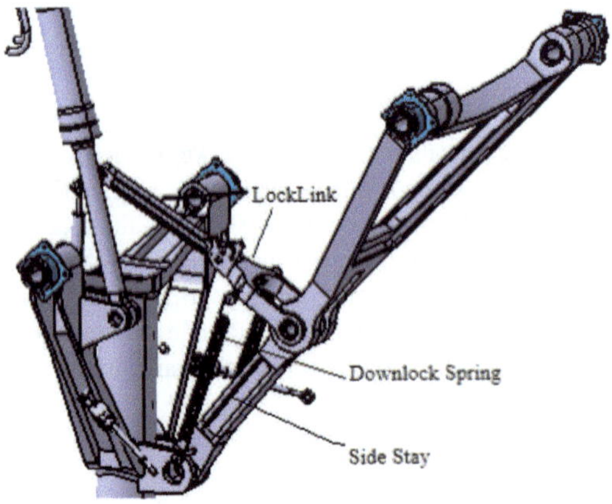

Fig. 1. Common arrangement schemes for downlock springs

2.2 Downlock Spring Design Method

The design of the downlock spring is a crucial part in the design of the downlock component and even the entire retractable system, and this parameter will affect the design of the downlock actuator as well as the performance of the entire retractable system. When the preliminary design of the downlock spring is completed, the design of the downlock actuator is started, and then the landing gear retraction actuator is designed using the obtained parameters, and finally the landing gear retraction performance results are obtained. If the landing gear retraction performance cannot meet the target requirements, the parameters of the downlock spring need to be adjusted, thus affecting the design of the downlock actuator and retraction actuator. The design parameters of the downlock spring should be finalized after several iterations of calculations with the design parameters of the downlock actuator, the design parameters of the retraction actuator and the landing gear retraction performance indexes.

In this paper, the following two operating conditions are considered when designing spring loads:

a) Prevents ground personnel from unlocking the lock by misuse, i.e. it cannot be unlocked when a vertical load of 400 N is applied to the center of the lock link (see Fig. 2).

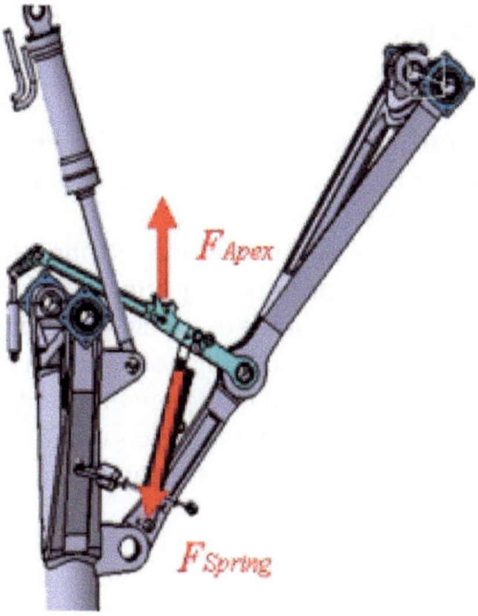

Fig. 2. Preventing accidental unlocking by ground personnel

b) Prevent the downlock from unlocking due to the inertial force of the lock link being greater than the spring tension caused by the 20 g overload, i.e., unable to unlock when subjected to a vertical overload of 20 g at the center of gravity of the lock link (see Fig. 3).

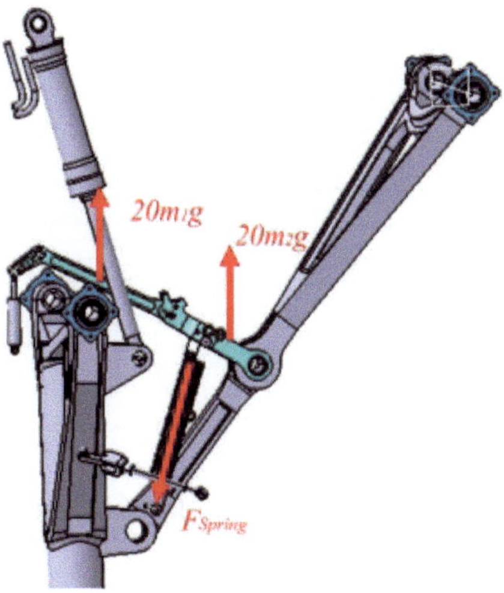

Fig. 3. Preventing accidental unlocking due to upward overloading

Case a) requires two springs to prevent unlocking, while case b) requires one spring to prevent unlocking. This requirement ensures that if one spring fails, the other spring will prevent the lock link from unlocking due to an upward overload of 20 g during takeoff and landing. In addition, it must be ensured that one spring will keep the landing gear down and locked during alternative extension.

The design process of the downlock spring of the landing gear system is shown in Fig. 4. During the design process, the downlock spring must ensure that the landing gear is lowered and locked in the above two scenarios.

After the preliminary design of the downlock spring, the unlocking actuator is designed, and then the landing gear retracting actuator is designed according to all parameters, and the retracting performance of the landing gear is designed.

If the retracting performance of the landing gear does not meet the requirements of the indicators, it is necessary to return to the design of the downlock spring and adjust the parameters. Combined with the design parameters of the unlocking actuator and the related parameters of the retracting actuator, the parameter optimization is completed through several iterations.

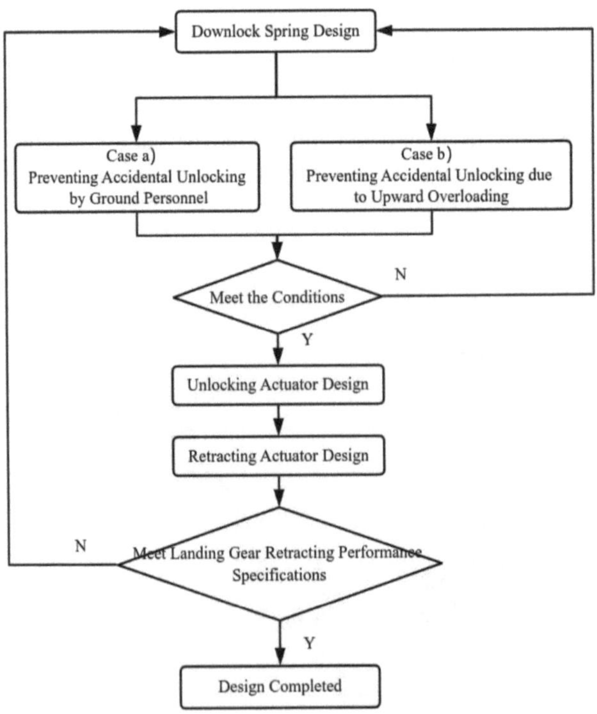

Fig. 4. Design process for downlock springs

2.3 Downlock Spring Tensile Characteristics

The downlock spring is not allowed to have permanent deformation, and its working stress should be ensured within the elastic limit range when designing the downlock spring. The downlock spring working in this range will produce corresponding elastic deformation when subjected to tensile load, as shown in Fig. 5, where H0 is the free length of the lock spring, H1 is the length of the downlock spring mounted to the landing gear and lowered into a locked state for the landing gear, and H2 is the maximum working tensile length of the downlock spring. Both sides of the downlock spring are earring joints, which are respectively installed on the lock link and the side stay.

Under the maximum working tension length, the working stress of the downlock spring should be within the allowable stress range of the material.

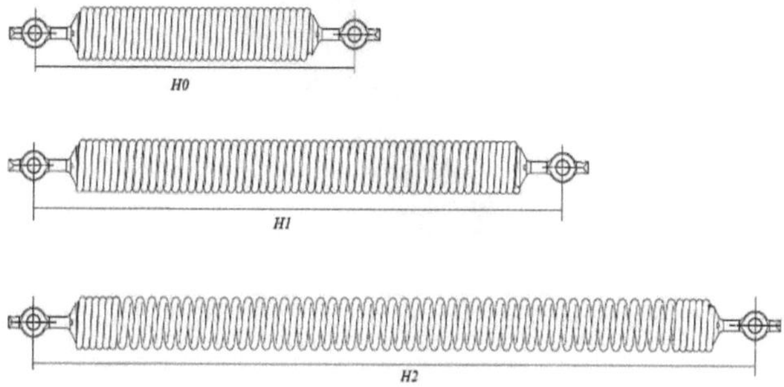

Fig. 5. Length of the downlock spring at different states

3 Example of Downlock Spring Design

Take the landing gear of a certain aircraft as an example, the uplock link of the landing gear has a length of 567.4 mm and a mass of 3 kg, the downlock link has a length of 390 mm and a mass of 5.4 kg, and the materials of both the uplock link and the downlock link are aluminum alloy. The downlock spring and lock link arrangement information is shown in Fig. 6.

A spring load analysis model is established based on ADAMS, and The external loads applied by the model are respectively 400 N due to personnel misoperation in case a) and inertia loads caused by vertical overload in case b). Friction, gravity, and unlocking actuator cylinder forces are not considered in the model. The spring is modeled as an equivalent strip, constrained at both ends with the lock link and drag strut respectively, so that the load on the equivalent strip after loading is the magnitude of the load that needs to be provided by the spring. The simulation model is shown in Fig. 7. The calculation results are shown in Table 1.

The specific steps for the design of the landing gear downlock spring are as follows:

First of all, according to the spatial arrangement of the position of the spring in the process of putting away can not be interference, in the process of movement between the upper and lower spring mounting point displacement change as small as possible, the spring mounting point is initially determined in combination with the position of other components.

Then select the spring material 17-7PH stainless steel and determine the spring material properties.

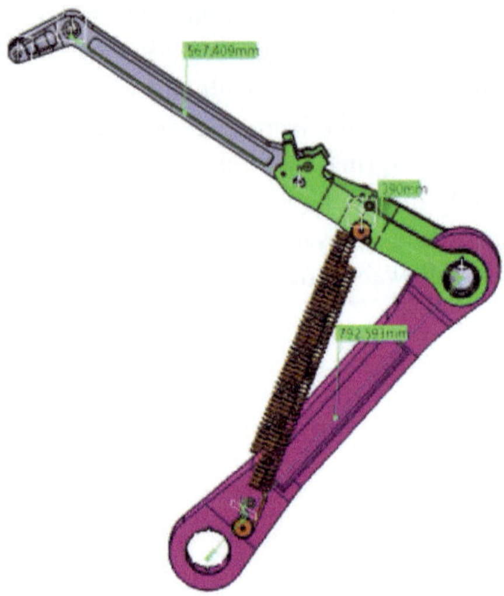

Fig. 6. Dimensional information on the arrangement of the downlocking spring and lock link

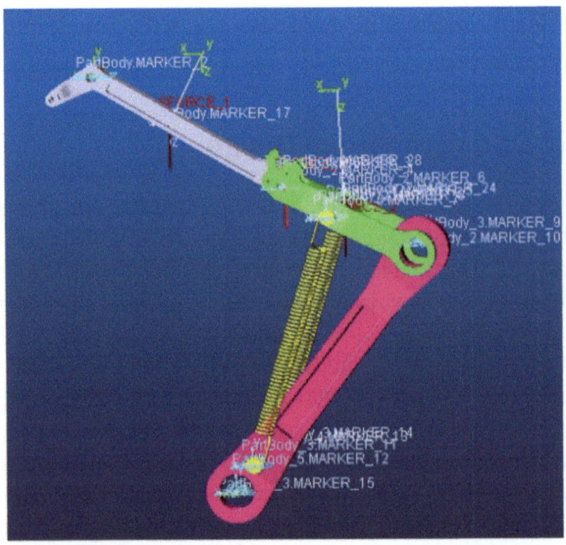

Fig. 7. Simulation model for spring force calculation of downlock spring

According to the installation space, the initial mean diameter of coil is D = 42 mm, estimate the diameter of the spring wire as d = 6 mm based on the winding ratio C value (range of 5–8), and check the allowable stress of the spring wire.

Table 1. Calculation of Required Spring Force

Type	Downlock spring	unit
Overloads to be resisted	20 g	g
Man-made external forces to be resisted	400	N
Load to be provided by a single spring to avoid accidental human unlocking	294	N
To avoid overload unlocking, the force that a single spring needs to provide	789	N

For a spring wire with a circular cross-section, the value of the curvature coefficient K is calculated [3]:

$$K \approx \frac{4C - 1}{4C - 4} + \frac{0.615}{C} \tag{1}$$

Calculated from$K = 1.21$

The initial stress $\tau_0^{'} = 95$ MPa is chosen according to the winding ratio C and the initial spring tension is calculated for the spring.

According to the Eq. (2) for calculating the initial spring tension [4]:

$$F_0 = \frac{\pi d_3 \tau_0^{'}}{8KD} \tag{2}$$

Among them:

$\tau_0^{'}$—Initial stress of the tension spring.

From Eq. (2), we get $F_0 = 158.19N$.

The initial working length of the spring H1 = 500 mm is estimated based on the position of the spring mounting points and the length of the locking hook.

Based on the initial working length H1 an estimate of the number of spring working turns is n = 50, and according to the formula $H_0 = nd$ the initial spring length H0 = 300 mm.

The load K_f required to produce a unit deformation in a spring is called the spring stiffness [5], i.e.:

$$K_f = \frac{\Delta F}{\Delta \lambda} = \frac{Gd}{8C^3 n} = \frac{Gd^4}{8D^3 n} \tag{3}$$

Bringing in the parameters and get $K_f = 3.47N/mm.$

Combined with mathematical model to analyze the maximum elongation of the spring during motion $\lambda_{\max} = H_2 - H_0$, where the maximum elongation of the spring $H_2 = 753.98$ mm, to get $\lambda_{\max} = 453.98$ mm, which is brought into Eq. (4) to calculate the maximum spring load$F_{\max}$:

$$\lambda_{\max} = \frac{F_{\max} - F_0}{k_f} \tag{4}$$

Among them:

F_0—The initial tension of the tension spring.

According to Eq. (4) we get $F_{\max} = 1733.5$ N.

Combined with mathematical model to analyze the spring working extension length during the initial process $\lambda_{\min} = H_1 - H_0$, where the spring is lowered to the locked position $H_1 = 500$ mm, to get $\lambda_{\min} = 200$ mm, which is brought into Eq. (5) to calculate the initial spring working load$F_{\min}$:

$$\lambda_{\min} = \frac{F_{\min} - F_0}{k_f} \tag{5}$$

Among them:

F_0—The initial tension of the tension spring.

According to Eq. (5) to get $F_{\min}$ =851.2 N, the load of putting down the locking position $F_{\min}$ are greater than the force required by the single spring in the case of human accidental unlocking and overload unlocking, which is eligible.

Based on the arrangement scheme of the landing gear, when the landing gear is in the lowered locking position, the length of the downlock spring $H_1 = 500$ mm, according to the change of the downlock spring elongation length with the retracting angle, as shown in Fig. 8, the downlock spring is at the maximum elongation $H_2 = 753.98$ mm when the retracting angle of the landing gear reaches 64°.

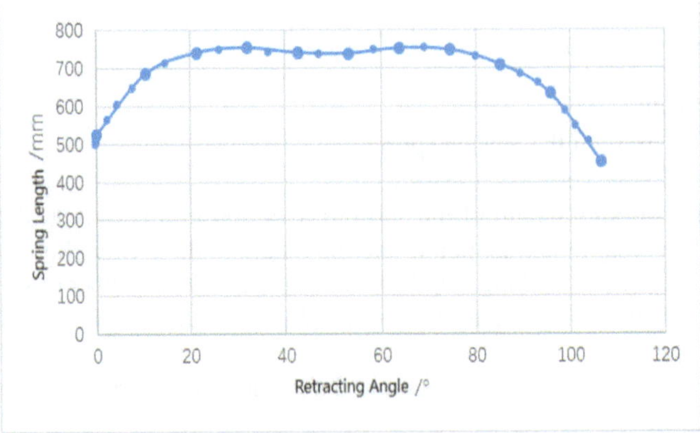

Fig. 8. The extension length of the downlock spring varies with the retracting angle

17-7PH stainless steel material tensile strength is $\sigma_b = 1720MPa$, then the permissible shear stress $[\tau] = 0.8 \times 0.6 \times \sigma_b \approx 827Mpa$. Calculate the maximum tangential stress of the spring during use Eq. (6) [4]:

$$\tau_{\max} = \frac{8F_{\max}DK}{\pi d^3} = 795.3Mpa \leq [\tau] \tag{6}$$

This result satisfies the conditions for use, the specific parameters of the spring obtained after calculation are shown in Table 2.

Table 2. Downlock Spring Design Parameters

Spring parameters	20 g overload
Medium Diameter D (mm)	42
Spring wire diameter d (mm)	6
Number of working turns n	50
Spring stiffness (N/mm)	3.47
Initial Stress $\tau_0^{'}$ (MPa)	95
Spring free length H_0 (mm)	300
Initial spring tension F_0 (N)	158.19
Drop Lock Position Length H_1 (mm)	500
Drop Lock Position Load F_{min} (N)	851.2
Maximum working length H_2 (mm)	753.98
Maximum working load F_{max} (N)	1733.5

When designing the unlocking actuator, it is necessary to consider that the downlock must be unlocked under the conditions of maximum locking force, maximum spring tension, maximum aerodynamic force, maximum friction force, and maximum load of the retractable actuator. At the same time, it is also necessary to consider that the downlock cannot be unlocked under the conditions of minimum locking force, minimum spring tension, no aerodynamic load, minimum friction force, and no load of the retracting actuator. When designing the retracting actuator, it is necessary to consider various factors such as overload coefficient, aerodynamic load, friction, and hydraulic system performance. The retracting performance of the landing gear must meet the system retraction time, and a comprehensive analysis should be conducted in conjunction with cabin doors, aerodynamic loads, overload coefficients, retracting actuators.

This paper omits the subsequent design and calculation process of unlocking actuator, retracting actuator and landing gear retracting performance. After the preliminary design of the downlocking spring, it is necessary to determine whether this parameter meets the requirements of the landing gear retracting performance indicators, if not, it is necessary to optimize the downlocking spring parameters, in the optimization process must meet the prevention of accidental unlocking and overloaded to prevent unlocking of these two scenarios [6–8].

4 Summary

This paper introduces a calculation method that is different from the traditional force analysis to determine the load of the lock spring, and completes the design of the spring by analyzing the principle of the lock linkage mechanism and the specific working conditions, and gives an example of the design and calculation process of the downlock spring in the landing gear of a certain type of aircraft, and with the help of Adams, the load required for the downlock spring to avoid unlocking under different working conditions is obtained. The spring design is then completed through theoretical calculation, and the required load conditions for the subsequent unlocking actuator, retracting actuator, and landing gear retracting performance design are provided. A reasonable and effective design process is formed, which is more operable and of higher value for engineering practice.

References

1. Civil Aviation Administration of China: China civil airworthiness regulation, part 25 airworthiness standards: transport category airplanes, civil aviation administration of China order: No.209: CCAR-25-4.Beijing: Civil Aviation Administration of China, 2011.
2. Knowles James, A.C., Lowenberg Mark, H., et al.: A bifurcation study to guide the design of a landing gear with a combined uplock/downlock mechanism. Proc. Math. Phys. Eng Sci. **470**(2172), 20140332 (2014)
3. Committee of Aircraft Design Manual editorial board: Aircraft Design Manual Section 14: Aircraft Landing System Design. Aviation Industry Press, Beijing (2002)
4. Committee of Mechanical Design Manual: Mechanical Design Manual Part 2, pp. 7–9. China Machine Press, Beijing (2004)
5. Committee of China Aviation Materials Handbook: China Aviation Materials Handbook Part 1 Structural Steel and Stainless Steel, pp. 759–763. Standards Press of China, Beijing (1988)
6. Li, Y., Zhao, J.: Dynamic analysis and optimization of aircraft landing gear retraction/extension system. J. Aerosp. Eng. **33**(4), 04020034 (2020)
7. Yuan, L., He, Y., Wu, L., et al.: Classification and analysis of aircraft landing gear deployable mechanism[J]. Mech. Eng. **9**, 23–26 (2014)
8. Yuan, L., Jiang, B.: Landing gear spatial retractable mechanism analysis and simulation. Mach. Des. Manuf. **4**, 143–149 (2014)

Design and Analysis of Bump Configuration in a Hypersonic Inlet

Shangcheng Xu(✉), Siyi Peng, and Yi Wang

National University of Defense Technology, Changsha, China
shangchengxu@nudt.edu.cn

Abstract. In this study, a new bump configuration that can be applied to hypersonic inlets to improve aerodynamic performances, is proposed. The new design contains bump-shaped surfaces only near the sidewalls and not in the center, unlike conventional designs. The performance parameters of the designed bump inlet are evaluated by numerical simulations. The results reveal that the bump exhibits good displacement capability for the boundary layer, which reaches an average thickness of 5.53 mm, showing a 5.6% decrease compared with a traditional bump. Additionally, because the proposed configuration does not contain a central bump-shaped surface, the unwanted total pressure loss and drag are avoided, increasing the pressure recovery by 5.0% and reducing the drag by 2.5%. Finally, a mass flow correction is conducted, which increases the mass flow ratio to 0.997.

Keywords: Boundary Layer · Bump · Drag · Hypersonic Inlet · Mass Flow Correction · Total Pressure Recovery

1 Introduction

Owing to the viscosity of airflow, a thick boundary layer (BL) is formed near the aircraft surface during the actual flight [1, 2]. When the BL airflow enters the inlet of a supersonic/hypersonic vehicle powered by a ramjet/scramjet, it ultimately reduces the captured airflow quality and engine efficiency [3, 4]. Moreover, the flow separation caused by the BL increases the thermal load and unsteady behavior and may lead to inlet unstart in certain cases [5–7]. Therefore, it is necessary to reduce the amount of BL airflow captured by the inlet.

To divert the BL flow, Lockheed Martin Corporation introduced a bump into the design of supersonic inlets, proposing a bump inlet concept (also known as the diverterless supersonic inlet, DSI) [8]. Such an inlet has a three-dimensional (3D) bump-shaped profile that diverts the BL to the two sides of the inlet, thereby improving the captured airflow quality [9–11]. Compared with the BL bleed scheme [12–14], the bump inlet features no diverter, bleed, or bypass, significantly reducing inlet weight and complexity. In addition, because only the wall surface near the inlet entrance needs to be modified to form a bump configuration, it is favorable for the integrated design of aircraft/inlet [15]. Bump inlets have attracted much attention from researchers and have been successfully applied in many supersonic vehicles [16–19]. Several numerical and experimental

Y. Zhao (Ed.): ICAMM 2025, SPM 96, pp. 23–39, 2025.
https://doi.org/10.1007/978-3-032-09862-7_3

studies have indicated that bumps could effectively displace the BL airflow of supersonic inlets, within which the generated transverse pressure gradient (TPG) on the bump surface plays a key role [20, 21].

The design methods of supersonic bump inlets have been studied extensively [22, 23]. Nevertheless, only a few works have focused on the design and analysis of hypersonic bump inlets [24]. The application of bumps in the hypersonic flow is a promising approach to improving inlet performance; however, there are several challenges, the most severe of which are the large total pressure loss and drag caused by the bump surface in the hypersonic flow [25]. Because hypersonic flow is faster than supersonic flow, the effective displacement for BL airflow requires a larger TPG and a higher bump configuration, which ultimately leads to large total pressure loss and drag in the hypersonic flow, thus deteriorating the inlet performance. This problem demands even stricter bump design requirements under hypersonic conditions. Generally, the ideal bump design applied in hypersonic flow should satisfy the following conditions. (a) It should be integrated into the inlet compression wall to reduce the flow loss caused by the bump. (b) It should generate a large TPG to effectively displace the hypersonic BL airflow. (c) The bump should not be too high to avoid large total pressure loss. Yu et al. [26] and Eiman et al. [27] proposed a novel hypersonic bump design method based on a pressure ridge concept. Although the designed bump generated a shock wave, the total pressure recovery was improved due to a good BL displacement capability. Xu et al. [28] designed a hypersonic bump based on the method of characteristics, which was fully integrated with the inlet without generating any bump-induced shock waves. However, its capability for diverting the BL flow was relatively low. These results indicate that although the introduction of a bump to a hypersonic inlet can improve the inlet performance, the bump configuration should be further optimized.

In this study, a new bump configuration for a hypersonic chin inlet was designed based on the understanding of the working characteristics of the bump applied in the hypersonic flow. Numerical simulations were performed to evaluate the BL displacement capabilities of the traditional and proposed bumps as well as their influence on the inlet total pressure recovery and drag. The obtained results revealed that the proposed bump configuration had a good BL displacement capability, which significantly improved the total pressure recovery and reduced the drag. Subsequently, the flow structures were analyzed to determine the underlying flow mechanism of the observed improvements of the inlet performance. Finally, a mass flow correction was conducted for the bump inlet.

2 Bump Inlet Design Applied in the Hypersonic Flow

2.1 Researched Model

The bump configuration design proposed in this study is based on a typical hypersonic chin inlet with a flight altitude of 26 km, Mach number of 6, and an angle of attack of 0°. Both the body and cowl are axisymmetric, and the cowl is located at the chin of the body. Figure 1(a) shows the profile of the inlet in the symmetry plane. The body radius at the throat section is R, and the key geometric parameters of the inlet are nondimensionalized using R as marked in the figure. The body is divided into a shock-dependent and compression area (Fig. 1(b)). The shock-dependent area is conical with a half-cone angle of 9.34° for the generation of an oblique forebody shock. Under the design point, the forebody shock wave impacts the cowl lip to meet a shock-on-lip (SOL) condition. The airflow is isentropically compressed by the external compression wall and then captured by the cowl and further compressed in the duct. The total contraction ratio (CR_t) and internal contraction ratio (CR_{in}) of the inlet are equal to 6.0 and 1.9, respectively. The angle between the symmetry and osculating planes is denoted by θ, as shown in Fig. 1(c) and (d). Two forward-sweeping sidewalls located in the $\theta = \pm 45°$ osculating planes are adopted to form side spillage windows for the airflow. The central angle of the inlet is 90°, which allows an airflow capture range of $-45° \le \theta \le 45°$.

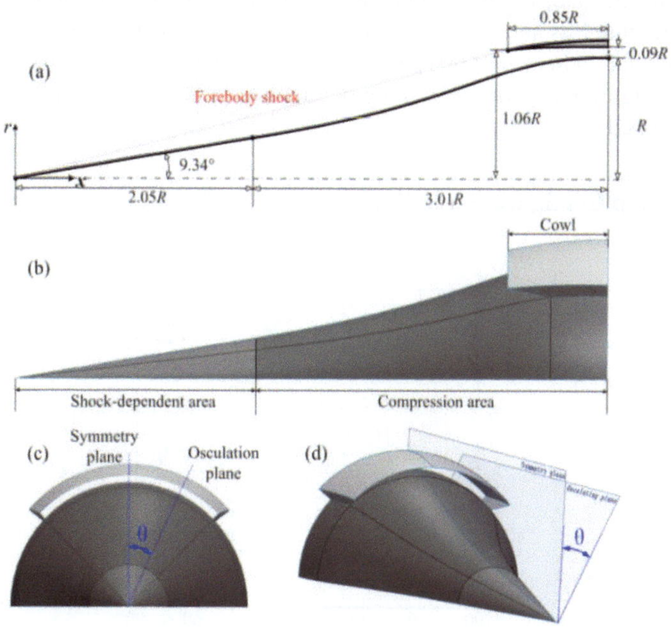

Fig. 1. Hypersonic inlet: (a) profile in the symmetry plane, (b) side view, (c) front view, and (d) 3D view.

2.2 Design Constraints and Requirements

To satisfy the requirements of the integrated design of the hypersonic inlet, the proposed bump configuration must be subjected to certain constraints. To meet the SOL condition under the design point, the shock-dependent area and cowl are to remain intact. To achieve a smooth transition between the inlet and the isolator, the location of the throat is unchanged. Moreover, the bump should be integrated with the inlet to minimize the total pressure loss and drag. At the same time, the BL diversion should be maximized to improve the quality of the captured airflow. Therefore, the design of the bump in the hypersonic inlet can be considered a multi-objective design problem under certain constraints (geometric and aerodynamic).

The height of the traditional bump configuration gradually decreases from the center to the two sides. When the airflow passes through the bump, the pressure at the central region is higher than that on both sides, which forms a pressure gradient along the transverse direction. The BL airflow driven by this gradient produces a transverse flow, which is ultimately excluded from the inlet. Therefore, an external compression wall with a TPG can be regarded as a bump configuration. There are two major problems related to the application of traditional bumps in hypersonic inlets. First, even though a transverse BL flow is generated in the bump central region, it is located far away from the spillage windows; as a result, the transverse BL airflow is not excluded as expected (in other words, the bump configuration in the central region weakly contributes to the BL diversion). Second, the bump is relatively high in the central region, resulting in large total pressure loss and drag in the hypersonic flow. These two problems are further discussed in detail in Sect. 4 of this paper. This implies that the bump design (especially the bump configuration in the central region) applied in the hypersonic inlet should be further developed.

2.3 New Bump Configuration Design

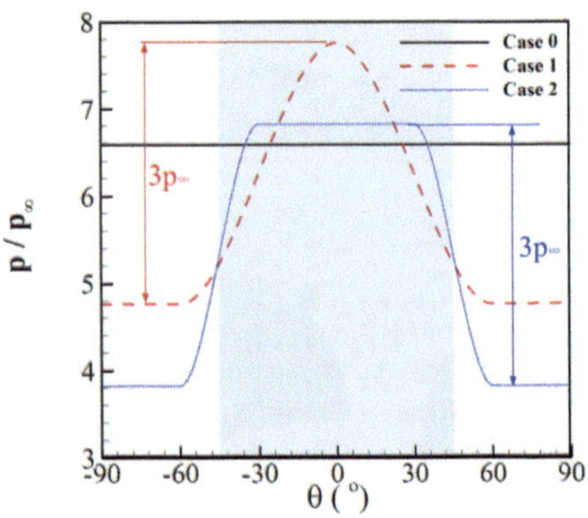

Fig. 2. Preset transverse pressure curves on the compression wall.

Based on the understanding of the working characteristics of the traditional bump in the hypersonic inlet, a new bump configuration is proposed. In this configuration, TPGs are only formed on the compression surface near both sidewalls and not in the central region. In other words, the bump is only placed on both sides and not in the central region. To evaluate its performance, a bump inlet was designed using the method based on the TPG proposed in a previous work. The bump design method is described in detail in [29], and its accuracy has been validated. Additionally, a traditional bump inlet was designed via the same method for comparison. For the convenience of description, the original inlet with an axisymmetric body is referred to as Case 0, the inlet with the traditional bump configuration is named Case 1, and the inlet with the new bump configuration is denoted as Case 2. Figure 2 shows the design curves of the transverse pressure on the compression wall, where the airflow capture range is marked in blue. The pressure does not change along the transverse direction in Case 0; however, it decreases from the center to the two sides within the range of $-60^\circ \leq \theta \leq 60^\circ$ in Case 1, indicating the presence of a TPG on the entire compression wall. In Case 2, the pressure is constant in the range of $-30^\circ \leq \theta \leq 30^\circ$ but decreases sharply in the ranges of $-60^\circ \leq \theta \leq -30^\circ$ and $30^\circ \leq \theta \leq 60^\circ$, forming large local TPGs near the sidewalls. For consistency, the transverse pressure drops in Cases 1 and 2 are set to $3p_\infty$. In addition, the internal contraction ratios of the designed bump inlets are set to 1.9 by adjusting the heights of the pressure curves. Table 1 lists the key parameters of the three inlets.

Table 1. Parameters of The Three Inlets.

Inlet	Bump range	Transverse pressure drop on bump	CR_{in}	CR_t
Case 0	–	0	1.9	6.0
Case 1	$-60^\circ \leq \theta \leq 60^\circ$	$3p_\infty$	1.9	6.0
Case 2	$-60^\circ \leq \theta \leq -30^\circ$, $30^\circ \leq \theta \leq 60^\circ$	$3p_\infty$	1.9	6.0

Figure 3 shows the three inlet configurations with the bumps marked in orange. The body of the original inlet is axisymmetric. The compression area in Case 1 forms an integrated bump in the range of $-60^\circ \leq \theta \leq 60^\circ$ that is high in the center and low on the sides. The compression area in Case 2 produces an axisymmetric profile without a bump in the range of $-30^\circ \leq \theta \leq 30^\circ$. Nevertheless, the wall height near the sidewalls ($-60^\circ \leq \theta \leq -30^\circ$ and $30^\circ \leq \theta \leq 60^\circ$) decreases sharply along the transverse direction, forming two distributed bumps in Case 2.

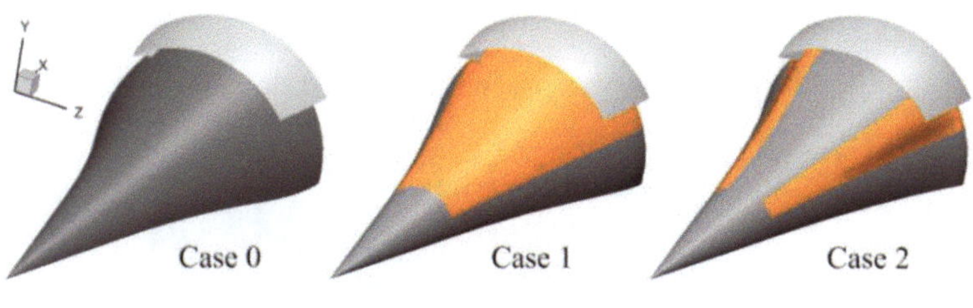

Fig. 3. Various inlet configurations.

3 Calculation Method and Wind Tunnel Test

3.1 Numerical Simulation Method

The 3D Reynolds-averaged Navier–Stokes equations are utilized to solve the inlet flow fields at the design point. Considering complex flow phenomena such as shock/BL interactions and separated flows in the calculated flow field, the k-ω SST model is utilized to simulate the turbulence flow, which has the advantages of both the k-ω model for calculating the near-wall flow and k-ε model for calculating the main flow. The air is simulated by adopting the ideal gas model, while its viscosity is calculated using the Sutherland law. Owing to the symmetry, only half of the model is simulated to save the computational cost. The computational domain is meshed with a structured grid as shown in Fig. 4; here, the grid points are clustered in the cone tip, inlet entrance, and duct to simulate the complex flow structures. The BL grid is generated to accurately model the near-wall BL flow and ensure that the y+ of the wall region is less than 1. A grid independence was undertaken in the author's previous work, and the grid number for the 3D computational region is 3,240,000 [29]. A pressure far-field condition is utilized in the inlet of the computational domain, a pressure outlet condition is set for the outlet, symmetric boundary conditions are adopted at the two sides, and the wall and cowl are assumed to be non-slip adiabatic walls.

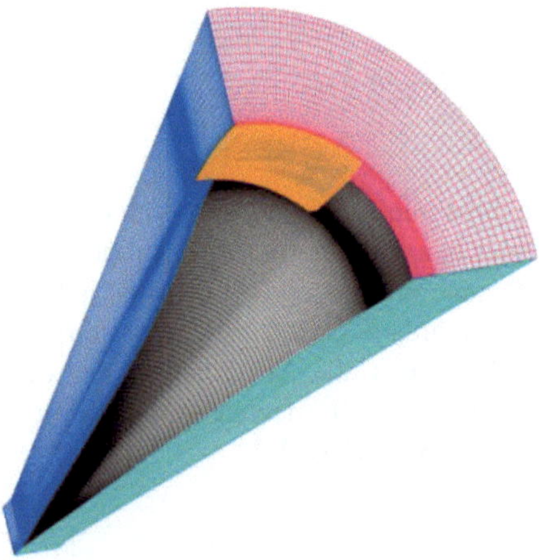

(a) 3D computational domain mesh distribution

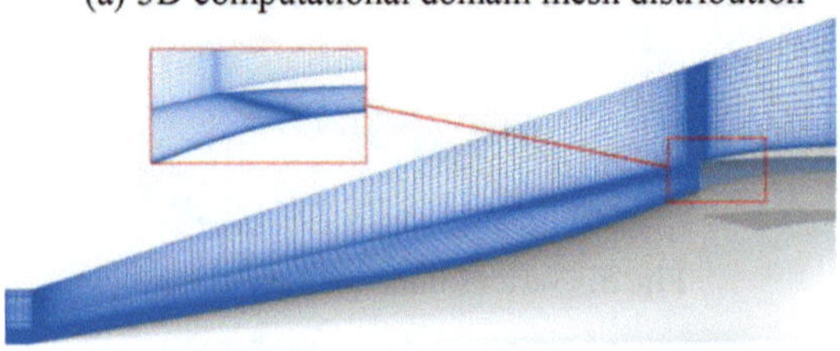

(b) Symmetry plane mesh distribution

Fig. 4. Various inlet configurations.

3.2 Wind Tunnel Test

An experiment was conducted in the LF-220 free-jet wind tunnel of the scramjet laboratory of the National University of Defense Technology. The nozzle outlet diameter was 220 mm, the inflow Mach number was 5.0, and the Reynolds number was 1.78 ×

107 m^{-1}. A typical mixed-compression inlet was employed as the test model. The surface static pressures along the flow direction were recorded during the test process. In addition, the experimental test was simulated employing the proposed numerical method, and the flow conditions of the wind tunnel were adopted as the boundary conditions of the calculating domain.

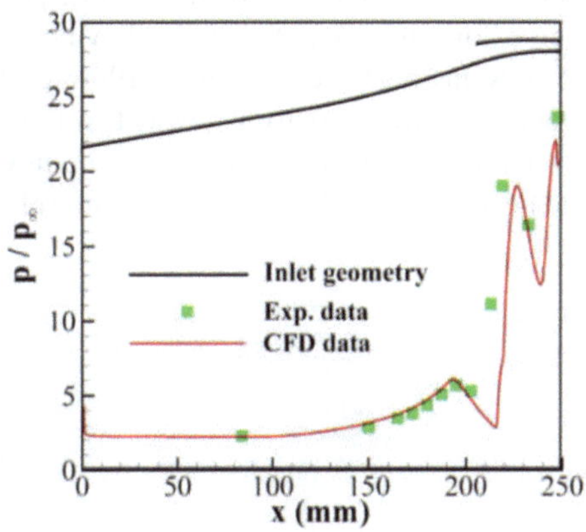

Fig. 5. Wall pressure ratios obtained by the computational fluid dynamics (CFD) and the wind tunnel experiment.

Figure 5 shows the wall pressure distributions obtained both experimentally and numerically. The results indicate that the pressure remains unchanged in the shock-dependent area and then gradually rises in the external compression area. After the pressure drops slightly, it increases rapidly due to the cowl shock/BL interactions. The obtained numerical data are basically consistent with the numerical results both on the external and internal compression regions, indicating that the numerical method adopted in this study can accurately simulate the hypersonic inlet flow field.

4 Results and Discussion

4.1 Boundary Layer Displacement

Figure 6 displays the distributions of the limiting streamlines of the inlets (front views). The streamlines in Case 0 maintain excellent two-dimensional (2D) characteristics and are not deflected. In contrast, the streamlines of the two bump inlets exhibit apparent 3D characteristics. The streamlines in Case 1 demonstrate different degrees of deflection on the bump surface. In the central area (Zone A) near the symmetry plane, although the streamlines are deflected towards the two sides, they nevertheless enter the inlet because of the long distance to the spillage windows. In the area near the sidewalls (Zone B), the streamlines are slightly deflected outwards and are partially displaced out of the inlet. This means that in Case 1, only the streamlines near the sidewalls are diverted, while the airflow in the central region is not excluded despite the generated transverse flow. In Case

2, the profile in Zone A is axisymmetric; as a result, the streamlines in this region form a 2D flow without apparent deflections. In contrast, under the influence of the bump, the streamlines in Zone B exhibit a large degree of deflection. Since their locations are close to the spillage windows, the deflected streamlines are easily displaced out of the inlet, leading to the large displacement of the BL airflow.

Figure 7 displays the transverse velocity (vt) contours near the inlet entrance. Herein, the transverse velocity refers to the velocity from the symmetry plane to the two sides, which reflects the transverse flow characteristics in the flow field. The flow field in Case 0 exhibits no transverse flow in most areas. In the airflow capture range of Case 1, transverse flow with a relatively low velocity is present in most regions. Meanwhile, there is almost no transverse flow near the symmetry plane in Case 2, but significant transverse flow is detected near the sidewalls with a maximum velocity exceeding 80 m/s.

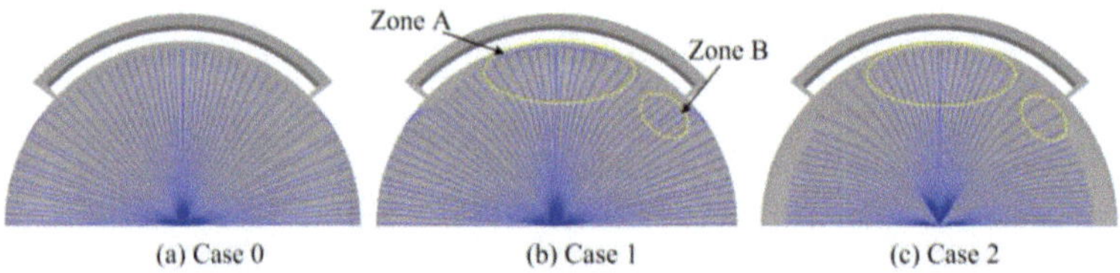

Fig. 6. Distributions of the limiting streamlines in various inlets.

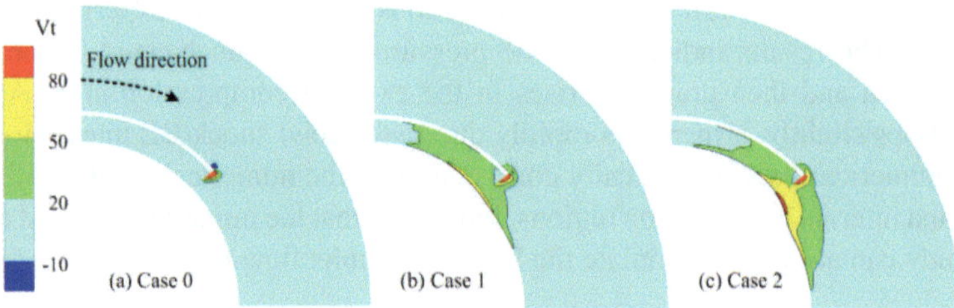

Fig. 7. Transverse velocity distributions at the cross-sections (x = 1.4 m) located upstream from the inlet port.

Figure 8(a) displays the BL distributions near the inlet entrance. Here, the red regions represent the main flows and the green regions refer to the BL flows. Considering that the inlet flow has strong non-uniform characteristics, the airflow with a total pressure recovery coefficient lower than 0.3 is defined as the BL in this study. As shown, the BL flows are formed near the compression surfaces as well as near the inner and outer cowl surfaces. Because the body is long, the BL near the compression wall is well developed. The low-energy BL airflow entering the inlet significantly reduces the engine efficiency; hence, it is necessary to decrease its intake by adding the bump. Figure 8(b) compares the thicknesses of the BLs on the compression walls to examine the diversion effect of the two types of bump configurations on the BL airflow. The BL in Case 0 exhibits an almost uniform distribution along the transverse direction. The BL in Case 1 is slightly thinner than that of Case 0 in the range of approximately $0° \leq \theta \leq 40°$ but obviously thicker outside the sidewall, indicating that part of the BL is displaced and accumulates near the bump side edge. The BL thickness in Case 2 fluctuates slightly in the range of $0° \leq \theta \leq 15°$; however, it sharply decreases along the transverse direction and becomes less than 5 mm at approximately $\theta = 40°$. Similarly, its thickness significantly increases outside the sidewall. The results show that within the airflow capture range, the average BL thickness is 6.04 mm in Case 0, 5.86 mm in Case 1 (slightly lower), and 5.53 mm in Case 2, which is the lowest. This implies that the bump configuration in Case 2 has a greater BL diversion capability than that in Case 1.

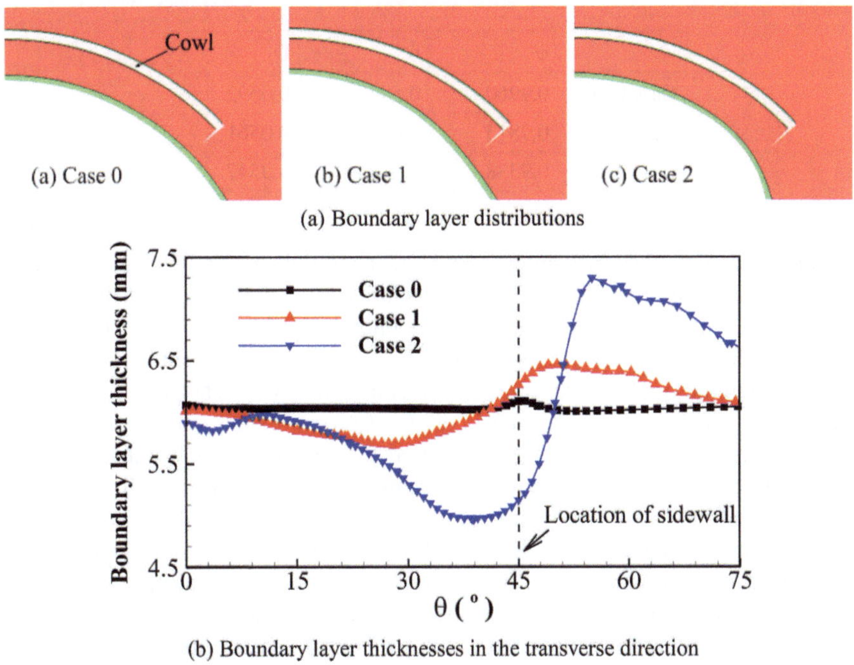

Fig. 8. Boundary layers at the cross-sections (x = 1.4 m) located upstream from the inlet ports.

4.2 Performance Under the Design Point

The mass flow ratios and mass-averaged total pressure recovery coefficients at the throats of various inlets are listed in Table 2. Case 0 demonstrates the highest mass flow ratio (0.996) but the lowest total pressure recovery coefficient (0.5907). Case 1 exhibits a slightly lower mass flow, while its total pressure recovery coefficient is increased to 0.6138. Case 2 has a significantly reduced mass flow ratio of 0.9338. However, its total pressure recovery coefficient is raised significantly to 0.6444. Compared with Case 0, the mass flow ratios in Cases 1 and 2 are reduced by 1.5 and 6.2%, respectively, while the total pressure recovery coefficients are increased by 3.9 and 9.1%, respectively, which indicates that the bump decreases the captured airflow and increases the total pressure recovery. In addition, the bump in Case 2 shows greater decreases in both the mass flow and total pressure loss compared with those in Case 1. Moreover, the drag coefficients calculated using the bottom area of the model as a reference area are also listed in Table 2. Case 0 exhibits the largest drag coefficient of 0.0592, while the drag coefficients in Cases 1 and 2 are decreased. In particular, the drag is significantly decreased in Case 2 as the corresponding drag coefficient is 2.5% lower than that in Case 1. The reasons for the observed differences in the aerodynamic performances of these inlets are determined by investigating the flow characteristics in the following sections.

Table 2. Performance Characteristics of Various Inlets Under the Design Point.

	φ	σ	D
Case 0	0.9960	0.5907	0.0592
Case 1	0.9813	0.6138	0.0561
Case 2	0.9338	0.6444	0.0547

Figure 9 displays the Mach number contours in multiple osculating planes of the flow fields with Mach numbers below 5.2 for better clarity. First, the flow field in the symmetry plane of Case 0 is analyzed to understand the inlet flow structure. A forebody shock is generated when the airflow passes through the forebody. Subsequently, as the airflow passes through the cowl leading edge, the outer and inner cowl shocks are generated as well, and the inner cowl shock impacts the BL to form a small-scale separation zone at the compression side. It is noteworthy that the forebody shock impacts the cowl leading edge to meet the SOL condition, indicating that there is no cowl-lip spillage and additional drag in Case 0. Owing to the axisymmetry of Case 0, the flow structure in each osculating plane is the same as that in the symmetry plane. The SOL conditions are also realized in both bump inlets. In addition, the two bumps isentropically compress the airflow without producing any shock wave in the flow field. This result demonstrates the following two characteristics of the designed bumps. (a) The bump does not change the SOL condition, which allows the inlet to capture the maximum possible amount of the compressed airflow while avoiding additional drag. (b) No shock wave is produced by the bump. In the hypersonic flow, a shock increases the flow loss; thus, the unwanted shock system should be avoided. The adopted bump design method enables the complete integration of the bump and inlet compression surfaces, thereby avoiding the bump shock wave.

The streamlines traced from the osculating plane of $\theta = 45°$ in the freestream field are displayed in Fig. 9. Because no transverse flow is formed in Case 0, the streamlines intersect the sidewall, indicating the absence of the transverse spillage. Therefore, a full mass flow capture is achieved in Case 0. The streamlines in Case 1 are deflected outwards at an increasing angle as they approach the wall. Ultimately, parts of the streamlines close to the wall are displaced outside the inlet, forming a small-scale transverse spillage (marked in blue). The streamlines in Case 2 (especially the near-wall streamlines) are severely deflected, leading to a large-scale transverse spillage on the outside of the sidewall; for this reason, Case 2. has a significantly lower mass flow ratio. The spillage regions in Cases 1 and 2 are large near the wall and small in the main flow, indicating that the displaced airflow is mostly the near-wall low-energy BL.

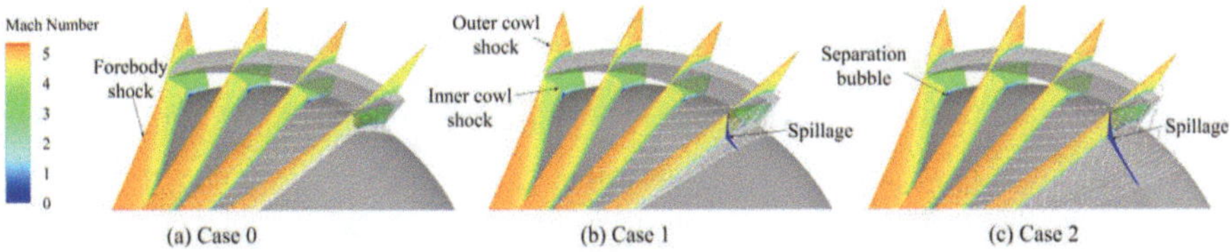

Fig. 9. Mach contour maps obtained for various osculation planes and streamline distributions originating from the far field on the $\theta = 45°$ osculation plane.

Figure 10 displays the distributions of the total pressure recovery in the inlet throat. The total pressure loss mainly originates from two sources: the loss of the BL airflow and the loss in the main flow caused by shock wave. The BL exists in the near-wall regions of the compression wall, two sidewalls, and inner cowl. In particular, the BL at the compression side is thicker than the BLs located near the cowl and sidewalls, which is the main factor contributing to the total pressure loss. Numerical results reveal that the mass flow ratio of the BL at the compression side in Case 0 is 0.0338; however, its values obtained for Cases 1 and 2 are equal to 0.0332 and 0.0313, which are 4.7% and 7.4% lower than that in Case 0, respectively. This indicates that due to the BL diversions, the BL airflows entering the bump inlets are considerably reduced. Furthermore, the BL airflow in Case 2 is significantly smaller than that in Case 1, suggesting that the BL-caused total pressure loss can be greatly reduced by adopting the bump configuration in Case 2. In addition, the total pressure loss in the main flow is also non-negligible. The total pressure recovery coefficient in Case 0 is uniform along the transverse direction. In comparison, the total pressure recovery in Case 1 is high on both sides and forms two high total pressure areas, but is significantly lower at the center. The total pressure recovery in Case 2 exhibits no significant reduction at the center and a relatively large high total pressure area exists in the flow field. Therefore, Case 2 possesses a high total pressure recovery performance.

The reasons for the observed total pressure loss in the central area of the throat in Case 1 are examined in the following sections by analyzing the inlet flow field structure. Figure 11 shows the distributions of the static pressure ratio, flow angle, and total pressure recovery coefficient near the inlet ports on the symmetry planes, where the static pressure ratio reflects the degree of airflow compression; the flow angle reflects the airflow direction; and total pressure recovery coefficient reflects the flow loss. To achieve a better understanding of the flow process at the inlet, the flow field structure in Case 0 is analyzed first. After the forebody shock, the airflow is compressed on the outer compression wall, which is then captured by the cowl. Because the cowl profile is not oriented in the same direction as the airflow, a cowl shock is generated to force the airflow to change its direction. Subsequently, the cowl shock is incident onto the compression walls to cause a shock/boundary layer interference. After that, the airflow is reflected several times until it reaches the throat. Figure 11 demonstrates that the inlet

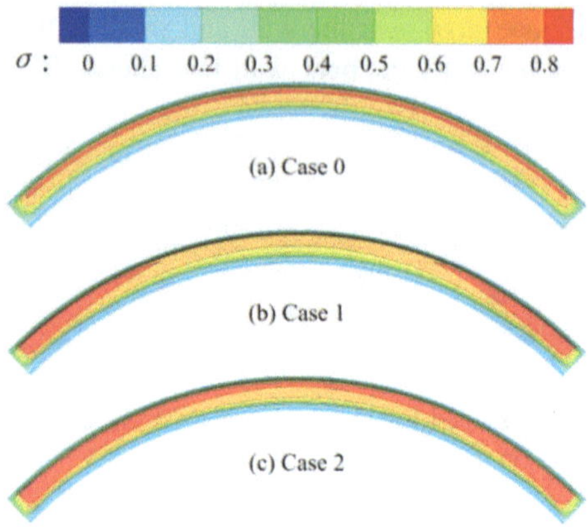

Fig. 10. Distributions of the total pressure recovery coefficient at the inlet throats.

outer compression walls isentropically compress the airflow without causing a total pressure loss. The total pressure loss in the inlet mainstream is caused by the forebody shock, cowl shock, and reflected shock. The flow field structures on the symmetry planes of the two bump inlets were investigated as well. Because the bump is set up on the symmetry plane in Case 1, its compression wall is higher than that in Case 2; as a result, the airflow compression on the outer compression wall surface in Case 1 is stronger than that in Case 2, and the flow angle at the inlet port in Case 1 is also greater than that in Case 2. As the angle between the airflow and the cowl increases, the cowl shock becomes stronger. The obtained flow field structure indicates that the average angle of the cowl shock in Case 2 is only 6.88°, while in Case 1, it equals 8.44°. The total pressure loss is increased by not only the stronger cowl shock, but also the stronger reflected shock. Therefore, the total pressure loss in the compression section of the symmetry plane flow field in Case 1 is relatively large.

The influence of the bump on the inlet drag is also studied. According to Table 2, the largest drag is obtained in Case 0 followed by Cases 1 and 2. This result indicates that adopting the Case 2 bump configuration a better drag-reduction effect than in Case 1 can be achieved. To analyze the drag differences between various inlets, the pressure and frictional drag coefficients of the compression wall and the cowl are summarized in Fig. 12, which shows that the frictional drags of the three inlets are similar, but the corresponding pressure drags are significantly different. In particular, the pressure drag generated by the cowl is low, which has a limited influence on the total drag. In contrast, large pressure drags are generated on the walls of the three inlets, which have significantly different magnitudes. The largest wall pressure drag coefficient of 0.2068 is obtained in Case 0, a slightly smaller (by 5.2%) value is observed in Case 1, and the smallest wall pressure drag coefficient is achieved in Case 2, which is 3.7% smaller than that in Case 1. Therefore, it can be concluded that the wall pressure drag is the main reason for the observed difference in the total drag of the three inlets.

Because the angle of attack is 0° under the design point, the pressure drag represents the horizontal component of the integral surface pressure. Figure 13 shows the distributions of the pressure ratios on the inlet surfaces. As can be seen, high-pressure areas with

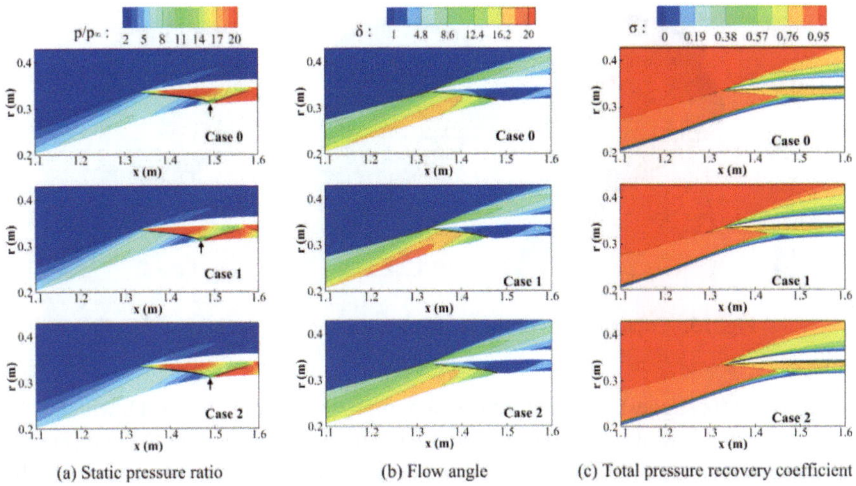

Fig. 11. Flow field structures present near the inlet ports on the symmetry planes.

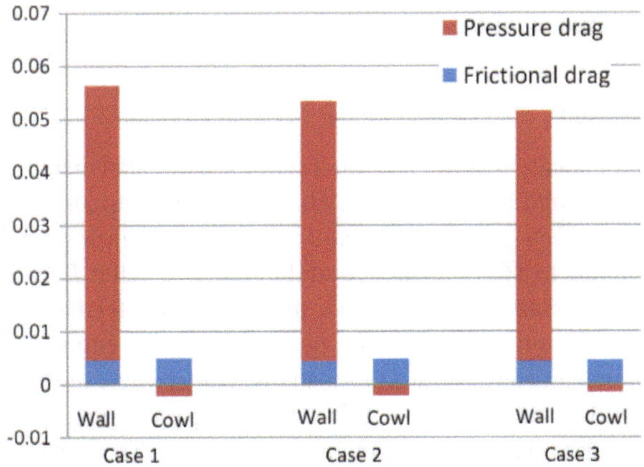

Fig. 12. Pressure and frictional drags of various inlet components.

different sizes and ranges are generated on the external compression walls for all three inlets. The high-pressure area in Case 0 is uniformly distributed along the transverse direction. Meanwhile, the high-pressure area is not observed at the two sides of Case 1; instead, it is formed at the center because the bump is relatively high, and the maximum pressure exceeds that in Case 0. Therefore, although the drag in Case 1 is lower than that in Case 0, this difference is not significant. Under the influence of the new bump, the high-pressure area at the two sides of Case 2 almost completely disappears. Meanwhile, because no bump configuration is set at the center of Case 2, the high-pressure area near the symmetry plane is inconspicuous. The bump configuration utilized in Case 2 can more effectively reduce the pressure drag compared with that in Case 1.

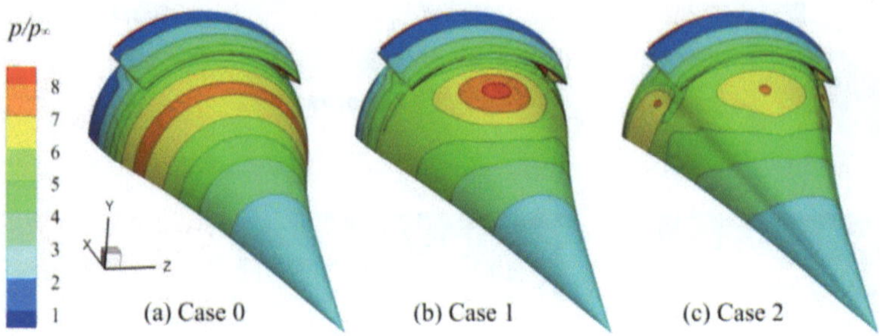

Fig. 13. Contour maps of the static pressure ratio on the inlet surfaces.

These results suggest that the traditional bump configuration (Case 1), which is high at the center and low at both sides, cannot displace the BL airflow in the central region and causes unwanted total pressure loss and drag as the bump is positioned too high. In contrast, the new bump configuration proposed in this study (Case 2) is relatively low at the center, effectively minimizing the total pressure loss and drag. Furthermore, setting up the bump near the spillage windows can generate a remarkable transverse airflow to promote the displacement of the BL airflow.

4.3 Flow Correction

The above results indicate that the introduction of a bump can displace the BL airflow and improve the captured airflow quality. However, the bump also decreases the inlet mass flow due to the diversion effect. In particular, the mass flow ratio in Case 2 is only 0.9338, corresponding to a 6.6% reduction of the designed value. Because the mass flow is one of the most important performance parameters of the inlet that directly determines the engine thrust, a mass flow correction is conducted for the bump inlet in this work to increase the captured airflow. For this purpose, the center angle of the inlet is increased to expand the airflow capture area, thereby increasing the mass flow ratio. This mass flow correction method has the following advantages. (a) No changes are made to the bump configuration, ensuring that the inlet still produces a diversion effect on the BL after the correction. (b) No changes are made to the SOL condition; hence, the inlet can capture all the compressed airflow without generating additional drag. Moreover, curved sidewalls are adopted to ensure that the inlet throat is not changed. By properly designing the shapes of the sidewalls, the airflow is slightly side compressed without obvious total pressure loss. When the center angle in Case 2 is increased to 95.6°, the inlet can reach the desired mass flow. Figure 14 shows the front views of the inlet configurations before and after the correction. The same method is adopted for the mass flow correction of Case 1, and its center angle increases to 91.4°. The models in Cases 1 and 2 obtained after the correction are named Cases 1.1 and 2.1, respectively.

Table 3 shows the center angle and performance parameters of each inlet under the design point. After the correction, full airflow captures are achieved in Cases 1.1 and 2.1, demonstrating that the correction method can effectively increase the captured airflow. In addition, the advantages of the bump inlets on the total pressure recovery and the drag performance are maintained. Moreover, the total pressure recovery coefficient in

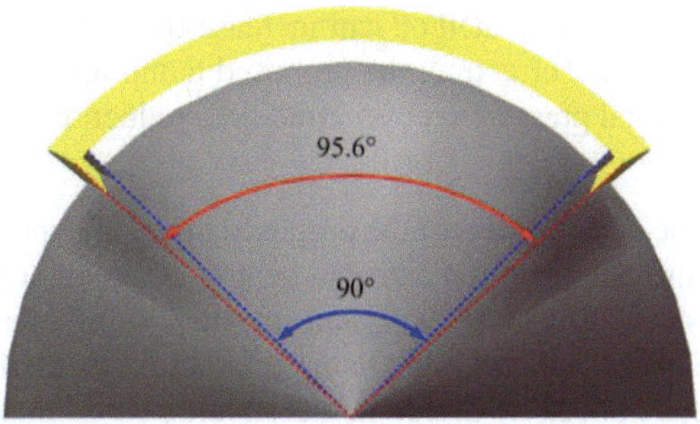

Fig. 14. Inlet configuration in Case 2 before and after the flow correction (front view).

Case 2.1 is 4.9% higher than that in Case 1.1, while the drag coefficient in Case 2.1 is 0.9% lower than in Case 1.1, indicating that at the same airflow capture ability, the bump configuration in Case 2.1 can improve the inlet performance more effectively.

Table 3. Key Design and Performance Parameters of Various Inlets Under the Design point.

Inlet	Center angle	φ	σ	D
Case 0	90°	0.9960	0.5907	0.0592
Case 1.1	91.4°	0.9993	0.6122	0.0563
Case 2.1	95.6°	0.9950	0.6424	0.0558

5 Conclusion

By analyzing the working characteristics of the bump, a distributed bump configuration applied to hypersonic inlets is proposed in this work. The influence of this configuration on the inlet performance is investigated using the numerical simulation. The following conclusions have been drawn.

(1) The proposed bump configuration produces a better BL diversion performance than that of the traditional bump. Although the traditional bump is capable of generating a transverse BL flow in the central region, this part of the BL airflow is not displaced outside the inlet, leading to a poor BL diversion. In contrast, the TPG on the new bump is concentrated near the sidewalls, which facilitates the sideways movement of the BL airflow and its ultimate displacement through the spillage windows, thus effectively promoting the BL diversion.

(2) The new bump inlet demonstrates better performances than those of the traditional bump inlet in terms of the total pressure recovery and drag reduction. The traditional bump configuration is too high at the center, causing large total pressure loss and drag. In contrast, the new bump is located on the two sides, thus avoiding the unwanted total pressure loss and drag caused by the high bump surface.

(3) The utilized mass correction method can increase the bump inlet mass flow to the desired value. After the correction, the proposed bump inlet still exhibits apparent advantages on the total pressure recovery and drag reduction at the same high mass flow ratio.

This study confirms the effectiveness of the proposed bump in hypersonic flow based on a chin inlet. This bump configuration can also be applied to other types of hypersonic inlets, such as 2D, sidewall-compression, and inward turning inlets. In addition, this work focuses on the performances of the bump inlet under the design point, while the performance parameters under off-design points should be examined in future studies.

References

1. Johnson, C.B., Lawing, P.L.: Mach 6 flowfield survey at the engine inlet of a research airplane. J. Aircr. **14**(4), 412–414 (1977)
2. Lawing, P.L., Johnson, C.B.: Inlet boundary-layer shapes on four aircraft forebodies at Mach 6. J. Aircr. **15**(1), 62–63 (1977)
3. Sziroczak, D., Smith, H.: A review of design issues specific to hypersonic flight vehicles. Prog. Aerosp. Sci. **84**, 1–28 (2016)
4. Lau, K.Y.: Hypersonic boundary-layer transition: application to high-speed vehicle design. J. Spacecr. Rocket. **45**(2), 176–183 (2018)
5. Bhanderi, H., Babinsky, H.: Improved boundary layer quantities in the shock wave boundary layer interaction region on bumps. In: 35th AIAA Fluid Dynamics Conference and Exhibit. AIAA, Toronto, Canada (2005)
6. Zhang, Y., Tan, H.J., Tian, F.C.: Control of incident shock/boundary-layer interaction by a two-dimensional bump. AIAA J. **52**(4), 767–776 (2014)
7. Datta, V.G.: Progress in shock wave/boundary layer interactions. Prog. Aerosp. Sci. **72**, 80–99 (2015)
8. Simon, P.C., Brown, D.W., Huff, R.G.: Performance of External-Compression Bump Inlet at Mach Numbers of 1.5 and 2.0, 2nd edn. National Advisory Committee for Aeronautics, Washington, USA (1957)
9. Lo, K.H., Kontis, K.: Flow characteristics of various three-dimensional rounded contour bumps in a Mach 1.3 freestream. Exp. Thermal Fluid Sci. **80**, 228–243 (2016)
10. Sang, D.K.: Aerodynamic design of a supersonic inlet with a parametric bump. J. Aircr. **46**(1), 198–202 (2015)
11. Tillotson, B.J., Loth, E., Dutton, J.C.: Experimental study of a Mach 3 bump compression flowfield. J. Propuls. Power. **25**(3), 545–554 (2009)
12. Schwartz, M.J., Gaitonde, D.V.: Effects of bleed on supersonic turbulence boundary layers. In: AIAA Aviation 2021 Forum. AIAA, USA (2021)
13. Yuan, Y.C., Liu, F.Z., Wang, X.: Design and analysis of a supersonic axisymmetric inlet based on controllable bleed slots. Sci. Technol. **118**, 107008 (2021)
14. Liu, J., Yuan, H.C., Wang, Y.F.: Unsteady supercritical/critical dual flowpath inlet flow and its control methods, Chinese. Aeronaut. **30**(6), 1877–1884 (2017)
15. Huang, G.P., Zuo, F.Y., Qiao, W.Y.: Design method of internal waverider inlet under non-uniform upstream for inlet/forebody integration. Sci. Technol. **74**, 160–172 (2018)
16. Sang, D.K., Song, D.J.: A numerical analysis on three-dimensional flow field in a supersonic bump-type inlet. J. Mech. Sci. Technol. **21**(2), 327–335 (2007)
17. Sang, D.K.: A numerical study on the effect of the flow control using bleeding systems in a bump-type inlet. J. Mech. Sci. Technol. **31**, 5821–5828 (2017)

18. Sang, D.K.: A computational study on the influence of a bleed slot on the flowfield of a bump-type inlet. J. Mech. Sci. Technol. **29**, 5245–5250 (2015)
19. McFarland, J.D.: Joint Strike Fighter Diverterless Supersonic Inlet, 2nd edn. National Press Club, USA (2000)
20. Xu, S.C., Wang, Y., Wang, Z.G.: The design and analysis of bump in supersonic flow. In: 21st AIAA International Space Planes and Hypersonics Technologies Conference, Xiamen, China (2017)
21. Svensson, M.: A CFD Investigation of a Generic Bump and its Application to a Diverterless Supersonic Inlet. Master's Dissertation, Swedish Defense Research Agency (2008)
22. Joongnyon, K., Hyung, J.S.: Wall pressure fluctuations in a turbulent boundary layer over a bump. AIAA J. **44**(7), 1393–1401 (2006)
23. Colliss, S.P., Babinsky, H., Nuebler, K.: Joint experimental and numerical approach to three-dimensional shock control bump research. AIAA J. **52**(2), 436–446 (2014)
24. Xu, S.C.: Design and Analysis of Hypersonic Inlet with Integrated Bump/Forebody. Master's Dissertation, National University of Defense Technology (2018)
25. Yu, Z.H., Huang, G.P., Chen, X.: Inverse design and Mach 6 experimental investigation of a pressure controllable bump. Aero. Sci. Technol. **81**, 204–212 (2018)
26. Yu, Z.H., Huang, G.P., Chen, X.: A pressure-controllable bump based on the pressure-ridge concept. Aero. Sci. Technol. **87**, 133–140 (2019)
27. Eiman, B.S., Huang, G.P., Qiao, W.Y.: Highly integrated inlet design based on the ridge concept. J. Propuls. Power. **32**(6), 1505–1515 (2016)
28. Xu, S.C., Wang, Y., Wang, Z.G.: Design and analysis of a hypersonic inlet with an integrated bump/forebody. Chin, J. Aeronaut. **32**(10), 2267–2274 (2019)
29. Xu, S.C., Wang, Y., Wang, Z.G.: Design method for hypersonic bump inlet based on transverse pressure gradient. J. Zhejiang Univ. Sci. **23**(6), 479–494 (2022)

Analysis of Cabin Floor Load Requirements for Civil Aircraft

Xuanzhe Tian(✉) and Congcong Li

State Key Laboratory of Airliner Integration Technology and Flight Simulation, Shanghai 200126, China
tianxuanzhe@comac.cc

Abstract. This paper analyzes the definition process of the cabin floor load requirements in the early aircraft design stage. Starting from introducing the current typical cabin floor structures, it analyzes the different load types on the floor structures. It introduces the scenarios of different load types, and gives the calculation process of some load types. The definition process of cabin floor load requirement discussed in this paper gives a reference to aircraft design engineers, helping them to better understand the load types and impact on the cabin floor structures.

Keywords: Cabin Floor Structure · Civil Aircraft · Requirement Definition

1 Introduction

The definition of cabin design requirements is a key aspect in the initial stage of aircraft design process. The key load bearing structures of cabin floor include cabin floor panels, seat rails, crossbeams, pillars, stringers and other supporting structures. The cabin floor shall withstand the load during normal flight conditions and emergency conditions. During the aircraft operation process, the objects inside the cabin, including passengers, passenger seats, galleys, lavatories, etc., will exert load on the floor structures. Therefore the scenarios of extreme loads must be analyzed to define an appropriate cabin floor load requirement to ensure structural safety and minimized weight. In addition, the cabin floor is part of the main load bearing structure of the entire aircraft. The load of the entire aircraft will be transferred to the floor cabin in different flight conditions. Since this type of load is not caused by objects inside the cabin, this paper will not analyze its impact on the cabin floor requirements in detail. However, the full aircraft load is a key aspect, and therefore must be taken into consideration when designing the cabin floor structures.

Weiwen has done research on the typical cabin floor structures and the load transfer analysis, which introduces the typical floor structure and different load types exerting on the cabin floor [1]. Much research has been conducted on the individual structure of the cabin floor. For example, Mingjing has conducted research on the application of composite material in aircraft [2]. Jun has conducted research on the seat rail structure [3] and Jieqiong has conducted research on the crossbeam structure [4]. However, there is still a lack of research on how to define appropriate load requirements of the whole

Y. Zhao (Ed.): ICAMM 2025, SPM 96, pp. 40–47, 2025.
https://doi.org/10.1007/978-3-032-09862-7_4

cabin floor structure in the early design phase to ensure structural safety and minimized weight. This paper will first introduce the cabin floor structures and typical load types. It will then analyze different cabin floor load scenarios and their impact on the cabin floor structures. Finally, this paper will give the process of how to define appropriate cabin floor load requirements.

2 Introduction of Cabin Floor Structures and Load Types

The main load bearing structures of the cabin floor includes floor panels, crossbeams, stringers, seat rails, pillar, etc. The typical cabin floor structure is shown in Fig. 1 [1].

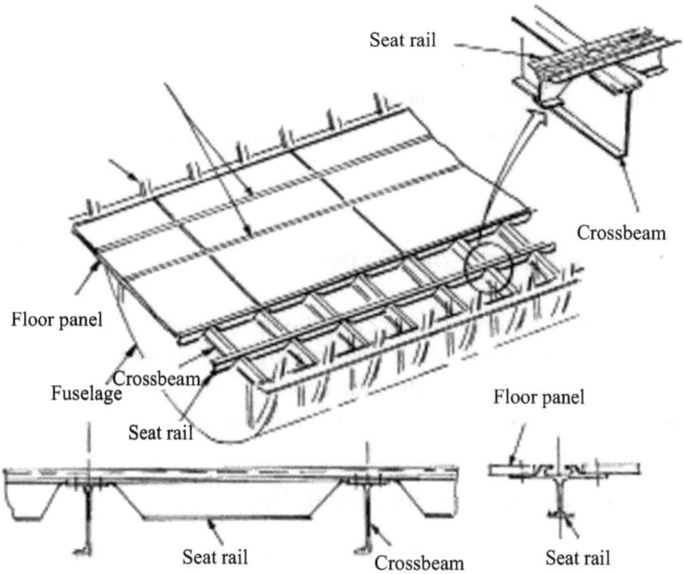

Fig. 1. Typical cabin floor structure [1]

Cabin floor panels are typically sandwich honeycomb structures using composite materials [2]. Floor panels shall withstand the local concentrated loads of items on the cabin floor, the rolling loads of luggage and food cart. The floor panels transfer these loads to the main load bearing structure. Seat rails are used to support facilities that are mounted on them such as seats, lavatories, galleys, etc. Seat rails shall withstand the loads from objects installed on it in extreme conditions. It shall also withstand uniform line loads in extreme conditions. Seat rails are usually made of aluminum or titanium alloys, and the cross section shapes varies with different aircraft [3]. Crossbeams are the major load bearing structure. They shall withstand the downward load caused by seats, passengers and food carts. They shall also withstand the pressurization load which stretches the crossbeams. Crossbeams are usually made of 7000 or 2000 series aluminum alloy, aluminum-lithium alloy and other composite materials [4]. The cross section shape mainly includes C-shape, Z-shape and I-shape [4].

The typical load types on cabin floor structures are shown in Table 1. These load types must be taken into consideration when defining appropriate cabin floor load requirements.

Table 1. Typical load types on cabin floor

Load Type	Caused By
Seat inertia loads	Inertia load in extreme conditions such as emergency landing
Uniform line loads	Seat and passenger vertical load in extreme conditions
Pressurization loads	Pressure difference between cabin and non-pressurized cabins
Decompression loads	Pressure difference due to sudden loss of pressure in pressurized cabins
Local concentrated loads	Interactions between occupants in the cabin and the cabin floor
Local rolling loads	Rolling objects such as food carts and luggage
Whole aircraft loads	Whole aircraft loads transferred to the cabin floor structure

3 Analysis of Different Load Types

When defining the load requirements of cabin floors, the possible scenarios of different load types must be analyzed first. The corresponding loads can be calculated or analyzed from the scenario analysis results. Then the appropriate requirement can be defined so that the floor structures satisfy the needs while minimizing structural weight.

3.1 Seat Inertia Loads in Emergency Conditions

Passenger seats are directly mounted on the seat rails and are not connected to fuselage frames. Besides passenger seats, other facilities such as closets, galleys, lavatories, etc. are also mounted on seat rails. However, they usually have smaller weight than the combined weight of passenger seats and passengers. Therefore, their loads on the cabin floor structures are usually smaller than that produced by passenger seats.

The seat and passenger exert a large inertia load in emergency conditions. According to CCAR25.785(f), each seat and its supporting structure must consider the maximum load factors, inertia forces, and reactions for each relevant flight and ground load condition including the emergency landing conditions prescribed in 25.561. The main scenario for extreme seat load is emergency landing. The ultimate inertia load factors required by CCAR 25.561 are shown in Table 2. Besides the load factors required in CCAR25.561, other extreme scenarios such as extreme maneuver, gusts, ground loads, etc. must also be taken into consideration, which depend on the design envelope of the aircraft. The seat loads should be the largest among all extreme conditions. In addition, according to CCAR25.785(f)(3), the inertia forces specified in 25.561 must be multiplied by a factor of 1.33 in determining the strength of the attachment of each seat to the structure.

Table 2. Typical Load types on cabin floor

Direction	Load factors in CCAR 25.561
Upward	3.0g
Forward	9.0g
Sideward	4.0g
Downward	6.0g
Rearward	1.5g

Passenger seats are mounted directly on seat rails, which is shown in Fig. 2. One typical three-passenger seat has four mounting points on two seat rails. The total mass of the system is the combination of passengers, passenger seats, passenger belongings and emergency equipment installed on the seats. The coordinates of passenger center of gravity and seat center of gravity can be chosen using a typical commercial aircraft seat. T

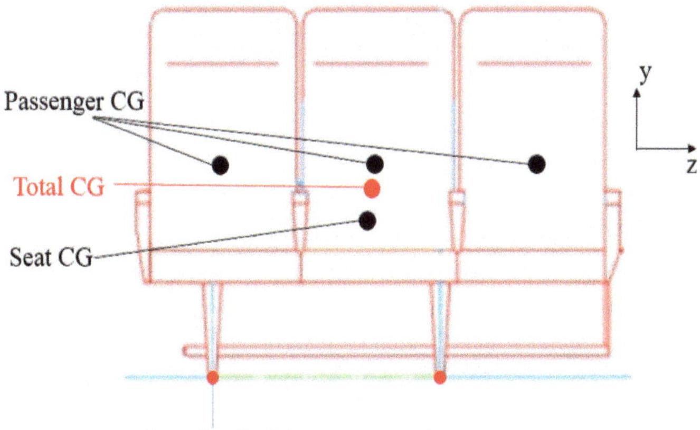

Fig. 2. Simplified seat mounting demonstration

The loads on mounting points can be calculated using the balance of force and moment equations. To give an example, loads under upward and downward load factors can be calculated using Eq. (1) to Eq. (4).

$$F_{a_{front}} = W_{total} \cdot n \cdot \frac{z_{cg} - z_b}{z_a - z_b} \cdot \frac{x_{cg} - x_{front}}{x_{aft} - x_{front}} \tag{1}$$

$$F_{a_{aft}} = W_{total} \cdot n \cdot \frac{z_{cg} - z_b}{z_a - z_b} \cdot \frac{x_{aft} - x_{cg}}{x_{aft} - x_{front}} \tag{2}$$

$$F_{b_{front}} = W_{total} \cdot n \cdot \frac{z_a - z_{cg}}{z_a - z_b} \cdot \frac{x_{cg} - x_{front}}{x_{aft} - x_{front}} \tag{3}$$

$$F_{b_{aft}} = W_{total} \cdot n \cdot \frac{z_a - z_{cg}}{z_a - z_b} \cdot \frac{x_{aft} - x_{cg}}{x_{aft} - x_{front}} \tag{4}$$

In the equations, F_a and F_b are the vertical loads on the two seat rails, W_{total} is the total weight of the system including passengers, seats, magazines, passenger belongings, etc. Passenger weight is 170 lbs. because CCAR 25.785(f) states that "Each seat or berth, and its supporting structure must be designed for an occupant weight of 170 pounds". cg is the center of gravity of the system, n is the most extreme load factor, x and z are the coordinates of the four mounting points and the center of gravity.

Similar equations can be deduced using the force and moment balance equations and therefore are not illustrated in detail in this paper. Note that the most extreme conditions are not necessarily when all three passenger are seated. All possible eight seating scenarios shown in Table 3 must be analyzed, where "1" stands for passenger seated and "0" stands for no passenger on that seat. Usually the extreme conditions are Case 1 to Case 4.

Table 3. Seating scenarios

	Seat A	Seat B	Seat C
Case 1	1	1	1
Case 2	1	1	0
Case 3	1	0	1
Case 4	0	1	1
Case 5	0	0	1
Case 6	0	1	0
Case 7	1	0	0
Case 8	0	0	0

3.2 Uniform Line Load

The seat load described in the previous section is usually exerted on the mounting point whereas the uniform line load is distributed along the seat rail. Aircraft cabin layouts are customized by airliners, and different airliners often prefer different layouts. As a result, the location of seats varies with different airliners. In the early design phase, a uniform line load requirement is defined to make sure the cabin floor structure can withstand loads in different layouts.

The equation to calculate line loads in shown in Eq. (5). P_L is the line load, $W_{passenger}$ is passenger weight. W_{seat} is the total seat weight in one row. L is the seat pitch.

$$P_L = \left(W_{passenger} \cdot n + W_{seat}\right)/L \tag{5}$$

The maximum line load occur when all 6 passenger are seated and the seat pitch is the minimum of all cabin layouts. CAAC gives the average passenger weight in its advisory circular AC-121-FS-135, which is shown in Table 4 [5]. An average passenger weight of 78 kg is used to calculate line loads. Additional weight margin is selected by engineers in case that the actual passenger weight exceeds the average weight. Seat weight varies largely with the seat configuration. In the early stage of aircraft design, the passenger seats are not selected, thus the design engineers must estimate the seat weight. The minimum seat pitch depends on the cabin layout selected by airliners. Typically for a single-aisle commercial aircraft, the minimum seat pitch is set to be 28 in. Note that the line load is in the vertical direction, thus the maximum vertical load factor used in the previous section must also be taken into consideration.

Table 4. Average passenger weight(luggage included [5]

Passenger type	Weight (lb)
Summer	
Male	192
Female	144
Average	168
Winter	
Male	197
Female	149
Average	173

3.3 Local Loads on Floor Panel

The cabin floor panel is the supporting structure in the cabin on the surface. In the aircraft operation process, passengers, luggage, food carts and other objects operate on the floor panels. The cabin floor panel is a sandwich honeycomb structure. If the local concentrated load is too large, the cabin floor panel might break.

EASA provides a guideline to select appropriate concentrated loads in its certification memorandum CM-S-009. The concentrated loads are shown in Fig. 3. The stepping force of occupants is 2200N and the application area is 10 cm*20 cm [6], and the resulting concentrated load is 110 kPa. Besides the guideline in the EASA memorandum, which is the minimum requirement for passenger safety, some more extreme scenarios might happen during flight. For example, a female whose weight is 70 kg steps on the floor panel with high heel shoes of 2 cm diameter. The resulting local concentrated load is 2.18Mpa, which is much larger than the EASA guideline. Therefore, the design engineer should analyze the possible scenarios of extreme concentrated loads and then define the appropriate concentrated load requirement.

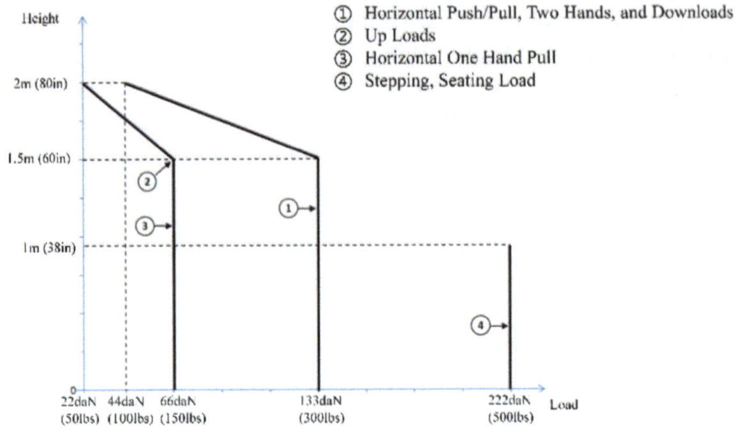

Fig. 3. Concentrated load types [6]

In addition, the cabin floor panel shall withstand the load of rolling items such as food carts and suitcases, which exert compression and shear loads onto the floor panels. The rolling load of a food cart is the largest since other rolling items normally have a much smaller weight. The floor panel shall withstand a rolling load of 100 pounds which is the standard food cart weight.

3.4 Pressurization and Decompression Loads

The pressurization load is caused by the pressure difference between the ambient pressure and the cabin pressure. During cruise stage, the aircraft is at the altitude between 35,000 ft. to 40,000 ft. The cabin pressure is kept at a lower pressure altitude level to keep passengers safe and comfortable. The pressure difference causes a pressurization load on the aircraft entire structure including the cabin floor structure.

The decompression load is caused by a sudden pressure drop in a pressurized cabin. The pressure difference causes the decompression load on cabin floor structures. For example, if the cargo cabin suddenly loses pressure due to penetration, there will be a sudden pressure difference between cabin and cargo bay, which will lead to a decompression load on the cabin floor structure.

The pressurization and decompression load are loads on the entire aircraft structure. It is very difficult in the early stage of aircraft design to calculate the correct pressurization and decompression loads on the cabin floor structures. Therefore, it is almost impossible to define the requirement load with exact numbers in the early design phase. However, the decompression and pressurization load must be taken into consideration when designing the cabin floor structures.

3.5 Other Whole Aircraft Loads

Besides the loads discussed in the previous chapters, the whole aircraft load also has a big impact on cabin floor structures. It is impossible to set a requirement with exact loads of the cabin floor structure since the structures of the entire aircraft are not designed completely in the early design phase. The cabin floor structures are usually considered to be part of the aircraft structure and are designed together with other structures.

4 Conclusion

This paper discusses the process of defining requirements for aircraft cabin floor structures in the early design phase. It first introduces the structures used in cabin floors. The different load types and their corresponding scenarios are then discussed. From the scenario analysis, the calculation processes of some load types are illustrated. Some load types are nearly impossible to define requirements with exact loads in the early design phase but they must be taken into consideration in the structural design process. This paper provides a reference for aircraft design engineers to make appropriate cabin floor load requirements in early design phase.

References

1. Zhang,W., SonglinZhang.: typical structures of aircraft floor and force transfer analysis. China Sci. Technol. Inf. **13**, 112 (2013)
2. Liu, M., Wu, Z.: Application of composite honeycomb sandwich structure in aircraft. Sci. Technol. Rev. **34**(3), 21–25 (2016)
3. Ma, J., Wang, S., Zhu, J.: Research on the structure of passenger cabin floor seat rail for civil aircraft. Civ. Aircr. Des. Res. **1**, 55–61 (2023)
4. Sun, J.: Research on cabin floor crossbeam structure for commercial aircraft. Civ. Aircr. Des. Res. **1**, 1–4 (2016)
5. CAAC AC-121-FS-135.: https://www.caac.gov.cn/XXGK/XXGK/GFXWJ/201911/t20191104_199320.html. Accessed 18 June 2025
6. EASA CM-S-009 Issue 01.: https://www.easa.europa.eu/en/document-library/product-certification-consultations/proposed-cm-s-009. Accessed 18 June 2025

Research on Drainage Design of Main Cargo Compartment of Civil Aircraft Passenger-to-Freighter

Yuhui Wang(✉), Jianjun Zhang, Zhou Zhou, Yang Wu, Chenlizi Zhang, Yanqiao Zhang, and Xueping Zhang

State Key Laboratory of Airliner Integration Technology and Flight Simulation, Commercial Aircraft Corporation of China Limited Company, Shanghai 200126, China
{wangyuhui,zhangjianjun,zhouzhou1,wuyang1,zhangchenlizi,zhangyanqiao,zhangxueping}@comac.cc

Abstract. The drainage design of main cargo compartment is of great significance to the safety of cargo aircraft. Therefore, the drainage design method of main cargo compartment is studied in this paper. First, through the analysis of the drainage scenario, the factors that need to be considered in the drainage design of the main cargo compartment are comprehensively identified. Secondly, the drainage path and drainage holes are optimized. The operation scenario of cargo aircraft determines the layout design of main cargo compartment drainage path. Rain/snow entering the main cargo compartment, water on the ULD (Unite Load Device) and water seepage of the cargo determine the drainage volume of the main cargo compartment. The drainage rate of the main cargo compartment is determined by the requirements of the cargo aircraft in different operating stages. In summary, combining with the engineering experience, this paper proposes a novel overall design method of the main cargo compartment drainage scheme of large cargo aircraft through theoretical analysis and test. The results show that the design method in this paper can meet the design requirements, and can provide a reference for the drainage design of cargo aircraft.

Keywords: cargo aircraft · main cargo compartment · overall design method · drainage design

1 Introduction

Water accumulation on aircrafts can cause corrosion to structures, equipment failure, and even catastrophic accidents [1–3]. How to quickly and reliably drain accumulated water is of great significance for improving aircraft operation safety [4–7]. For example, in the design of passenger aircraft, drainage holes will be arranged near the cabin door, and waterproof design will be carried out in the kitchen and lavatory area, etc. The cargo aircraft will design drainage facilities in the main cargo compartment. Through the analysis of the drainage schemes of passenger aircraft and cargo aircraft, three mainstream drainage methods are identified (as shown in Table 1), and it can be seen that the current

Y. Zhao (Ed.): ICAMM 2025, SPM 96, pp. 48–59, 2025.
https://doi.org/10.1007/978-3-032-09862-7_5

schemes cannot simultaneously take into account the requirements of high reliability, low leakage area and large displacement.

Table 1. Existing or potential drainage schemes.

No.	Scheme	Advantage	Weakness	Application
1	The drainage method of the gray water interface valve is installed.	(1) Reduce leakage area; (2) Improve cabin comfort	(1) Easy to block and leak; (2) Large space occupation	Aircraft washroom
2	The Drainage method of connecting to the outside of the aircraft directly	(1) Simple form; (2) High reliability, not easy to plug and leak	Increase the leakage area	Main cargo compartment
3	The drainage method of establishing a storage tank in the belly compartment.	(1) Reduce the leakage area of the aircraft; (2) Improve the comfort of the cabin.	(1) High requirements for structural anti-corrosion; (2) Low drainage volume.	Electronic cabin

Therefore, through the analysis of the drainage scenario, this study identifies that there is no requirement of high-altitude drainage in the front of the main cargo compartment. By installing a suitable check valve in the front pipeline of the main cargo compartment, the increase of leakage area can be avoided and the comfort of the cabin can be improved. By raising the height of the bottom of the ventilation grille on both sides of the main cargo compartment, the problems of water leakage prevention on both sides of the main cargo compartment are solved. By installing a structural check valve under the main cargo compartment door, the problems of drainage and high-altitude gas leakage in the main cargo compartment door area are solved.

There are currently no airworthiness regulations regarding the requirements for aircraft drainage volume and drainage rate, and there is very little research on this aspect. The drainage volume and drainage rate of the main cargo compartment are the key factors affecting the number, size and layout design of the drainage holes. At the same time, the design of the drainage holes also needs to consider the design characteristics of the aircraft structure and system. In order to solve the design problem of the drainage holes, this study identified the rainwater/snow water entering the main cargo compartment door area, water accumulation on the surface of the ULD and water seepage of the cargo as the main sources of water in the main cargo compartment. Define the drainage requirements of the cargo aircraft on the ground and in the air, and proposed a calculation model of the drainage volume and drainage rate of the main cargo compartment. According to the design characteristics of aircraft, a method of layout design of drainage holes is presented. The size, quantity and layout of the drainage holes are optimized, which has the characteristics of easy design, easy modification and good economy. The feasibility of the method is proved by the test and engineering application, and it can meet the design requirements.

2 Main Cargo Compartment Drainage Design Considerations

2.1 Analysis of Main Cargo Compartment Drainage Scenarios

From the analysis of operation stage, there will be water in the main cargo compartment on the ground and in the air. The main cargo compartment shall be capable of ground and air drainage ability.

When on the ground condition, the aircraft ground angle is generally negative. Affected by gravity, the water inside the main cargo compartment flows to the front through the floor of the main cargo compartment and the diversion devices on both sides of the floor. In order to ensure the effective drainage of the water, a drainage path should be designed in the front of the main cargo compartment. When in the air, the aircraft usually has an elevation angle, and due to the influence of gravity, the water inside the main cargo compartment flows to the rear through the main cargo compartment floor. In order to ensure effective drainage of water, drainage channels should be designed in the rear of the main cargo compartment.

The drainage conditions under different posture are shown in the Fig. 1.

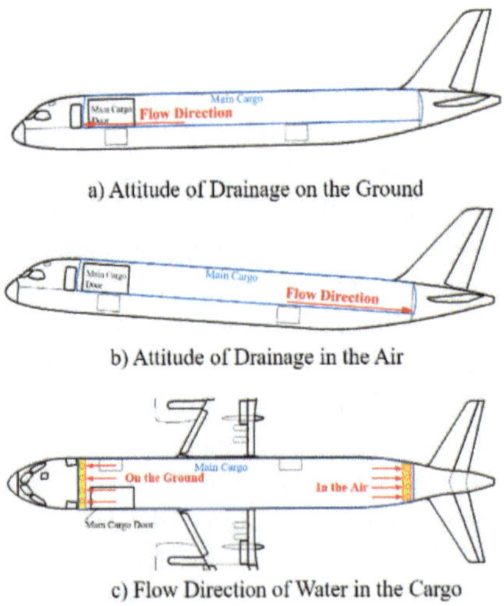

Fig. 1. Aircraft drainage attitude on the ground and in the air.

2.2 Requirements for Drainage Design of Cargo Aircraft

Main compartment drainage design shall meet relevant airworthiness requirements such as FAR 25.609 "Protection of structure", FAR25.841 "Pressurized cabins", FAR25.843

"Tests for pressurized cabins", FAR25.1455 "Draining of fluids subject to freezing" etc. [4]. In addition, when carrying out scheme design, factors such as drainage rate, aircraft attitude, equipment protection and waterproofing [3], and wet area design need to be comprehensively considered. The drainage design requirements for the main cargo compartment of cargo aircraft are shown in Table 2.

Table 2. Drainage design requirements.

No.	Design requirements	Source
1	Corrosion protection requirements of the structure	Airworthiness requirement
2	Leakage area requirements	Airworthiness requirement
3	Discharge requirements for easily frozen liquids.	Airworthiness requirement
4	Requirements for drainage rate	Aircraft design requirements
5	Accessibility, reliability and maintainability requirements	Aircraft design requirements
6	Requirements for wet area protection	Aircraft design requirements
7	Water leakage prevention requirements on both sides of main cargo compartment floor	Aircraft design requirements

The design requirements in Table 2 are described as follows:

- It shall be able to meet the corrosion prevention requirements of the structure along the drainage path [4].
- It shall be able to meet the requirement that the cabin height change rate does not exceed 2000 ft./min during emergency pressure relief [4, 8].
- It shall be able to avoid dangerous amounts of ice formation on the aircraft during drainage [4].
- It should be able to meet the requirements of drainage rate on the ground and in the air.
- It should have high reliability, good accessibility and maintainability.
- The main cargo compartment should be designed according to the waterproofing requirements of the wet area, and the drain pan shall be provided in the electronic compartment.
- In order to reduce water flow into the lower part of the floor, the lower height of the ventilation grille on both sides of the main cargo compartment should be raised.

3 Drainage Design Method for the Main Cargo Compartment of a Large Cargo Aircraft

The layout design of drainage scheme of the main cargo compartment needs to consider five aspects: drainage path design, drainage hole design (including size, quantity, layout, etc.), drainage of the main cargo door area, water leakage prevention on both sides of the main cargo compartment and waterproof of the electronic compartment. The waterproof design of the electronic compartment is generally to install a drain pan above the equipment and drain the water to the bottom of the fuselage through the drainage pipe.

Considering the above factors, a drainage design scheme for the main cargo compartment is proposed in this study. The drainage path is designed. A calculation model of water drainage volume and drainage rate is proposed, and the optimal design is carried out by calculating and analyzing the diameter and arrangement of drainage holes. At the same time, combined with the design characteristics of the aircraft, the size, quantity and layout of the drainage holes are determined, and the verification is carried out on the aircraft.

Figure 2 shows the schematic of the front/rear drainage path of the main cargo compartment, drainage of the main cargo door area, water leakage prevention on both sides, and drainage holes layout.

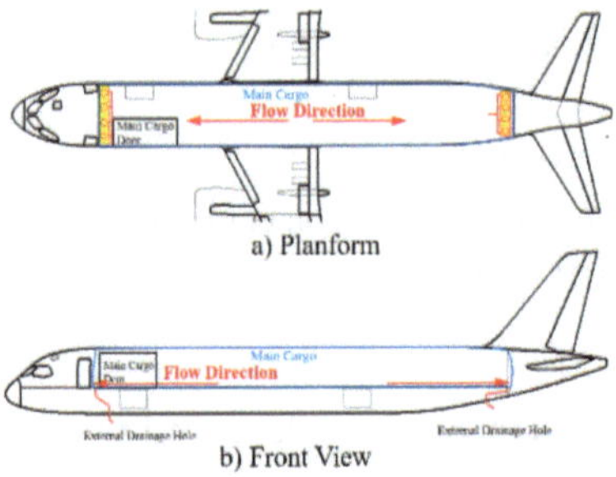

Fig. 2. Main cargo compartment drainage schematic.

3.1 Study on Design Method of Drainage Path in Main Cargo Compartment Area

The water in the main cargo compartment flows under the action of gravity and is discharged from the drainage pipe. In order to achieve a good drainage effect, the drainage path of the main cargo compartment should have the following characteristics.

- Drainage channels should be arranged in front and rear of the main cargo compartment. The drainage holes should be arranged as close as possible to the front and rear walls of the main cargo compartment.
- Considering the arrangement of freight rail, the drainage hole is arranged between the two rails. Generally, at least one drainage hole is arranged between the two freight rails.
- In the area of the front and rear drainage holes, a collection tank should be designed to collect and guide the accumulated water in the main cargo compartment for easy drainage.
- Considering that the water in the middle area of the main cargo compartment is discharged from the aircraft through the front/rear water holes, there should be a certain gap between the side rails of the cargo system and the interior side panels of

the main cargo compartment to facilitate the water flow to the front/rear of the main cargo compartment.

- Considering the waterproof requirements of the device, drainpipes and connectors should not pass above the device or adequate protective measures should be taken to prevent water dripping on the device.

In order to reduce the increase in the leakage area of the aircraft and improve the comfort, a check valve is installed in the front drain pipe of the main cargo compartment to ensure that the front drain pipe has the drainage capacity on the ground and at low altitude, and is closed at high altitude. If there is a device under the drainpipe, set a drain pan to prevent leaking water from dripping onto the device. The rear drainage pipe should ensure that it is constantly connected to the outside of the aircraft and has the ability to discharge water at high altitude. The drainage path in the front and rear of the main cargo compartment is shown in Figs. 3 and 4.

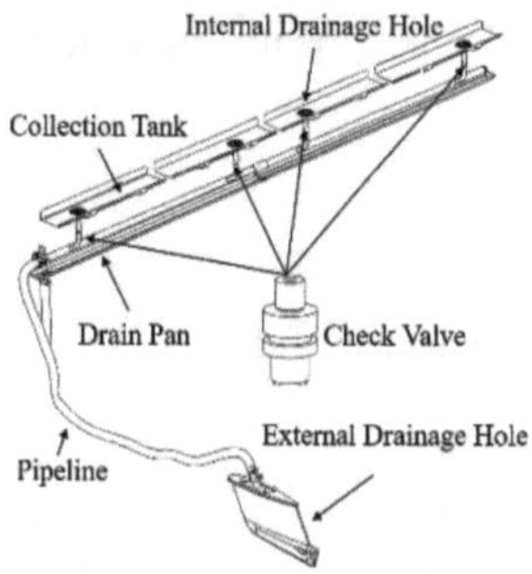

Fig. 3. Schematic of drainage path in front of main cargo compartment.

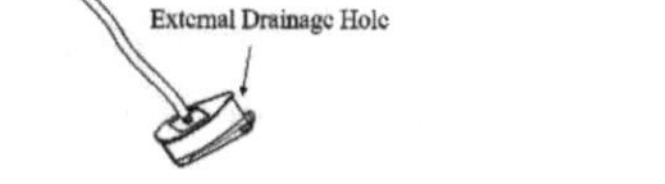

Fig. 4. Schematic of drainage path in the rear of main cargo compartment.

3.2 Design Method for Layout of Drainage Holes

The water in the main cargo compartment flows under the action of gravity and is discharged from the drainage pipe. In order to achieve a good drainage effect, the drainage path of the main cargo compartment should have the following characteristics.

Main Cargo Compartment Drainage Volume Calculation Model. The water in the main cargo compartment mainly comes from the water on the surface of the ULD, the rainwater/snow water entering from the main cargo door area, and the water seepage of the cargo itself. The drainage volume of the main cargo compartment is shown in Eq. (1).

$$V = V_1 + V_2 + V_3 \tag{1}$$

In the equation
V -Total drainage volume.
V_1 – Volume of water on the surface of the ULD.
V_2 - Volume of Rain and snow water entering from the main cargo door area.
V_3 - Volume of water seepage from cargo.

The calculation of the volume of water on the surface of the ULD is shown in Eq. (2). The water accumulation area on the surface of the ULD is determined based on the type of the ULD. As defined in the Unite Load Device Regulations [9], the maximum top area of AAY (a type of ULD) is 2.76 m^2, and the water accumulation height can generally be selected as 1 mm.

$$V_1 = S_1 h_1 \tag{2}$$

In the equation
V_1 - Volume of water on the surface of the ULD.
S_1 - Top area of the ULD.
h_1 - Water accumulation height.

The rain and snow entering from the main cargo door area is calculated by the equivalent precipitation coverage area and rainfall in the main cargo door area. The equivalent precipitation coverage area is mainly related to the size, the opening angle and the closing time of the main cargo door. Rainfall is selected in extreme cases. According to the records of MIL-HDBK-310 [10], the extreme rainfall rate is 7.25 mm /min, and the rainfall time is selected for the duration of closing the main cargo door. Usually, the closing time of the main cargo door is 1 min. The calculation of rain and snow water volume entering from the main cargo compartment area is shown in Eqs. (3), (4)

$$V_2 = S_2 h_2 \tag{3}$$

In the equation
V_2 - Volume of rain and snow water entering from the main cargo door area.
S_2 - Equivalent precipitation coverage area of main cargo door area.
h_2 - Rainfall.

$$h_2 = rt \tag{4}$$

In the equation
h_2 - Rainfall.
r - Rainfall rate.
t - Rainfall time.

The amount of water leaking from cargo during flight is related to the cargo itself. The maximum water seepage of each ULD can generally be selected as 10 L.

Drainage Rate Calculation Model. When the aircraft is on the ground, the maximum amount of liquid that may occur in the main cargo compartment is required to be drained through the forward drainage path during loading time. The drainage rate in the front of the main cargo compartment is shown in Eq. (5).

$$Q_1 = \frac{V_1 + V_2}{t_1} \tag{5}$$

In the equation

Q_1 – Drainage rate on the ground.

V_1 - Volume of water on the surface of the ULD.

V_2 - Volume of rain and snow water entering from the main cargo door area.

t_1 – Loading time.

In the air, the aircraft is required to discharge the maximum amount of liquid that may occur in the main cargo compartment through the rear drainage path during the time between takeoff and enter cruising altitude. The drainage rate at the rear of the main cargo compartment is shown in Eq. (6).

$$Q_2 = \frac{V_1 + V_3}{t_2} \tag{6}$$

In the equation

Q_2 – Drainage rate in the air.

V_1 - Volume of water on the surface of the ULD.

V_3 - Volume of water seepage from cargo.

t_2 – The time between takeoff and enter cruising altitude.

In the initial stage of scheme design, Bernoulli equation can be used for preliminary estimation. However, it should be noted that in actual drainage, the drainage rate will be reduced due to different pipe joints, filters, pipe layout paths, and sizes, et al.

Bernoulli's equation is shown in Eq. (7).

$$\frac{1}{2}mv_1^2 + mgh_1 + P_1\Delta V_1 = \frac{1}{2}mv_2^2 + mgh_2 + P_2\Delta V_2 \tag{7}$$

Assuming that the velocity of water at the inlet is approximately 0 and the pressure at the inlet and outlet is equal. Ignoring the viscosity of the liquid, the outlet water drainage volume rate is shown in Eq. (8).

$$Q = Sv = \frac{1}{4}\pi d^2\sqrt{2g\Delta h} \tag{8}$$

In the equation

Q – Drainage volume rate.

d - Diameter of the drainpipe.

g - Gravitational acceleration.

Δh - Equivalent height difference.

For the design with structural check valve installed in the drainage pipe, the inner diameter of the check valve should be selected as "d". When calculating the height difference, select the distance from the floor surface to the outlet of the check valve as "Δh".

For example, the inner diameter of a check valve installed on an aircraft is 5 mm, i.e., the "d" is 5 mm. The distance between the floor surface and the check valve outlet is the equivalent height difference, i.e., the "Δh" is 100 mm. The drainage rate is 1.72 L/min calculated by Eq. (8). For the drainage rate test, a drainage hole is selected, and 1 L of water is injected at a constant speed, and the time of complete discharge is measured. The test shows that the drainage rate of a single check valve is 0.58 L/min. The test operation on the aircraft is shown in Fig. 5. The difference between the theoretically calculated rate and the measured rate is corrected by the loss coefficient. In this model, the coefficient can be approximated as $C = 0.58/1.72 = 0.34$. The causes of the loss are the flow loss caused by the main hole dispersed to a number of small holes inside the check valve, the resistance along the pipeline, etc.

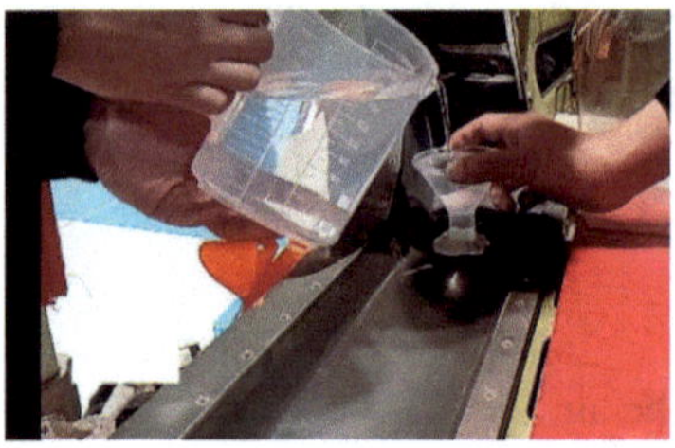

Fig. 5. Drainage rate test of check valve

Layout Design of Drainage Holes in the Front and Rear of Main Cargo Compartment. Based on the aforementioned computational model, assume that the main cargo compartment of the aircraft is loaded with 5 AAY containers, the main cargo door's closing time is 1 min, the equivalent precipitation area of the main cargo door area is 3.85 m^2, and the drainage amount is calculated and analyzed as follows.

- On the ground condition. According to Eq. (2), the maximum amount of water on the top of a ULD is 13.8 L. Suppose that heavy rain suddenly falls before the main cargo door begins to close after the ULD is loaded. According to Eq. (3), the rainfall during the door closing time is 27.9 L. According to Eq. (5), the maximum drainage volume of the aircraft on the ground is 41.7 L.
- In the air condition. The total water leakage of the cargo is 50 L. According to Eq. (6), the maximum drainage volume of the aircraft in the air is 63.8 L.

The drainage rate is analyzed as follows.

- On the ground, the aircraft is required to discharge the maximum amount of water that may occur in the main cargo compartment through the front drainage path during loading and unloading time (generally within 30 min). According to Eq. (5), the drainage rate is required to be 1.39 L/min.

- In the air, the aircraft is required to discharge the maximum amount of water that may appear in the main cargo compartment through the rear drainage path during the time between takeoff and cruising (generally within 20 min). According to Eq. (6), the drainage rate is required to be 3.19 L/min.

The CV92-14 check valve is installed in the front drainage pipe. According to the test data of the drainage rate of the CV92-14 check valve, the number of front drainage holes should be set at least 3. Considering the number of rails on the floor of the main cargo compartment, set 4 drainage holes in the front and the maximum drainage rate is 2.32 L/min.

For the drainage pipe in the rear of the main cargo compartment, the diameter of the drainage pipe is about twice that of the front for the check valve is not added. According to (8), the drainage rate is about 4 times as that in the front. Considering the number of freight rails, 4 drainage holes are arranged in the rear, and the maximum drainage rate is about 9.28 L/ min. The schematic of layout and quantity of thc drainage holes are shown in Figs. 3 and 4.

3.3 Other Considerations

Main Cargo Door Area Drainage Design. Due to the structural design features, water in this area cannot flow through the floor to the front/rear drainage pipe.

In order to solve the drainage problem in this area, the drainage scheme in the main cargo door area should be designed. Considering there's no need for drainage at high altitude and the requirement of reducing the leakage area of the aircraft, the structural drainage valve is installed in the drainage hole. The structural check valve is capable of draining at ground and low altitude and closing at high altitude. To ensure the drainage effect, arrange two drainage holes at least in this area. Fig. 6 shows the drainage diagram in the main cargo door area of the aircraft.

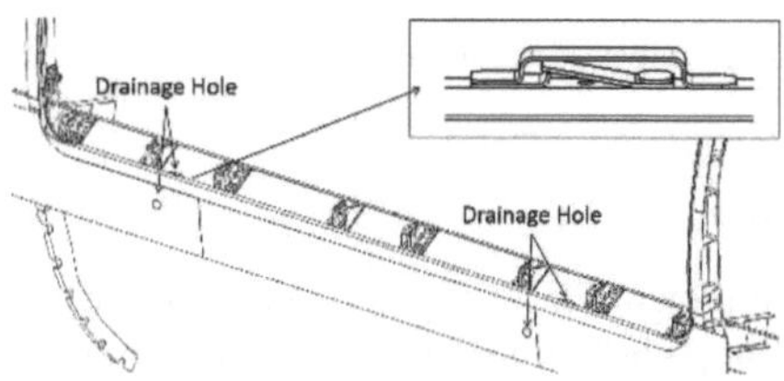

Fig. 6. Schematic of the drainage in the main cargo door area

Analysis of Water Leakage Prevention Design on Both Sides of Main Cargo Compartment. To solve the problem that water in the main cargo compartment flows into the lower part of the floor through the holes on both sides of the main cargo compartment, which may cause equipment failure and structural corrosion, et al. A drainage path should be designed between the side rail of the cargo system and the ventilation grille of the main cargo compartment to facilitate the water to the front/rear drainage holes and discharge out of the aircraft. At the same time by raising the height of the ventilation grille. The distance between the bottom of the ventilation grille and the floor is generally not less than 2 in. to prevent water from flowing into the belly compartment when the aircraft rolls/bumps.

3.4 Test Method for Drainage Rate

In order to verify the feasibility of the scheme, a test method is proposed and tested on the aircraft. Figure 7 is the schematic of drainage rate test and Fig. 8 is the real site of the drainage rate test on the aircraft.

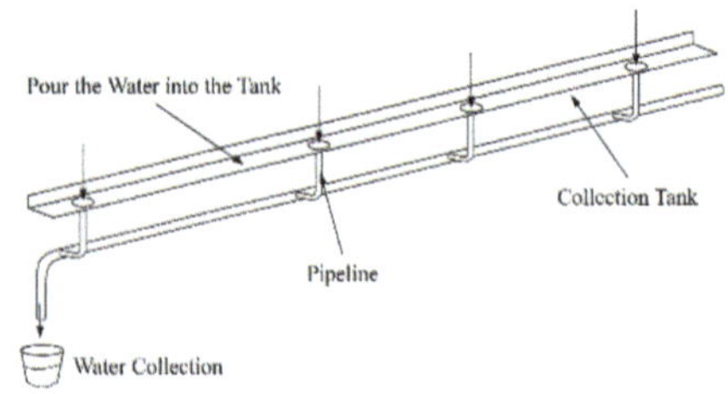

Fig. 7. Schematic of drainage rate test

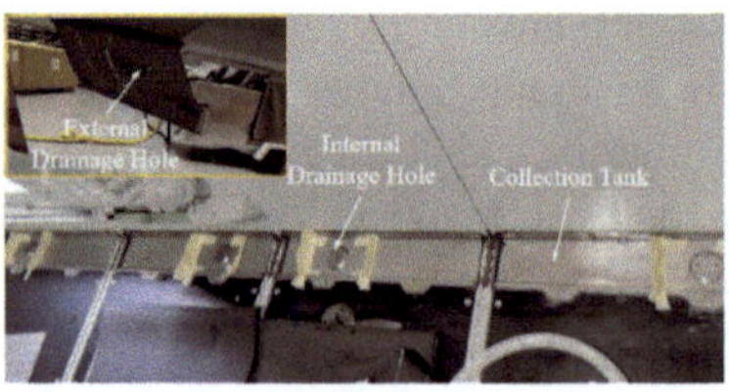

Fig. 8. Drainage rate test of the main cargo compartment drainage system

The test method for drainage rate is as follows.

- Pour a sufficient amount of water into each row of drainage holes, and the water level is higher than the drainage hole filter in the collection tank.
- During the drainage process, maintain a sufficient amount of water in the collection tank area.
- The water receiving device is set under the drainage hole of the exterior of the aircraft, and the timing starts when the water flows out.
- Record the drainage volume within 5 min.

The drainage rate test results show that the front drainage rate is greater than 1.39 L/min, and the rear drainage rate is greater than 3.19 L/min. The sealing and drainage rate meet the requirements and the drainage scheme is feasible.

4 Summary

There are many factors involved in the study of the drainage volume and drainage rate of the main cargo compartment, and the design of the drainage scheme of the main cargo compartment is also a comprehensive design problem under multiple constraints, which needs to take into account the factors such as aircraft operation scenario, drainage path, leakage area of the aircraft, structural drainage design, pipeline design and waterproof of equipment, etc. Under the condition of fully taking into account these factors, the design characteristics of the modified platform itself should also be combined.

Based on the analysis and study of the design of the main cargo compartment drainage scheme, this paper proposes a drainage design method for the main cargo compartment, as well as the calculation model of the drainage volume and drainage rate, and verifies the rationality of the method through experiments, aiming at providing reference and calculation ideas for the drainage design of cargo aircraft, which has important application value.

References

1. Kant, R., Chauhan, P.S., Bhatt, G., Bhattacharya, S.: Corrosion monitoring and control in aircraft: a review. Sens. Automot. Aerosp. Appl., 39–53 (2018)
2. Wang, Z.: Research on ‘GJB5431-2005 the Design Requirements for Waterproofing and Drainage of Aircraft Structures’ GJB5431-2005, vol. 2, pp. 41–51 (2013)
3. DO-160G Environmental Conditions and Test Procedures for Airborne Equipment, RTCA, Washington, DC, the Unite States (2010)
4. FAR Part 25 Airworthiness Standards Transport Category Airplanes, FAA
5. AC 25.863-1 Draft Flammable Fluid Fire Protection, FAA (2002)
6. Yue, N., Li, Y., Bai, M., Hou, S.P.: Flammable fluid fire protection airworthiness de-sign and verification method of civil transport aircraft. Procedia Eng. **80**, 110–118 (2014)
7. Zhang, X.P., Lei, M.L., Zhu, C.: The method of civil aircraft water and waste system icing-protection design. Adv. Aeronaut. Sci. Eng. **7**, 106–111 (2016)
8. ARP 1270B Aircraft Cabin Pressurization Criteria, SAE (2010)
9. Unite Load Device Regulations (ULDR), IATA (2024)
10. MIL-HDBK-310 Global Climatic Data for Developing Military Products, Department of Defense, Unite States (1997)

Civil Aircraft Dismantling Process

Jiayu Gu(✉)

COMAC Shanghai Aircraft Design and Research Institute, Shanghai, China
gujiayu1@comac.cc

Abstract. With the rapid development of civil aviation industry, the number of aircraft decommissioning continues to rise. With the gradual industrial development and large-scale operation of COMAC C919, the demand for aircraft dismantling business in China will grow rapidly. Aircraft dismantling industry can produce huge economic benefits, but the domestic start is late and lack of experience. This paper comprehensively considers economic benefits, laws and regulations, safety requirements and environmental protection to establish a general aircraft dismantling process, in order to promote and standardize the development of domestic dismantling industry.

Keywords: Disassembly · Dismantling · Decommissioning · Recycling

1 Introduction

COMAC releases "Commercial Aircraft Corporation of China Market Forecast Annual Report (CMF) (2022–2041)". It forecasts 42,428 jetliners of all categories will be delivered worldwide, In addition, new freighters and passenger-to-freighter aircraft delivery will be 2,991. By 2041, China's fleet size will reach 10,007 aircraft, accounting for 21.1% of the global passenger aircraft fleet. China's aviation market will become the world's largest single aviation market. Besides, in the past two years, the development speed of the domestic large aircraft C919 has far exceeded expectations, from development and production to delivery and mass production, and then to fleet operation and multi-airline operation. In particular, the three major airlines, China Eastern Airlines, Air China and China Southern Airlines, soon appeared to operate the 919 at the same time. According to IATA statistics on retired airplanes from 1980–2017, the proportion of passenger airplanes was 82%, and the average age of retirement was 25.9 [1]. This means that twenty years later, a large number of domestic C919 airplanes will face the problem of disposal after retirement.

Foreign aircraft dismantling industry is developing rapidly and the technology is mature. Boeing led the establishment of AFRA (Aircraft Fleet Recycling Association) and released the Best Management Practice (BMP) for guiding the aircraft dismantling [2]. Airbus established a complete aircraft dismantling process and standards through the actual implementation of a series of aircraft dismantling projects. The domestic dismantling industry started late, and there is no common aircraft dismantling process specification to provide operational guidance. The CAAC issued the first AC-145-FS2019-017

Y. Zhao (Ed.): ICAMM 2025, SPM 96, pp. 60–68, 2025.
https://doi.org/10.1007/978-3-032-09862-7_6

"Aircraft Dismantling" on aircraft dismantling in 2019 [3], which supplements the relevant content of aircraft dismantling on the basis of AC145, and another regulation related to aircraft dismantling is AC-120-FS-058R3 "Qualified Aircraft Material", which mainly covers the control requirements for dismantled parts of aircraft [4]. At present, aircraft dismantling work in China is mainly carried out according to CCAR-145 [5] and these two AC, but the detailed dismantling process is not given.

Throughout the domestic research, most of them also just stay in the general overview and experience summary of the Aircraft Fleet Recycling Association (AFRA) and Airbus PAMELA (Process for Advanced Management of End-of-Life of Aircraft) project [6], without forming a complete set of dismantling process, making a clear comb of the dismantling stage, and focusing on the key steps in the dismantling stage, so this paper comprehensively considers economic benefits, laws and regulations, safety requirements and environmental protection and proposes a referable aircraft dismantling process to disassemble recyclable parts as much as possible and increase the value of retired airplanes as well as minimize harm to people and the environment.

2 Factors and Principles

Through the study of AFRA BMP, Airbus' PAMELA, IATA BIPAD and CCAR-145, generally speaking, aircraft disassembly and dismantling refers to the use of professional equipment and technology to dismantle and divide retired aircraft to realize the whole into zero. After dismantling, the parts and aeronautical materials obtained after maintenance, testing and certification can be returned to the aviation materials market or recycled in other industries.

According to experience, aircraft dismantling follows the principle of first dismantling the whole aircraft into assemblies, dismantling the assemblies into components, and then dismantling the components into parts. Some equipment is sold directly. Some equipment and parts can be repaired and certified for sale. The metal, carbon fiber and other materials on the aircraft are utilized in a high-value way through new technologies.

Civil aircraft dismantling mainly involves four aspects: economic benefits, laws and regulations, safety requirements and environmental protection.

1) **Economic benefits**: Economic benefits relate to fleet planning, maintenance, aircraft value and parts supply and demand. An aircraft technology may not have reached the design life, but the operation is not as fuel-efficient as the new generation of models, and maintenance of important components such as engines, APU will produce high maintenance costs. So the benefit of continuing to operate the aircraft is less than the profit obtained by dismantling, then it is necessary to consider dismantling. When selecting the disassembly facilities, it is also necessary to consider transportation costs, storage and disassembly costs, import taxes, etc. Additional economic benefits can also be derived from the conversion of obsolete aircraft parts, which have zero value or even disposal costs, into new products with added value after all parts are dismantled for reuse for aerospace purposes. According to ARFA, the continued airworthiness of high-value components accounts for approximately 90% of the value of aircraft obsolescence. The weight of recyclable materials accounts for about 50% to 60% of the weight of the aircraft, but the recycling value only accounts for 10% of

the value of the aircraft scrap. Due to the high cost of disposal of hazardous substances and other wastes, additional costs are often required and have a negative effect on economic returns, as shown in Table 1.

Table 1. Weight and economic value of the parts of the aircraft when dismantled (Source: AFRA BIPAD)

Proportion	High-value components	Recyclable materials	Hazardous substances and waste
Aircraft empty weight	10–15%	50–60%	20–30%
Scrap value	90%–95%	10%	−10%–15%

2) **Laws and regulations:** Laws and regulations refer to standards and regulations established by national or international bodies, tax regulations and ownership rights protection regulations. The noise and carbon dioxide emission standards proposed by ICAO could speed up aircraft owners' decision to decommission. Other national civil aviation administrations have age restrictions on imported aircraft. In addition, when parts are dismantled for sale, it is easier to obtain maintenance and certification in the same aviation jurisdiction. Therefore, before choosing a dismantling site, it is also necessary to fully understand national and international tax policies. Reusable parts may only be removed by a maintenance organization approved by the national Civil aviation Authority, subject to compliance with applicable aviation regulations. In China, it must comply with the requirements of CCAR-145.
3) **Safety requirements:** Safety requirements include parts airworthiness, quality management, equipment certification, employee qualifications, health and safety requirements, etc. Regardless of whether the aircraft has been involved in a previous accident or not, before the disassembled parts can be sold, the engine or other parts must be cleared of the effects of the accident and can continue to be used safely. Unauthorized or counterfeit parts may not be re-entered into the aviation market. Airline operators, licensed aircraft maintenance engineers, maintenance facilities and other civil aviation participants are responsible for ensuring that components used on civil aircraft meet the airworthiness standards set out in Annex 8 to the Chicago Convention. At the same time, both the FAA and EASA have developed quality and safety guidelines and policies to ensure the safety of personnel and the level of quality during the dismantling process. AFRA also recommends the implementation of globally recognized health and safety Standards such as ISO 45001 and Aerospace Standards (AS) based on ISO9001.
4) **Environmental protection:** Environmental protection refers to the disposal and recycling of hazardous substances and materials during aircraft dismantling. AFRA estimates that about 700 aircraft a year have been dismantled and recycled after retirement. As more and more aircraft are taken apart, the growing amount of waste becomes a big problem if it is not done in a sustainable way to minimize the damage to the environment. Hazardous wastes (such as asbestos or depleted uranium ore),

noise, greenhouse gas emissions, wastewater, etc. during the dismantling process should be strictly controlled and discharged, managed, and disposed of according to the requirements of the local jurisdiction. Before disassembly, you can identify all hazardous substances on the aircraft through the OEM (Original Equipment Manufacturer) document, figure out whether risks such as overflow, fire, and explosion occur when using tools, and keep track of the final destination of hazardous substances. ISO14001 is the most widely used environmental standard, not only for the aviation industry, but also for airlines, airports, OEMs, MROs, dismantling companies, recycling companies and distributors.

From the above analysis, the basic principles of civil aircraft disassembly can be summarized as follows:

- The dismantling process follows industry practices and regulatory requirements, especially civil aviation regulations.
- In order to ensure safety and airworthiness, the dismantling of the aircraft (aviation-related parts) should be done by a maintenance organization or specialized dismantling enterprises recognized by CCAR Part 145.
- Safety must come first at all times.
- Compliance with the requirements of the quality management system ISO 9001.
- Comply with the requirements of the environmental management system ISO 14001.
- As far as possible, the dismantling process does not damage valuable used parts.
- Historical information such as delivery manuals and maintenance and utilization records of dismantled aircraft need to be complete.

3 Specific Steps for Civil Aircraft Dismantling

Combined with the dismantling of Boeing 747, the general process of civil aircraft dismantling was established in accordance with Best Management Practice of the Aircraft Fleet Recycling Association (AFRA) and Best Industry Practices for Aircraft Decommissioning of the International Air Transport Association (IATA). Each of these steps takes into account the four factors mentioned in Sect. 2: economic benefits, laws and regulations, safety requirements and environmental protection, as shown in Fig. 1.

Selection of Storage Sites and Dismantling Areas

Arid areas are usually chosen for facilities that store aircraft. Dry climates, low-salt air (far enough from the sea) and small temperature differences between day and night are ideal conditions for aircraft storage. The climate not only affects the structure of the aircraft (corrosion and delamination), but also the way in which the removed components are stored. Parts with magnetic properties, for example, need to be stored in a way that keeps the storage facility clean, well ventilated, and at a constant drying temperature to minimize the effects of condensation. Ideal storage temperatures for magnetic parts are between 15 °C and 25 °C and ideal storage humidity is between 40% and 60%. Also storage racks should be strong enough to provide adequate support for large aircraft parts so that the parts do not deform during storage. All aircraft components should follow the component manufacturer's guidelines and be packaged in protective material to minimize damage and corrosion. Moisture monitoring shall be performed using a

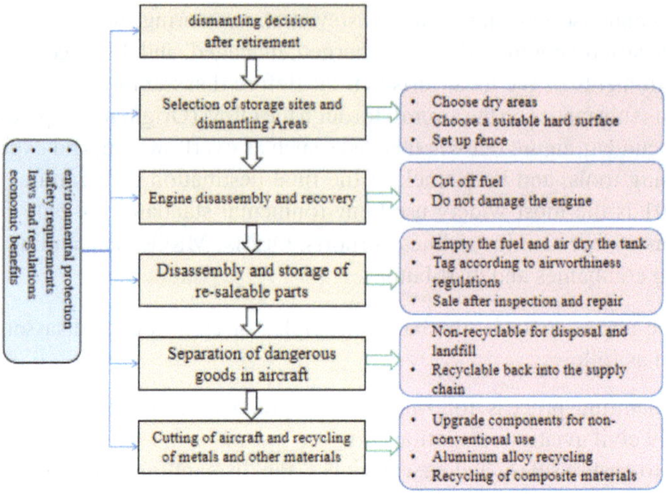

Fig. 1. General Process for Dismantling Civil Aircraft

hygrometer to record data on a daily basis. Rubber parts should not be exposed to bright light (e.g., direct sunlight) and storage rooms should be kept as dark as possible. Electronic turntables and circuits of electrostatic discharge sensitive equipment can be easily damaged by electrostatic discharge if they are not properly protected and handled, and therefore must be individually packed in anti-static bags. Electrostatic precipitators should not be stored on shelves covered with carpet, Styrofoam, vinyl, or any other material that can store or generate an electrical charge.

When using mobile equipment to dismantle an aircraft, a suitable dismantling area must be provided. At a minimum, this area needs to have an enclosed hard surface for dismantling activities. In most cases part of the operational airfield can be used as a dismantling area and therefore it is necessary to comply with aviation related regulations. After selecting a suitable hard surface, a sufficiently high movable fence should be provided around the dismantling area, taking into account the wind direction around the airport and the distance to the runway.

Engine Disassembly and Recovery

The goal of aircraft dismantling is to recover 98% of aircraft components. The engine is the largest and most expensive component, costing about $1.5 million for one engine that need to be evaluated for removal. For example, the recovered parts of the B747 worth $6.8 million.

In the case of the B747, the exact procedure for removing the engine is as follows: screw in the bolts to completely separate the fan blades and the core compressor, cut off the electrical connectors at all points, separate the three hydraulic hoses to restrict the main fuel supply and remove catches. Any mistake happened during this process could have resulted in death. Moving airplane parts is likely to knock someone over, and a mechanic can be swept into the machine or blown over by high winds while the engine is working. So the dismantling work had to be done with safety in mind before technique.

It is safe to remove the engine after loading the hoist, with customized slings holding the engine in place to ensure that the engine would not be damaged and the crew would not be injured during the removal process. The engine was removed and transported by forklift.

Disassembly and Storage of Re-saleable Parts

In general, high-value components can be defined as those components that individually account for more than 1% of the total aircraft scrap value, and components with a market value of more than $20,000 (under the condition of "removal") are considered high-value in IATA's BIPAD. The evaluation is based on market analysis and past experience. In addition to engines, landing gear, APU, electrical power units (e.g., generators), flight controls, and navigation systems are typically higher-value components than others. Component value is highly dependent on its condition and varies from condition to condition.

To prevent corrosion and the formation of explosive mixtures of air and fuel vapors, and to balance weight, the aircraft is stored intact. Prior to the dismantling of high-value components, the aircraft should be emptied of residual fuel. Aircraft fuel is stored in the main and reserve fuel tanks. Since aircraft fuel is highly flammable, it is first emptied before dismantling to prevent explosions and ensure that the tanks have enough time to dry out. The mechanic dismantling the aircraft closes the pumps and lateral valves in the cockpit, and the mechanic outside the aircraft locates the fuel lines, connects them to the fuel nozzles, and drains the stored fuel. After emptying the tank, the tank is forced ventilated with compressed air. To ensure that the air in the tank does not explode, the gas in the tank is analyzed during and after ventilation. The pressurized hydraulic oil system is then treated, the system is depressurized, and the liquid is stored in a special container.

After all fuel is removed and the aircraft is powered down, the removal of electronic equipment can be performed, which is an important step in the process of taking the aircraft to debris. There are about hundreds of instruments to be dismantled, and they must be dismantled as early as possible to reduce the chance of accidental damage due to their fragility. Recoverable avionics equipment is located in the electronics bay, which contains a lot of boxes controlling the fuel system, flight control and communications instruments, etc., which can be used in other aircraft after inspection. The black box (flight recorder, cockpit voice recorder) is located at the rear of the airplane, above the passenger cabin. Pipes, wires, fasteners, etc. located in the belly hatch space of the aircraft are also of recovery value.

Recyclable parts need to be properly sorted, removed and carefully identified with the part's assembly number, The removal label including registration number, filing number, part nomenclature, part number and serial number should be filled out [7], and then attached to the part and placed on a shelf for filing (in accordance with the requirements of CCAR-145 and AC 145). There are approximately 500–1000 labels on parts removed from the B747. Once labeled, the parts are sent to the aircraft inspector for confirmation. From complex electronics to plumbing, from flight controllers to hydraulic lines, even the most mundane items can have real value, and aircraft inspectors are required to sort through each of these parts for recycling.

Separation of Dangerous Goods in Aircraft

The hazardous goods contained in the aircraft must be handled with care. Special attention should be paid to the parts containing radioactive material, which include smoke detectors, fluorescent markers, and ballast made of depleted uranium ore (on old airplanes built in the early 1980s). These parts are disposed of under nuclear law (Europe: Atomic Energy Act) rather than national waste management regulations. For hazardous liquids marked with an asterisk in the European Waste Catalog, records of waste disposal are required. Records of waste disposal provide strong evidence of the orderly disposal of waste.

Ensuring that experienced and expert personnel handle hazardous materials is essential. For example, fiberglass materials produced before 1995 for cabin insulation have hazardous properties similar to asbestos. Employees working with such materials should wear the special equipment required, including protective clothing (anti-static), work boots (anti-static), gloves, special masks, helmets and safety glasses.

The goal of this phase is to dismantle the aircraft to a condition "free of liquids or other hazardous components". The next step is to segment the aircraft in the dismantling area in preparation for cutting and subsequent recovery.

Cutting of Aircraft and Recycling of Metals and Other Materials

Before the crushing process begins, a second gas check is carried out to ensure that there is no risk of explosion. Cutting begins when the analysis shows that the risk of fuel explosion has been eliminated. Once everything that can be recycled has been disassembled, the remaining work is usually done by a metal dismantling company. This procedure is performed after the aircraft is deregistered from the Aviation Authority, so it is not necessarily performed by a certified aircraft maintenance organization. The dismantling process needs to be methodical and purposeful. Heavy equipment specialists break all the windows, lift the roof and sort the metal into aluminum, steel, wire and copper as they rummage through the aircraft's cabin. Minimize landfills by recycling the most material.

1) *Aluminum alloy recycling.* Aircraft use a variety of aluminum alloys. Prior to dismantling operations, different types of aluminum alloys are identified by mapping to the documentation from OEM, resulting in a higher quality end product. The more materials are categorized, the more efficient it is to recycle the different types of alloys separately. To separate the aluminum from the other debris, a magnet with strong suction is used to cycle through all the metal. Aluminum that is not magnetized rolls off, while the iron is sucked onto a magnetic drum and sent to a smelter for recycling. The clean aluminum is sent to a smelter, where a gas-fired furnace heats the metal internally to 1,000°, the scrap is converted to aluminum water, and this liquid metal is filled with molds to form aluminum blocks for reuse, for example in the beer and beverage industries as beverage packaging (cans).
2) *Upgrade components for non-conventional use.* Passenger seats, emergency evacuation slides, etc., can be used in aviation academy training. Some decoration design companies or private collectors re-polished, shaped and design tires, hatch doors, portholes as the production of art works or household items, as shown in Fig. 2. The cockpit and airplane seats can also be used for movie sets. MROs sometimes repairs,

certifies and re-uses equipment such as seats, internal partitions, kitchens and luggage compartments.

a A lighting System Made of Portholes b A Table Made of Wheels

Fig. 2. The Production after Upgrading the Aircraft Components (SOURCE: boredpanda.com)

3) *The use of composite materials.* Modern aircraft increasingly use carbon fiber composite materials to reduce aircraft weight and reduce fuel consumption. Due to technical barriers, these composites were usually transported as waste to landfill or incinerated at a treatment plant after dismantling the aircraft. Now, composites companies continue to improve recycling methods, exploring new ways to recycle carbon fiber materials from aircraft for automotive applications, such as engine covers, car roofs and fender well component [8]. In the future, composites can be certified first for non-critical parts of the aircraft such as the cabin and interior, while their use on the fuselage requires further optimization of processes and methods to meet mechanical requirements.

4 Conclusion

Civil aircraft dismantling needs to consider the impacts of regulations, environment, safety, technology, quality, economy and other aspects, and the dismantling process should be standardized in order to ensure efficient and safe dismantling of the aircraft and the recycling value of retired aircraft. Stakeholders (CAAC, aircraft owners, dismantling companies, OEM, dismantled parts users, government agencies, etc.) need to pay attention to the best practices and advanced technologies of aircraft dismantling at home and abroad, and continuously improve the general dismantling process of civil aircraft.

Boeing and Airbus, the world's two largest aircraft manufacturers, have long been secretly competing in aircraft recycling programs. As the main manufacturer, COMAC should more actively participate in the industry management and practice of aircraft dismantling, and make efforts in the whole life cycle of the aircraft and the whole industrial chain to promote China's transformation from a civil aviation power to a civil aviation power.

References

1. International Air Transport Association (IATA), Best Industry Practices for Aircraft Decommissioning (BIPAD), Montreal—Geneva (2018)
2. Aircraft Fleet Recycling Association (AFRA): AFRA Best Management Practice (BMP), Guide, USA (2017)
3. Civil Aviation Administration of China (CAAC), AC-145-FS-2019-017 Aircraft Dismantling, Beijing: Civil Aviation Administration of China (2019)
4. Civil Aviation Administration of China (CAAC): AC-120-FS-058 R4 Qualified Aircraft Material. Civil Aviation Administration of China, Beijing (2023)
5. Civil Aviation Administration of China (CAAC): CCAR-145 Approval of Civil Aircraft Maintenance Organization. Civil Aviation Administration of China, Beijing (2022)
6. Airbus, S.A.S.: Process for Advanced Management of End-of-life of Aircraft (PAMELA). Airbus Academy, France (2008)
7. S, Y.U.H., T., H.E.: Research on airworthiness requirements of civil aircraft dismantling. Civ. Aircr. Des. Res. **1**, 157–161 (2022)
8. Asmatulu, E., Overcash, M., Twomey, J.: Recycling of aircraft: state of the art in 2011. J. Ind. Eng. **8** (2013)

Numerical Investigation on Aerodynamic Performance of Asymmetric Airfoil Under Oscillating Turbulence

Liming Long[1], Ji Zhang[1(✉)], Xu Zheng[1], Chuntao Yang[1,2], Yan Li[1], Han Yuan[1], and Ning Mei[1]

[1] College of Engineering, Ocean University of China, Qingdao 266100, China
zhangji@ouc.edu.cn
[2] Haier Group, Qingdao 266100, China

Abstract. In this study, the transient aerodynamic performance of an asymmetric airfoil under oscillating turbulence at different angles of attack was studied. Experimental data were used to validate the precision of the numerical simulation approach. The results were obtained by establishing single-and double-wing models of the NACA6412 airfoil with and without disturbance sources. The velocity, pressure, and lift-drag characteristics of the airfoil were investigated. Compared with the single-wing model, the unsteadiness of the double-wing model was more significant, and the turbulence generates larger vortex structures above the wing's upper-right region. The presence of oscillating turbulence led to an increase in the airfoil's peak lift force. The lift-to-drag ratio of the airfoils reached a maximum in the range of 5°–10° angle of attack. Improved deep stall behavior and lift generation at small angles of attack were more pronounced under turbulence with smaller oscillation amplitudes.

Keywords: Aerodynamic performance · Angle of attack · NACA6412 airfoil · Oscillating turbulence · Detached Eddy Simulation

1 Introduction

The wing is an important part of the aircraft, and its main role is to provide lift for the aircraft and control the flight attitude. NACA6412 [1] airfoil is a typical asymmetric airfoil with a complex structure, high machining difficulty, and large pitching moment. With the rapid developments in computers and numerical analysis, numerical calculation and analysis of the flow field and aerodynamic characteristics can be performed using computational fluid dynamics (CFD). The CFD method based on the Navier–Stokes Equations [2] is used to solve the wing flow field from the basic equations reflecting the nature of the fluid, and the viscosity and compressibility of the gas are considered. There are two difficulties in simulating the vortex structure evolution using the Navier–Stokes (N–S) equation [3]: (1) Accurate calculation of unsteady effects. The time discretisation accuracy has a noteworthy effect on the calculation results. (2) Calculation of

Y. Zhao (Ed.): ICAMM 2025, SPM 96, pp. 69–79, 2025.
https://doi.org/10.1007/978-3-032-09862-7_7

turbulent kinetic energy at high Reynolds numbers. In this study, the detached eddy simulation (DES) model [4] was used to calculate turbulence to obtain a finer eddy structure simulation junction.

Slowing down the boundary layer separation and improving the lift-drag characteristics are important factors to be considered in the design of wings with a large angle of attack [5]. The boundary layer separation can be effectively inhibited by adding a periodic disturbance to the boundary layer of the wing [6]. Compared to the turbulence amplitude, adjusting the turbulence oscillation frequency can improve the aerodynamic performance of the airfoil under a deep stall. In this study, the transient aerodynamic performance of the NACA6412 airfoil was studied by numerical simulation, which is a typical airfoil. The disturbance source was the same scaled-down NACA6412 airfoil placed in front of the main wing, and the wake of the small wing was introduced as the oscillation turbulence at a high angle of attack. The transient aerodynamic performance of the NACA6412 wing at different angles of attack was investigated, including its velocity, pressure, and lift-drag characteristics.

2 Numerical Case Description

2.1 Methodology

In both the RANS and LES approaches, the governing equations for the incompressible flow are written in a dimensionless form [7, 8].

$$\frac{\partial \overline{u}_j}{\partial x_j} = 0 \tag{1}$$

$$\frac{\partial \overline{u}_i}{\partial t} + \frac{\partial \left(\overline{u}_i \overline{u}_j\right)}{\partial x_j} = -\frac{\partial \overline{p}}{\partial x_i} - \frac{1}{\mathrm{Re}} \frac{\partial \tau_{ij}^{mol}}{\partial x_j} - \frac{\partial \tau_{ij}^{turb}}{\partial x_j} \tag{2}$$

$$\tau_{ij}^{mol} = -2\nu \overline{S}_{ij}, \tag{3}$$

$$\overline{S}_{ij} = \frac{1}{2}\left(\frac{\partial \overline{u}_i}{\partial x_j} + \frac{\partial \overline{u}_j}{\partial x_i}\right) \tag{4}$$

where u_i and u_j are the velocity and p is the pressure. The $^{-}$ indicates the resolved scales for LES and the time-averaged components for RANS, respectively. τ_{ij}^{mol} is the momentum transport of molecular motion, which can be defined by Eq. (3). The term τ_{ij}^{turb} is arouse due to the momentum transport of turbulence motion.

The dependent variable in the DES model is defined as $\tilde{\nu}$, the transport equation of $\tilde{\nu}$ is:

$$\frac{\partial \rho \tilde{\nu}}{\partial t} + \frac{\partial}{\partial x_j}(\rho \tilde{\nu} u_j) = c_{b1} \rho \tilde{S} \tilde{\nu} + \frac{1}{\sigma_\nu}\left[\frac{\partial}{\partial x_j}\left\{(\mu + \rho \tilde{\nu})\frac{\partial \tilde{\nu}}{\partial x}\right\} + C_{b2}\rho\left(\frac{\partial \tilde{\nu}}{\partial x_j}\right)^2\right] - c_{w1} f_w \rho \left(\frac{\tilde{\nu}}{\overline{d}}\right)^2 \tag{5}$$

The right side of the Eq. (5) is the generation term, diffusion term and wall damage term of eddy current viscosity respectively, and the values of each coefficient are as follows:

$c_{b1} = 0.1335,\ c_{b2} = 0.622,\ \sigma = 2/3,\ c_{w1} = \frac{c_{b1}}{k^2} + \frac{1+c_{b2}}{\sigma},\ c_{w2} = 0.3,\ c_{w3} = 2,$
$r = \frac{\tilde{\nu}}{\tilde{S}k^2d^2}, g = r + c_{w2}(r^6 - r), f_w = g\left(\frac{1+c_{w3}^6}{g^6+c_{w3}^6}\right)^{\frac{1}{6}}, \tilde{S} = S + \frac{\tilde{\nu}}{k^2d^2}f_{v2}, S = \sqrt{2S_{ij}S_{ij}},$
$\chi = \frac{\tilde{\nu}}{\nu}, f_{v1} = \frac{\chi^3}{\chi^3+c_{v1}^3}, f_{v2} = 1 - \frac{\chi}{1+\chi f_{v1}}, c_{v1} = 7.1, c_{des} = 0.65, \tilde{d} = \min(d, C_{des}\Delta)$

Where, d is the nearest distance. Δ is the filter scale in LES method.

2.2 Numerical Model

In this model, the Integrated Computer Engineering and Manufacturing software was used to perform structural mesh division for the NACA6412 wing; the mesh numbers of the single and double wings were 25000 and 55000, respectively. The mesh distributions are shown in Fig. 1. It should be noted that the smaller wing was scaled down to the larger wing with a ratio of 3:1 in the case of a double wing. The time steps of the single and double wings were set to 0.002 s and 0.0002 s, respectively. To simulate the state of high-speed flight, the inlet velocity was set to 200 m/s (corresponding to Re $= 1.33 \times 10^7$). Angles of attack of 0°, 10°, 20°, and 30° were considered to analyse the variation in the velocity and pressure profiles near the wing for the two cases. The outlet pressure was set to 1 atm, and the boundary temperature was 20 °C. The gas was compressible and treated as an ideal gas, and there was no slip on the wing wall. The coupled algorithm of FLUENT (2020) was used for the numerical calculations.

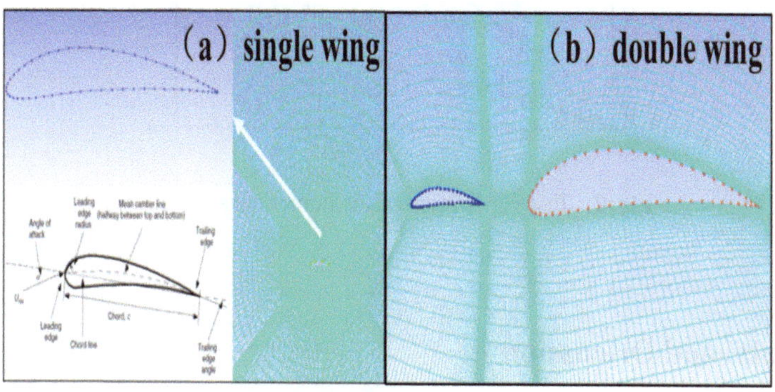

Fig. 1. Mesh distribution of the single and double wings

2.3 Mesh Validation and Test Verification

Table 1 lists the aerodynamic coefficients of the different grids. The difference between the aerodynamic coefficients of the 25000 and 50000 grids is less than 0.25% for the single-wing model. The difference between the aerodynamic coefficients of the 55000 and 80000 grids is less than 0.17% for the double-wing model, which indicates that grid independence has been achieved. It is verified that the numerical results obtained are independent of the grid density through an analysis of the grid independence.

Table 1. Grid independence verification for aerodynamic coefficient

Single-wing Model/ Double-wing model/	C_l (AOA = 0°)	C_l (AOA = 0°)
20000/50000	1.175	1.152
25000/55000	1.205	1.182
50000/80000	1.208	1.184

To verify the reliability of the calculation method, the viscous flow of the NACA6412 airfoil (angle of attack varies from 0° to 10°) without turbulence was simulated numerically. The calculation condition was Re = 10000. Figure 2 shows the airfoil lift drag coefficient results obtained by numerical calculation and the wind tunnel test results [6, 9] under the corresponding conditions. The consistency observed between the numerical and experimental results demonstrates the validity of the simulation method.

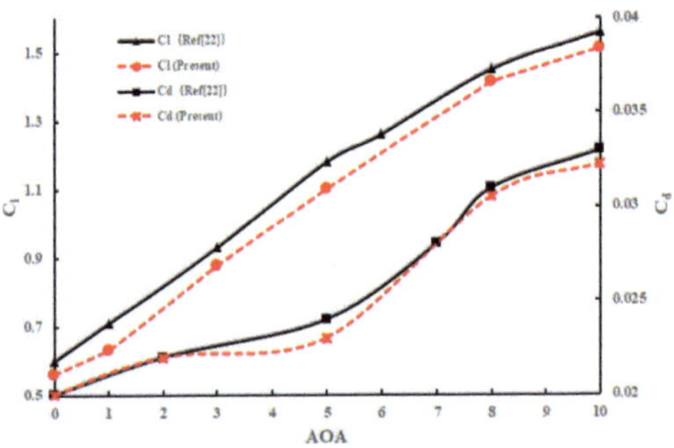

Fig. 2. Test verification of the numerical simulation

3 Results and Discussion

3.1 Velocity Characteristics

Figure 3 shows the velocity distributions of the single-wing model at the velocity attack angles of 0°, 10°, 20°, and 30°. For the single-wing model, the velocity field remains stable after the calculation. Variations in the angle of attack lead to noticeable alterations in velocity profiles. When the velocity angle of attack is 0°, an area with high velocity mainly exists in the upper part of the airfoil. The closer it is to the wing wall, the higher is the velocity. The maximum speed at the angle of attack of 0° is smaller than that at angles of attack of 10°, 20°, and 30°. When the angles of attack are 0° and 10°, a relatively slow stagnation zone is produced at the front of the airfoil. With an increase in the angle of attack, the wake area gradually increases. When the angles of attack are 20° and 30°, there is a lower velocity zone in the upper area of the airfoil. In addition, distinct shock waves are observed on the upper surface of the airfoil. Crossing the shock wave causes the airflow to experience elevated pressure and temperature, along with a reduction in velocity and a Mach number shift from above to below one. The location where shock waves generally occur also produces wave resistance. Wave resistance increases the resistance coefficient and consumes power, which makes a difference on the wing surface and a certain influence on the position of the pressure centre and lift centre.

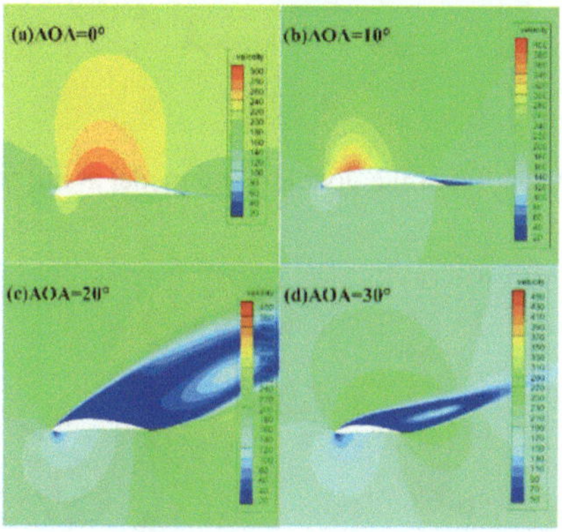

Fig. 3. Velocity distribution of the single-wing model at different angles of attack

Figure 4 shows the velocity distributions of the double-wing model at different angles of attack. Compared with the single-wing model, the velocity distribution of the double-wing model fluctuates greatly after the steady state, particularly when the angle of attack of the double-wing model is 10°, 20°, and 30°. It can be observed from the figure that the velocity distributions of the small and large airfoils are similar under different velocity angles of attack. With an increase in the angle of attack, the degree of flow disorder increases. In particular, when the angle of attack is 30°, the velocity distribution above the airfoil is more uneven, and the unsteadiness is significant.

Flow diagrams of the two models at an angle of attack of 30° are shown in Fig. 5. As can be seen from Fig. 5(a), at a large angle of attack, the flow situation of the single-wing model near the wing is very complex; thus, vortex structures of different scales are generated. The flow diagram of the double-wing model at an angle of attack of 30° is shown in Fig. 5(b). Many vortex structures appear above the airfoil. The largest-scaled vortex appears at the upper right of the large airfoil. The existence of the oscillating turbulence strikingly changes the streamline structure of the airfoil. The counter-rotating vortices continue to grow, which cancels part of the main vortex and reduces the lift force.

3.2 Pressure Characteristics

The pressure characteristics of the single-wing model, as illustrated in Fig. 6, vary significantly between 0° and higher angles of attack such as 10°, 20°, and 30°. When the airflow is aligned horizontally (0°), elevated pressure appears at the airfoil's front

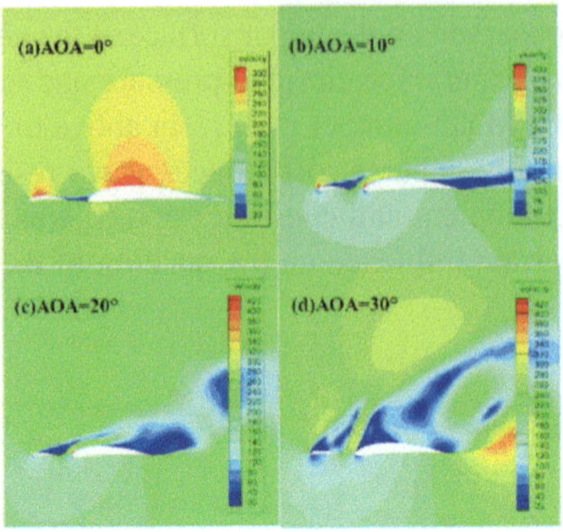

Fig. 4. Velocity distribution of the double-wing model at different angles of attack

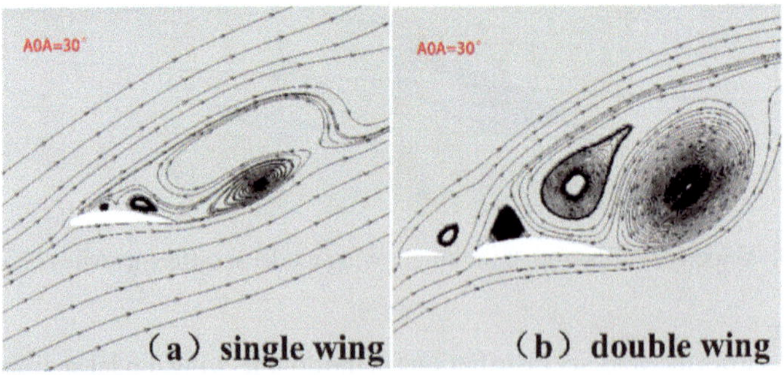

Fig. 5. Flow diagram at 30° Angle of attack

section, and a suction region forms above the airfoil. When the angles of attack are 10°, 20°, and 30°, the high-pressure area is located at the lower part of the airfoil.

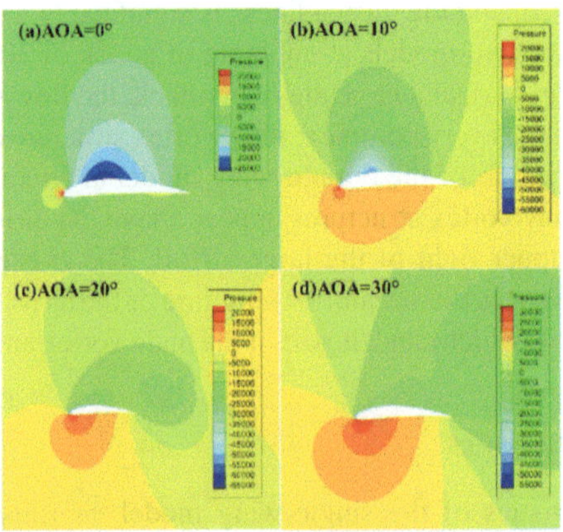

Fig. 6. Pressure distribution of the single-wing model

Figure 7 shows the pressure distribution of the double-wing model under different angles of attack. As opposed to the single-wing model, the velocity distributions of both wings fluctuate strikingly. When the angle of attack is 30°, there is a large negative pressure area at the rear and upper positions of the airfoil. NACA6412 is an asymmetric airfoil, and its upper surface area is larger than that of its lower surface. The airflow maintains symmetry across the airfoil, but due to higher velocity on the upper side, a pressure gradient develops, giving rise to aerodynamic lift. Compared to the single-wing model, the lift force is affected as the pressure difference diminishes somewhat once the angle of attack goes above 10°. The maximum-pressure areas for both airfoils are located at the tip of the airfoil.

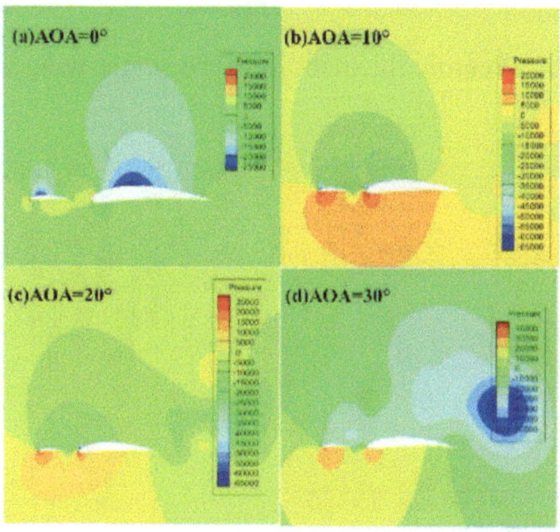

Fig. 7. Pressure distribution of the double-wing model

The pressure fluctuations at the pressure monitoring point of the single-wing model are shown in Fig. 8. When the angle of attack is 0°, the pressure at the steady-state monitoring point does not change with time, and its value is positive. When the angles of attack are 10°, 20°, and 30°, the pressure fluctuates over time. With a higher angle of attack, the pressure at the monitoring point remains negative and its absolute value increases.

Figure 9 shows the pressure fluctuations at Point 1 and Point 2 calculated by the double-wing model under a steady state. When the angle of attack is 0°, the pressure at the Point 1, which is the monitoring point behind the small airfoil, fluctuates periodically. Point 2, the rear point of the large airfoil, is relatively stable. When the angles of attack are 10°, 20°, and 30°, the pressure values at Points 1 and 2 fluctuate strikingly.

3.3 Characteristics of Lift-Drag Force

The lift response of the single-wing model at angles of attack from 0° to 35° is depicted in Fig. 10. It can be observed that the lift coefficient grows, then declines, and subsequently increases again as the angle of attack grows. When the angle of attack is 9°, the lift coefficient reaches a maximum of 2.2. As the angle of attack grows, the velocities on the airfoil's top and bottom surfaces change, with the lower surface experiencing slower

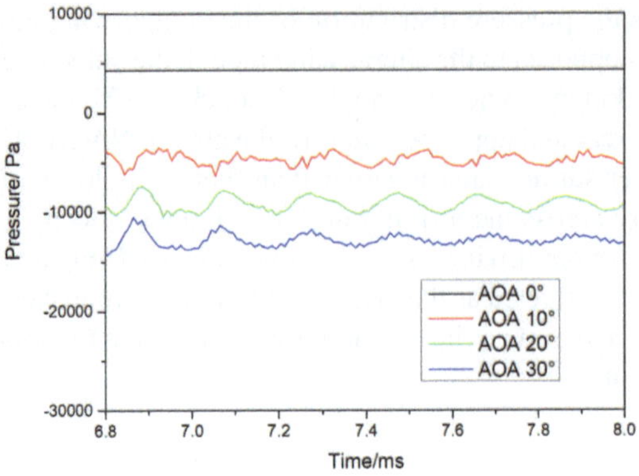

Fig. 8. Pressure fluctuations of the single-wing model

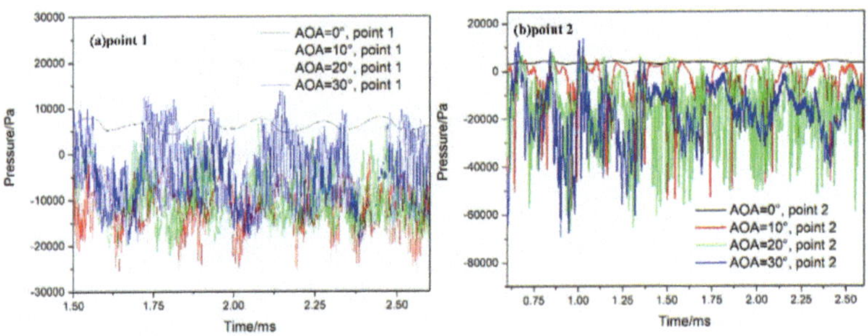

Fig. 9. The pressure fluctuations of the double-wing model

airflow compared to the upper surface. Therefore, the pressure on the upper surface is less than that on the lower surface, hence, the lift coefficient increases. From 9° to 25°, the lift coefficient of the airfoil shows a downward trend and then rises. The fundamental reason for this phenomenon is that local supersonic zones and shock waves appear on the upper and lower surfaces of the airfoil. In addition, the sequence and expansion speed of the local supersonic flow on the upper and lower surfaces of the wing are not the same. The pressure on the upper and lower surfaces of the wing decreases to different degrees, which results in obvious changes in the lift coefficient at this stage.

Figure 11 shows the lift characteristics of the double-wing model at different angles of attack. The lifting-drag characteristics of a large wing under different attack angles were studied by comparing with the single-wing model. The lift coefficient of the double-wing model fluctuates significantly with time. The lift coefficient varies little over time at 0° angle of attack but shows the highest variation at 35°. Compared to the single-wing model, the maximum lift coefficient increases at elevated angles, although such angles also lead to increased instability in lift behavior.

Figure 12(a) shows the lift-to-drag ratio characteristics of the single-wing model at different angles of attack. As shown in the figure, the lift-drag ratio of the airfoil diminishes gradually with increasing angle of attack. Figure 12(b) depicts the lift-drag

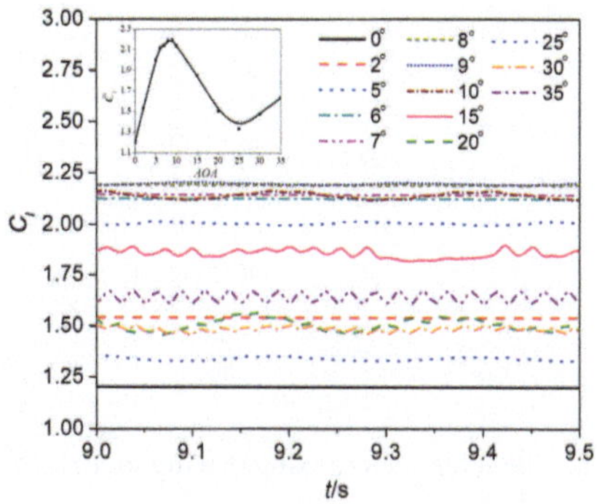

Fig. 10. Lift characteristics of the single-wing model

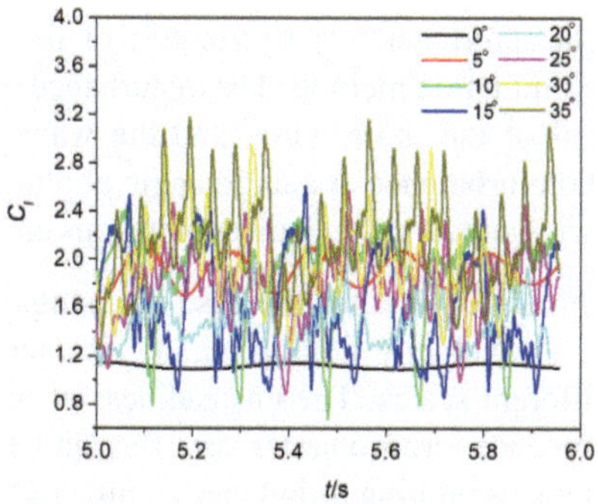

Fig. 11. Lift characteristics of the double-wing model

ratio trends of the double-wing model across multiple angles of attack. It can be observed that the lift-to-drag ratio of the airfoil is the largest in the range of 5°–10° angle of attack, and there are large fluctuations. At an angle of attack of 0°, the lift-drag ratio of the airfoil is stable at approximately 30. Compared with the single-wing model, the lift-drag ratio of the airfoil with the oscillating turbulence decreases when the angle of attack is 0°, but the lift-drag ratio increases significantly when the angle of attack is between 5°–10°, indicating that the oscillating turbulence attached to the airfoil head greatly improves the lift force and deep stall characteristics of the airfoil. As the angle of attack continues to increase, the airfoil lift does not improve further. Oscillation turbulence with a lower oscillation amplitude is more beneficial for improving the deep stall performance and lift of the airfoil at a small angle of attack.

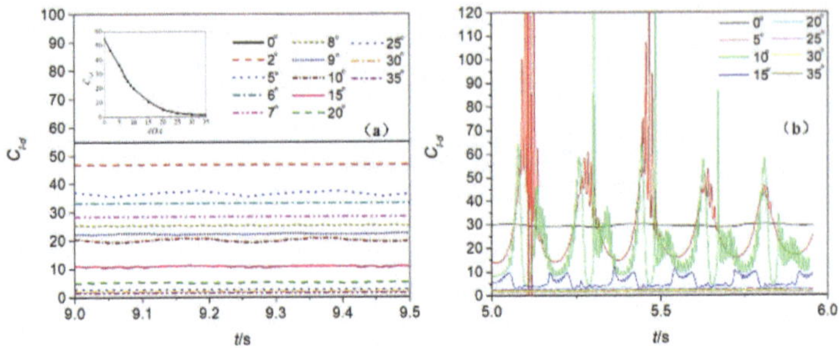

Fig. 12. Lift-drag ratio characteristics (a) single-wing model; (b) double-wing model

4 Conclusions

In this study, the transient aerodynamic performance of the NACA6412 airfoil was studied based on the DES simulation method. The disturbance source of the NACA6412 airfoil was placed in front of the main wing, and the wake of the small wing was introduced as the oscillation turbulence at a large angle of attack. Through the analysis of the single and double-wing models, the main conclusions are summarised as follows:

1. In terms of velocity characteristics, for the case of a single wing, with an increase in the angle of attack, the velocity field gradually becomes complex, resulting in vortex structures of different scales. The unsteadiness of the double-wing model is more obvious, and turbulence forms a larger vortex structure in the upper right of the wing. The existence of the oscillating turbulence significantly changes the streamline structure of the airfoil. The counter-rotating vortices continue to grow, which cancels part of the main vortex and reduces the lift force.
2. In the single-wing model, a high-pressure zone forms below the airfoil, and the tail area shows negative pressure values that intensify as the angle of attack rises. Pressure variations are more pronounced in the double-wing model than in the single-wing one. At a higher angle of attack, a large negative pressure area is formed at the upper right of the airfoil. The maximum-pressure areas of the two types of airfoils appear at the tip of the airfoil.
3. In terms of the lift-drag force characteristics, the lift-drag ratio of the airfoil decreases gradually with an increase in the angle of attack. The non-stationarity of the flow profile under the double-wing model is strong, and the lift and drag coefficients fluctuate significantly with time. Oscillation turbulence with a lower oscillation amplitude is more beneficial for improving the deep stall performance and lift of the airfoil at a small angle of attack.

Overall, the research on oscillating turbulence, which could analyze the real surroundings, can be applied in real-world airfoil design for further optimization. But it mainly focuses on 2D without analyzing 3D effects, which needs to be further analyzation.

Acknowledgments. The authors acknowledge the financial support provided by the National Natural Science Foundation of China (NO. 52401352), the Shandong Provincial Natural Science

Fund for Excellent Young Scientists Fund Program (Overseas)(2025HWYQ-035), Postdoctoral Fellowship Program of China Postdoctoral Science Foundation(GZC20252079),Qingdao Science and Technology Benefiting People Demonstration Project (25-1-5-cspz-4-nsh).

References

1. Vigneswaran, C.M., Vishnu Kumar, G.C.: Numerical analysis of co flow jet airfoil on enhancement of aerodynamic performance. Mater. Today Proc. **82**, 103–107 (2023)
2. Han, S., Song, W., Han, Z., Xu, J.: A high-resolution numerical investigation of unsteady wake vortices for coaxial rotors in hover. Chin. J. Aeronaut. **37**(11), 278–299 (2024)
3. Farhloul, M., Fall, N., Dione, I., Léger, S.: Finite element analysis of a pseudostress–pressure–velocity formulation of the stationary Navier–Stokes equations. J. Comput. Appl. Math. **469**, 116665 (2025)
4. Casadei, L., Könözsy, L., Lawson, N.J.: Unsteady Detached-Eddy Simulation (DES) of the Jetstream 31 aircraft in One Engine Inoperative (OEI) condition with propeller modelling. Aerosp. Sci. Technol. **91**, 287–300 (2019)
5. Bunge, U., Mockett, C., Thiele, F.: Guidelines for implementing Detached-Eddy Simulation using different models. Aerosp. Sci. Technol. **11**(5), 376–385 (2007)
6. Sharma, D., Goyal, R.: Methodologies to improve the performance of vertical axis wind turbine: a review on stall formation and mitigation. Sustain. Energy Technol. Assess. **60**, 103561 (2023)
7. Brockmeyer, L., Saini, N., Tentner, A., Fang, J., Merzari, E.: Evaluation of RANS vs. LES simulation of fluid flow through 3×3 rod bundle with a simple spacer grid as a precursor to coupled fluid–structure interaction simulations. Nucl. Eng. Des. **430**, 113662 (2024)
8. Yu, Y.Q., Bergeron, A., Licht, J.R.: Numerical simulation of involute-plate research reactor flow behavior using RANS, LES and DNS. Ann. Nucl. Energy. **207**, 110709 (2024)
9. Cheng, X., Han, Q., Huang, Y., Shi, M., Shi, H., Ji, M., Yang, C.: Bioinspired ultra-fine hybrid nanocoating for improving strength and damage tolerance of composite fan blades in flexible manufacturing. Compos. Sci. Technol. **259**, 110956 (2025)

Numerical Study on Damage Assessment of Aircraft Fuel Tank Cover Under Impact Load of Rim Debris

Zhou Changwei[1(✉)], Wang Zhaodong[2], Wu Xiaoyang[3(✉)], Wang Peiyan[4], Chen Yu[1], Chen Xuegang[1], Li Baofeng[1], Zhang Lijuan[1], Zhang Chenlizi[1], Liu Fu[5], and Wu Xihui[1]

[1] Shanghai Aircraft Design and Research Institute, COMAC, Shanghai, China
zhouchangwei@comac.cc
[2] Shanghai Aircraft Airworthiness Certification Center, CAAC, Shanghai, China
[3] Xi'an Institute of Modern Chemistry, NORINCO Group, Xi'an, China
wxyxia204@163.com
[4] Northwestern Polytechnical University, Xi'an, China
[5] Shanghai Polytechnic University, Shanghai, China

Abstract. It is a potential hazard mode of aircraft tire blasting that the wheel fails and throws out the rim fragments, which has a low probability but is extremely harmful. In order to evaluate the threat of rim debris to the typical double-layer foam-filled impact-resistant fuel tank cover of aircraft, the actual impact position and impact angle of rim debris were characterized by a digital prototype, and a numerical analysis model was established based on the ductile failure criterion, which could reflect the structural dynamic response. The damage degree under three impact postures, namely, outer-section, cross-section and sharp-corner, was evaluated from three dimensions, such as stress level, damage morphology and displacement response. The results show that: the rim debris of the nose landing gear has no effect on the aircraft, and the rim debris of the main landing gear impacting the fuel tank cover at an angle of 16 is the most dangerous condition; The rim debris has the risk of penetrating the fuel tank cover. The cross-sectional impact is the most dangerous impact posture, causing the most serious damage. The internal structure of the cover has reached the critical damage value, which requires special attention in the design and verification of the fuel tank cover.

Keywords: Wheel Failure · Rim Debris · Impact Posture · Damage Assessment · Numerical Simulation

1 Introduction

Wheel failure is a potential hazard mode of aircraft tire blasting. Under extreme conditions, wheel rim fragments with 60 arc segments will be thrown out, which may damage important structures such as wings and landing gear, destroy fuel tank cover and cause fire, seriously affecting aircraft safety [1, 2]. In February, 1999, a Boeing 747-200' s tire

Y. Zhao (Ed.): ICAMM 2025, SPM 96, pp. 80–90, 2025.
https://doi.org/10.1007/978-3-032-09862-7_8

burst during take-off, and the rim fragments were thrown out and penetrated the cabin, damaging the hydraulic systems of No.1 and No.4, and the plane returned to the field after dumping fuel urgently. The accident investigation report shows that it is a common aviation accident that the fuel tank structure is impacted by the rim debris, which will cause catastrophic accidents such as aircraft fire in serious cases. Therefore, it is of great significance to carry out research in this field to improve the safety of aircraft.

Wang [3] studied the impact response of tire debris to the cover and analyzed the relationship between the impact area, impact angle and impact energy. Yao [4] studied the impact resistance of two types of fuel tank covers, machine-added aluminum alloy and precision cast aluminum alloy, and found that the impact resistance of the cast part was weaker than that of machine-added parts; Johnson [5] studied the optimal design form of composite material covers under the impact of tire debris; Mines and Karagiozova [6, 7] conducted mechanical properties tests of tire debris, established an impact simulation model and analyzed the dynamic response of the fuel tank cover; Wang [8] studied the effects of tank pressure and oil sloshing on the tank structure based on the coupled Euler-Lagrange (CEL) method.

The related research mainly focuses on the impact of tire debris, and the damage assessment of fuel tank cover under the impact load of wheel rim debris has not been carried out. Based on the digital prototype, the impact angle of the rim debris on the aircraft fuel tank cover is systematically analyzed, and the damage assessment of the fuel tank cover under different impact postures of the rim debris is carried out, so as to provide guidance for the design and verification of the aircraft fuel tank cover.

2 Failure Model and Dangerous Conditions of Rim Debris Based on Digital Prototype

Although the probability of rim debris has been minimized according to the design and manufacturing specifications such as TSO-C135 and TSO-C62e, it is still required that the hazard of rim debris must be evaluated in aircraft airworthiness verification. The airworthiness regulations for transport aircraft, such as CCAR-25-R4, FAR-25(Amdt. 25-144), and CS-25(Amdt. 20), all set clear requirements for this [9–11].

Referring to Zhou [12]'s practical experience on airworthiness of civil aircraft, taking an aircraft as an example, a wheel rim debris model that meets the airworthiness requirements is established. The wheel rim debris is 1/6 of the wheel flange, the size is 47 mm*31 mm, the material is magnesium-zinc alloy, the mass is 0.86 kg and the impact speed is 100 m/s, as shown in Fig. 1.

Wheel rim debris occurs within 20° of the lateral direction of the wheel, as shown in Figure 2. When multiple wheels are installed on the same landing gear, only the lateral projection of the outermost wheel rim needs to be considered. If only one wheel is installed on a landing gear, the lateral projection of the rims on both sides of the wheel should be considered. Based on the digital prototype of the aircraft, a model of the debris impact area of the nose landing gear(NLG) and main landing gear(MLG) is established, as shown in Fig. 3.

It is found that rim debris of the NLG has no effect on the fuselage structure and system, while rim debris of the MLG will impact the fuselage, wing, landing gear and

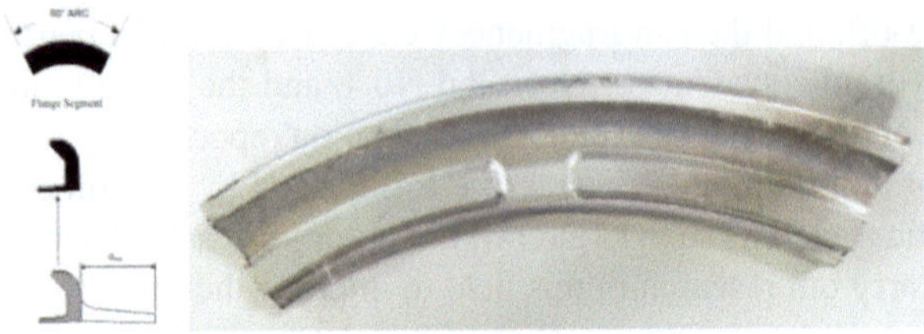

Fig. 1. Schematic diagram of rim debris

Fig. 2. Schematic diagram of failure model of rim debris

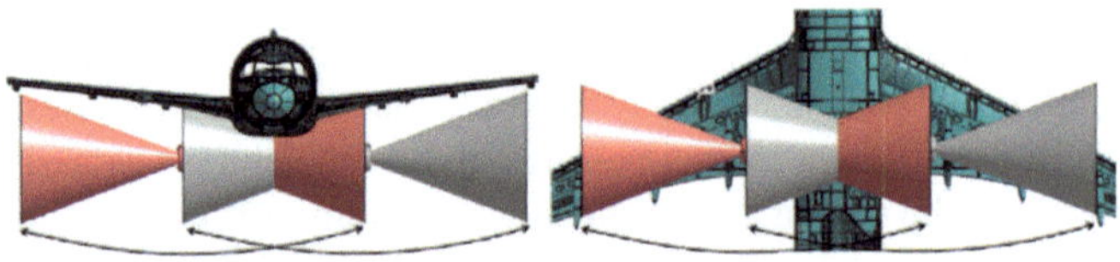

Fig. 3. Influence range of rim debris based on digital prototype

engine and other key parts, including the MLG cabin door, wing fairing, wing box from 5th ribs to wing tip, engine hanging joint and so on.

The fuel tank cover is the weakest part of the wing structure. When it is impacted by the rim debris, it is very likely to break through and cause dangerous fuel leakage, which will lead to serious accidents. According to the measurement of digital prototype, the most dangerous situation is that the rim fragments impact the wing fuel tank cover at 16° and the impact speed is 100 m/s. The velocity component of the wheel rim debris is 96.12 m/s in the heading direction and 27.56 m/s in the vertical direction.

3 Finite Element Modeling

The fuel tank cover of an aircraft consists of inner and outer panels, pressure rings, fasteners, filled foam and sealing rings, as shown in Fig. 4. The inner and outer panels are made of the same material as the pressure ring, and both are made of aluminum alloy 2024-T42; The filled foam is made of Rohacell 71WF material. Fasteners use shear-resistant flat-headed high-lock bolts made of titanium alloy, which are used to connect the outer cover panel, pressure ring and wing panel.

2024 aluminum alloy is a common aviation material, and Johnson-Cook is widely used in engineering to describe the mechanical behavior of aluminum alloy materials

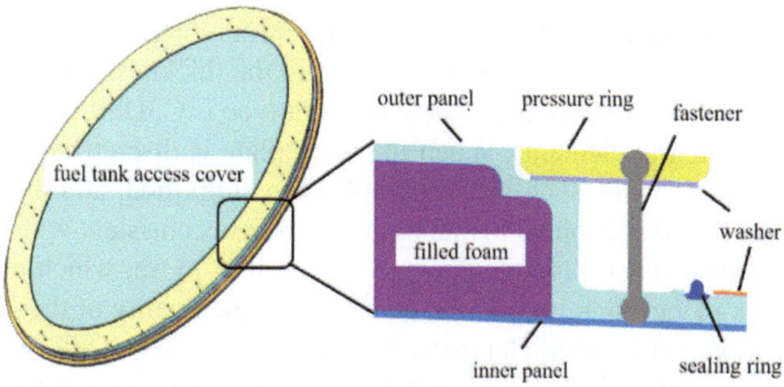

Fig. 4. Structural schematic diagram of a typical double-layer foam-filled impact-resistant fuel tank cover

under high-speed impact [13], and the expression is as follows:

$$\sigma = \left(A + B\varepsilon^{n}\right)\left(1 + C\ln\frac{\dot{\varepsilon}}{\dot{\varepsilon}_0}\right) \tag{1}$$

Where is the material strain; is the strain rate; is the reference strain rate; A, b, c, n, m are material parameters.

The material parameters of 2024-T42 are shown in Tables 1 and Tables 2.

Table 1. 2024-T42 Material Parameters [14]

material	Elastic modulus (GPa)	Shear modulus (GPa)	Poisson's ratio	Density (g/cm3)	A (MPa)	B (MPa)	C	n
2024 T42	71.08	48.42	0.33	2.768	293.7	737.4	0.00376	0.582

Table 2. Rohacell 71 WF Foam Material Parameters [15]

Elastic modulus (MPa)	Shear modulus (MPa)	Density (kg/m3)	Tensile strength (MPa)	Compressive strength (MPa)	Shear strength (MPa)
75	24	75	1.56	1.05	1

Geometric nonlinearity and material nonlinearity are considered, and the ABAQUS/Explicit universal explicit analysis module is used to solve the problem. In the finite element analysis, the structural strength of the cover is focused on, simplifying the influence of washers and sealing rings. In order to save computing resources, the cover inner and outer plates and pressure rings are modeled by shell elements, the element type is S4R (4-node quadrangular finite membrane strain linear reduction integral shell element), and the element integration points are 5 in the thickness direction. The filling foam is modeled by solid elements, and the element type is C3D8R (8-node hexahedral linear reduction integral element). When the wing plate is discretized, the connecting parts between components, such as rivets, bolts, are simplified, and the cross-section type of a full-shell grid is adopted to maintain a thickness consistent with the real structure. The main quadrilateral mixed elements are used in mesh, which not only meets the requirements of Jacobian matrix, but also ensures the quality of the mesh. The rim fragments are modeled by solid elements, the element type is C3D8R (8-node hexahedral linear reduced integral element), and the element size is determined according to the real structure size. The numerical model is shown in Fig. 5.

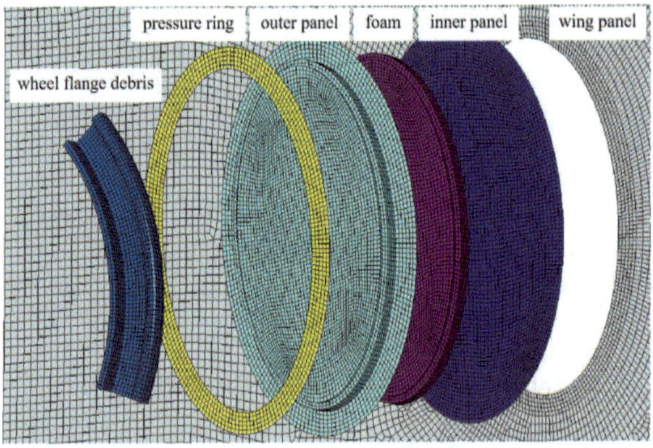

Fig. 5. Schematic diagram of discretization grid of rim and fuel tank cover.

Before analyzing, the model needs to be verified for grid independence to ensure that the calculation results are not limited by the grid size. Figure 6 compares structural displacement and grid number for different grid size (3–10 mm) at 2.5 ms. The results show that when the grid size is 5 mm, the displacement calculated by the model fluctuates around 91 mm, and the accuracy does not change as the number increases. Based on the above analysis, a cell size of 5 mm was selected for mesh division in subsequent simulation calculations. Specifically, the element number of rim fragment is 5670, the element number of wing plate is 55354, and the element number of cover is 37737, of which the element number of panel and filled foams is 31995 and 5742 respectively.

In order to study the harmfulness of impact posture, three typical impact postures are simulated and analyzed, namely, the outer-section, the cross-section and the sharp-corner, as shown in Fig. 7.

The difference between these three models lies in the impact posture of debris, while other simulation parameters such as impact velocity, boundary conditions and contact conditions remain unchanged. In the simulation, the ductile failure criterion based on effective plastic displacement is adopted to describe the damage evolution of materials and delete the failure units, so as to obtain clear observation results of damage morphology and ensure the accuracy of calculation results.

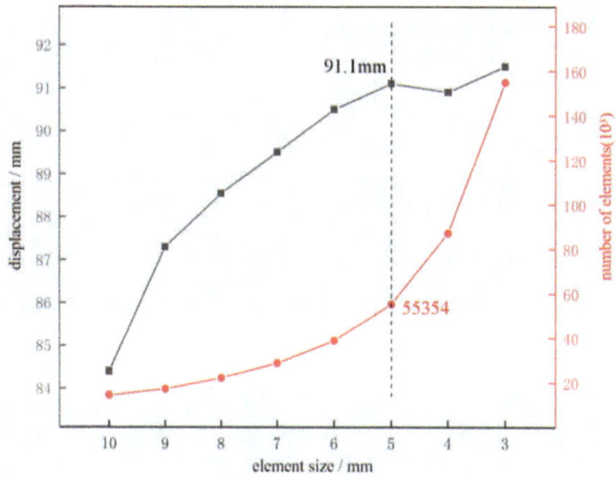

Fig. 6. Grid independence verification

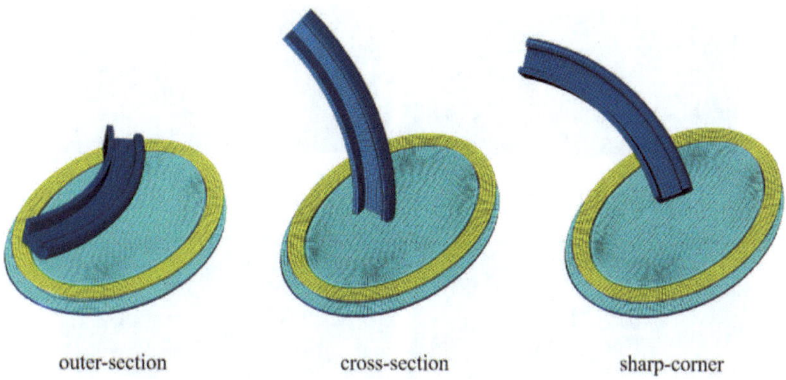

Fig. 7. Schematic diagram of different impact postures

4 Numerical Analysis and Result Discussion

4.1 Structural Stress Level

Figure 8 shows the impact process of rim debris with different postures. The analysis shows that at 0.25 ms, the rim debris touches the cover, causing obvious stress concentration; At 0.5 ms, the rim fragments continue to impact, which increases the stress level; At 1 ms, the stress level gradually decreased and the stress wave began to spread outward; At 4 ms, the rim fragments are separated from the cover.

Different from the structural response when the tire fragments impact [3], the rim fragments have great rigidity and hardly deform during the impact process, so the impact process of the rim fragments takes shorter time, the strain rate response is higher, and the structural damage is more serious.

Different from outer-sectional impact, cross-sectional impact and sharp-corner impact both cause embedded damage to the fuel tank cover outer panel, and the stress

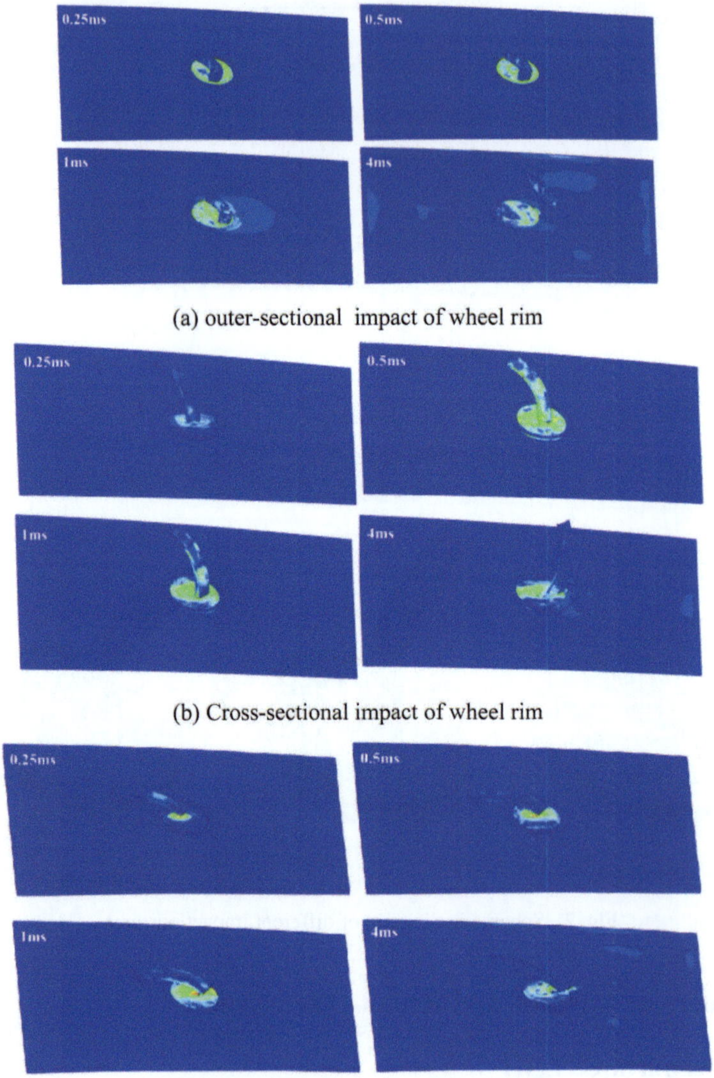

(c) sharp-corner impact of wheel rim

Fig. 8. Impact process of different postures

level of the cover caused by cross-sectional impact is the highest, so it can be inferred that cross-sectional impact is the most dangerous impact condition.

4.2 Structural Damage Morphology

The damage morphology of the components of the aircraft fuel tank cover at 4 ms is extracted, as shown in Table 3. DUCTCRT represents the material criterion of ductile failure. When the material reaches the equivalent plastic strain, it will be damaged, and when the ductile damage state variable is greater than or equal to 1, the material will fail.

Table 3. Summary of Damage Morphology of Fuel Tank cover Components

Impact posture	legend	pressure ring	Outer panel	foam	Inner panel
outer-sectional impact		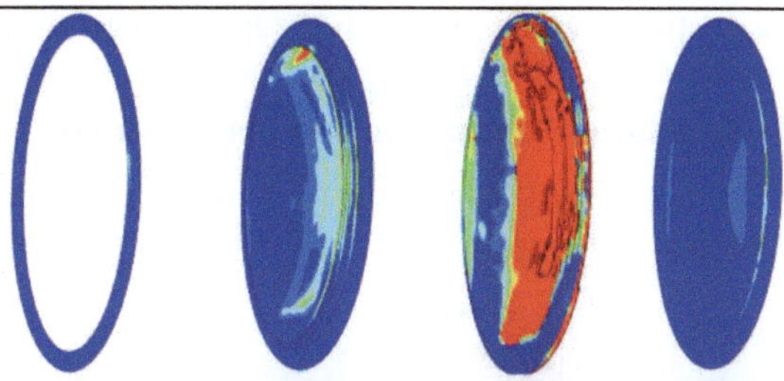			
Cross-sectional impact	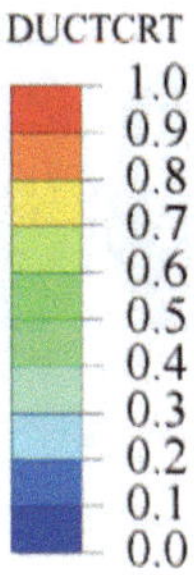	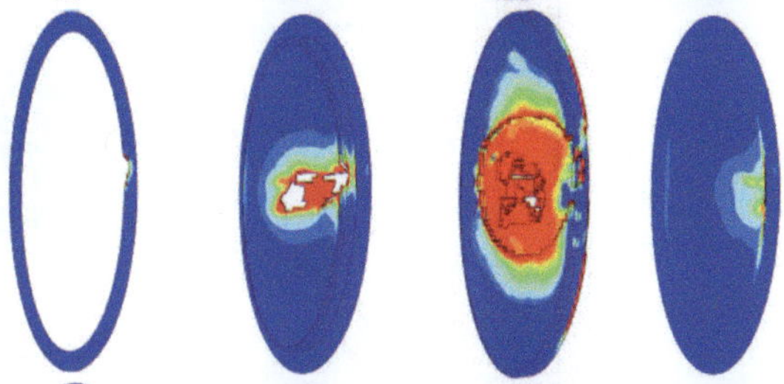			
sharp-corner impact		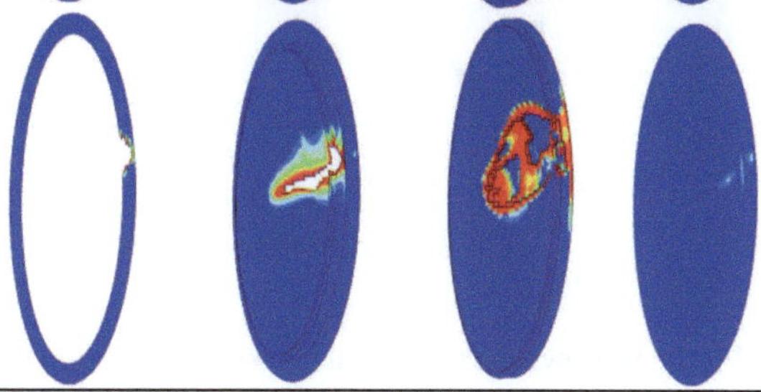			

The cover outer panel is the first layer structure of the fuel tank cover and the outer-sectional impact will cause a large range of deformation of the cover outer panel. Compared with the outer-sectional impact, the cross-sectional impact and sharp-corner impact will lead to the damage of the cover.

The filling foam inside the cover is the main energy-absorbing part. It is worth noting that the foam is penetrated during the cross-sectional impact, while the other two impacts only damage the foam, which indicates that although the sharp-corner impact is easier to scratch the surface of the cover outer panel, it will not damage the deep structure, while the cross-sectional impact will lead to penetrating damage.

The cover inner panel is the last layer structure of the fuel tank cover. Although the three impact postures did not lead to the failure of the cover inner panel, the damage of the cross-section impact is close to the critical value, and the equivalent plastic strain area is concentrated in the boundary, which may lead to the initiation and propagation of cracks and lead to leakage under the fuel tank pressure.

To sum up, the outer-sectional impact will lead to a large-scale deformation of the cover outer panel, but the deep structure is still intact; sharp-corner impact leads to the structural damage of the cover outer panel, but has little effect on the deep structure. In contrast, the structural damage caused by cross-sectional impact is the most serious, which may lead to penetrating damage and leakage. Therefore, it can be inferred that cross-sectional impact posture is the most dangerous impact condition.

4.3 Structural Deformation and Plastic Strain

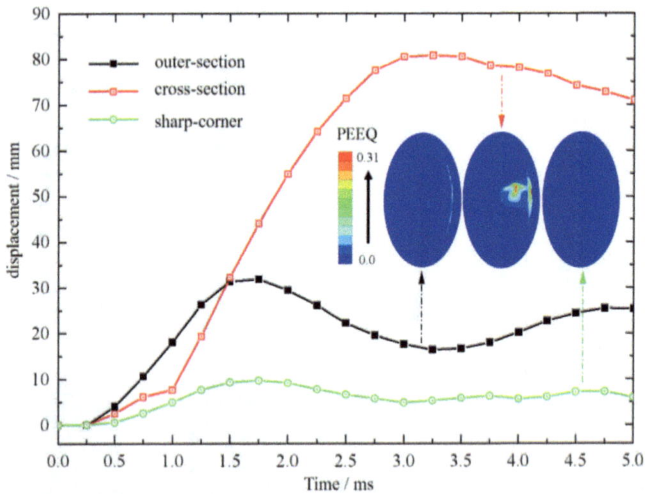

Fig. 9. Structural response of cover inner panel

Figure 9 shows the structural response of the cover inner panel. The maximum deformation of the impact point is 9.6 mm, 31.8 mm and 80.8 mm, respectively under three impact postures: sharp-corner, outer-section and cross-section. The maximum equivalent plastic strain of cross-section impact is 0.31, which is close to the failure threshold, which indicates that the cross-sectional impact posture has the risk of penetrating the fuel tank cover. The equivalent plastic strain under the other two impact postures is far less than the failure strain, and the structure has not been damaged. Cross-sectional impact posture is the most dangerous impact condition.

5 Summary

Based on the digital prototype, the wheel rim debris model that meets the airworthiness requirements is established, and the dynamic analysis model of a typical double-layer foam-filled impact-resistant fuel tank cover is established by using the ductile failure criterion, and the structural damage assessment under different impact postures is carried out. The results show that:

1) Taking an airplane as an example, the rim fragments of NLG have no effect on the fuselage and system, while rim fragments of MLG will impact the fuselage, wings, landing gear and engine and other key parts. The most dangerous condition that the rim debris hits the wing fuel tank cover at an inclination angle of 16;
2) The outer-sectional impact will lead to a large-scale deformation of the cover outer panel, but the deep structure is still intact; Sharp-corner impact leads to the structural damage of the cover outer panel, but has little effect on the deep structure;
3) Under the cross-sectional impact, the stress level and deformation are the highest, and the penetration damage of deep structure is more serious, the equivalent plastic strain of the cover inner panel is close to the failure threshold.
4) Rim debris has the risk of penetrating the fuel tank cover, especially the cross-sectional impact, which is considered to be the most dangerous impact posture, and needs to be paid special attention to in the design and verification of the fuel tank cover.

References

1. European Aviation Safety Agency: Certification Specifications for Large Airplanes: CS-25 Amendment 14, pp. 3–5 (2013)
2. European Aviation Safety Agency: Notice of Proposed Amendment, Protection from Debris Impacts: NPA 2013-02, pp. 1–7 (2013)
3. Wang, F.S., Jia, S.Q., Wang, Y.: Impact response study on covering cap of aircraft big-size integral fuel tank. Appl. Compos. Mater. **23**(5), 953–972 (2016)
4. Yao, S.L., Yue, Z.F., Geng, X.L.: Safety assessment of aircraft fuel tank access cover under the impact load by Tire fragments. Chin. J. Mech. Eng. **32**(1), 14 (2019)
5. Johnson, A.F., Kempe, G., Simon, J.: Design of composite wing access cover under impact loads. Appl. Compos. Mater. **7**(4), 219–229 (2000)
6. Mines, R., McKown, S., Birch, R.S.: Impact of aircraft rubber tyre fragments on aluminium alloy plates: I - experimental. Int. J. Impact Eng. **34**(4), 627–646 (2007)
7. Karagiozova, D., Mines, R.: Impact of aircraft rubber tyre fragments on aluminium alloy plates: II - numerical simulation using LS-DYNA. Int. J. Impact Eng. **34**(4), 647–667 (2007)
8. Wang, P.Y., Lian, C.C., Yue, C.X., et al.: Experimental and numerical study of tire debris impact on fuel tank cover based on coupled Eulerian-Lagrangian method. Int. J. Impact Eng. **157**, 103968 (2021)
9. Civil Aviation Administration of China: Airworthiness Standards of Transport Category Aircraft:CCAR-25-R4. Civil Aviation Administration of China, Beijing (2011)
10. USA Federal Aviation Administration: Airworthiness Standards: Transport Category Airplanes: 14 CFR, PART25, pp. 1–4 (2000)
11. European Aviation Safety Agency: Certification Specifications for Large Airplanes: CS -25, pp. 1–3 (2008)
12. Changwei, Z., Zhaodong, W.: Study on the airworthiness certification technical system and key contents of aircraft tire burst. Adv. Aeronaut. Sci. Eng. **16**(4), 148–154,180 (2025)
13. Changwei, Z., Lijuan, Z., Chenglizi, Z.: Structural damage analysis and experimental study of aviation structure under tire blasting jet load. J. Vib. Shock. **42**(8), 334–340 (2023)

14. Zhang, L.: A new fatigue cumulative damage model based on material parameters and stress interaction. Int. J. Fatigue. **179**, 108035 (2024)
15. Zaldivar, R.J., Marques, D., Barrie, C.: The effect of radiation exposure on carbon fiber-reinforced Rohacell core laminate structural composites. J. Compos. Mater. **54**(17), 2261–2269 (2020)

Correction of Modal Experiment Results for Hydraulic Pipeline Based on Finite Element Method

Jun Li(✉) and Zhi Niu

COMAC Shanghai Aircraft Design & Research Institute, Shanghai, China
lijun7@comac.cc

Abstract. Hydraulic system provides vital hydraulic energy for aircraft flight, and the pipeline vibration problem caused by pressure pulsation seriously affects the reliability of the hydraulic system. Conducting vibration and modal experiment of hydraulic pipeline is an important means to support hydraulic pipeline system design. However, thin-walled and slender hydraulic pipelines are often lightweight, their dynamic characteristics are easily affected by the additional mass of sensors during experiment. This article proposes a method to eliminate additional mass errors and correct the modal experiment results of pipelines. By establishing a elaborate and complete finite element model, Using additional mass elements to carry out numerical simulation, and finally the corrected modal parameters of pipelines are obtained. The data from the application case of the aircraft shows that this method can eliminate the influence of sensor mass on modal experiment, and the method provides references for finite element analysis of hydraulic pipelines and correction of modal experiment results of other thin-walled circular tube structures.

Keywords: Modal experiment · Hydraulic pipeline · finite element method

1 Introduction

Hydraulic system as an important aircraft onboard system, for flight control, landing gear, brakes and other users to provide power energy. Aircraft hydraulic system is mainly composed of hydraulic pump, pipeline, oil filter, tank, etc., hydraulic pipeline as a transport tube to transfer power to the fuselage all over the actuator. Hydraulic pump output hydraulic oil is a periodic pulsating fluid, pressure pulsation is the main cause of hydraulic pipeline vibration and noise, when the hydraulic pump pulsation frequency and hydraulic pipeline natural frequency close, fluid and pipe between the fluid-structure coupling resonance phenomenon, resonance response beyond a certain range will lead to pipes or joints, supports, clamps and other cracks, Make the hydraulic system fail and affect the flight safety of the aircraft [1–3].

Hydraulic pipeline failure is one of the most common failure forms of aircraft. It is very important to study the dynamic characteristics of hydraulic pipeline for pipeline vibration analysis, strength analysis and fatigue life analysis. Researchers at home and

Y. Zhao (Ed.): ICAMM 2025, SPM 96, pp. 91–101, 2025.
https://doi.org/10.1007/978-3-032-09862-7_9

abroad have carried out a lot of research on the dynamic characteristics of the pipeline, analyzed the vibration mechanism of the hydraulic pipeline, and studied the influence of different materials and fluid pulsation factors on the vibration of the pipeline by establishing the mathematical model and simulation model of the hydraulic pipeline, and applied the modal experiment in the analysis of the dynamic characteristics of the pipeline [4–8]. Modal experiment, as a method that can accurately obtain the dynamic characteristics of hydraulic pipelines, provides important basis for the design and analysis of hydraulic pipeline systems.

However, aviation hydraulic pipelines usually use thin-walled and slender pipelines, which have a weight of only a few tens of grams and are easily affected by the additional mass of sensors in modal experiment. Only after eliminating the errors caused by the additional mass can the modal parameter identification results be used for the study of the dynamic mechanical characteristics of hydraulic pipelines [10, 11].

To sum up, scholars at home and abroad have achieved certain achievements and experience in theoretical analysis and experiment of dynamic characteristics of hydraulic pipelines. In order to study the influence of the sensor additional mass in the pipeline system experiment, this paper established a elaborate finite element model of the pipeline system and calculated the modal parameters in two states including the concentrated mass and without the concentrated mass through modal analysis, corrected the modal experiment results of a section pipeline on a certain type of aircraft, and obtained the dynamic parameters of the section of the hydraulic pipeline in the real state.

2 Modal Parameter Identification

The main task of modal parameter identification is to measure the modal parameters of the structure or system from the input and output data, including modal frequency, vibration mode, damping, etc. Modal experiment will simplify the continuous infinite multi-degree of freedom system to multi-degree of freedom system, and transform the vibration differential equation of linear time-invariant system from physical space to modal space. The equations are decouple into an independent equation described by modal coordinates and modal parameters, and the modal parameters are obtained by solving the equation.

The motion differential equation of the multi-degree-of-freedom system is:

$$[M]\{\ddot{x}\} + C\{\dot{x}\} + K\{x\} = \{f\} \tag{1}$$

Where $[M]$, $[K]$, $[C]$ are respectively mass matrix, stiffness matrix and damping matrix.

The differential equation under modal coordinates can be obtained by using coordinate transformation:

$$[\varphi]^T[M][\varphi]\{\ddot{q}\} + [\varphi]^T[C][\varphi]\{\dot{q}\} + [\varphi]^T[K][\varphi]\{q\} = [\varphi]^T\{f\} \tag{2}$$

The decoupled differential equation of motion is obtained:

$$[m_r]\{\ddot{q}\} + [c_r]\{\dot{q}\} + [k_r]\{q\} = \{f_r\} \tag{3}$$

Thus the vibration response is obtained:

$$\{x\} = [H]\{f\} \tag{4}$$

Where the frequency response function matrix is:

$$[H] = \sum_{r=1}^{n} \frac{\{\varphi_r\}\{\varphi_r\}^T}{k_r - \omega^2 m_r + j\omega c_r}, H_{ij}(\omega) = \sum_{r=1}^{n} \frac{\varphi_{ir}\varphi_{jr}}{k_r - \omega^2 + j\omega c_r} \tag{5}$$

Where m_r, k_r, c_r, φ_r and are called the r order modal mass, modal stiffness, modal damping, modal vibration and modal frequency:

$$\omega_r = \sqrt{\frac{k_r}{m_r}} \tag{6}$$

The physical meaning of the frequency response function $[H]_{ij}$ acts on the unit force at the j point, and the response caused at the i point. Corresponds to the transfer function in the time domain. The modal experiment obtains the frequency response function by measuring the excitation force at the j point and the response at the i point at the same time, and then solves the modal frequency and modal shape from the frequency response function matrix.

3 Modal Experiment

3.1 Experiment Object

The modal experiment object is a section of hydraulic pipeline between the electric pump and the pressure oil filter actually installed by the aircraft. The experiment pipeline system consists of 8 parts, including the pipe joint, nut and bracket fixed for the electric pump end, the pipe joint and nut fixed for the oil filter end, the bracket and clamp fixed in the middle of the pipeline. There are four straight sections in this section of pipeline. In this experiment, a total of 11 measuring points are arranged on the pipeline road, and the layout of the pipeline measuring points is shown in Fig. 1.

3.2 Experiment Method

The mode parameter identification method of hammer method is adopted in this modal experiment. The experiment equipment includes impact hammer, acceleration sensor and dynamic signal analyzer, etc. By stimulating in the lateral and vertical directions of point No. 5, the acceleration response in three directions is picked up at measuring points No. 1-11 in turn, and the modal parameter identification is carried out according to the frequency response function. The experiment device is shown in Fig. 2.

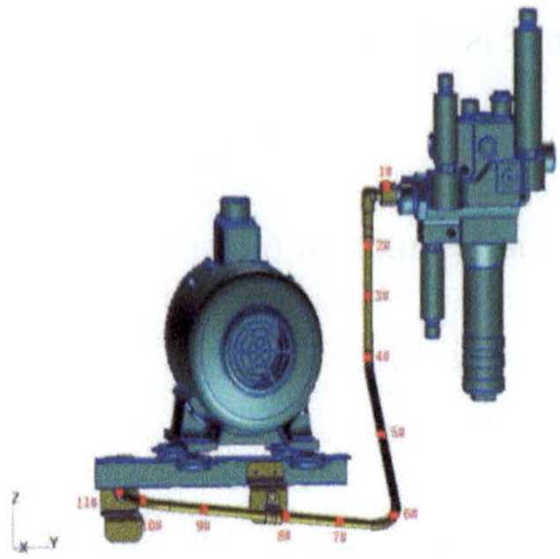

Fig. 1. Pipeline modal experiment point arrangement

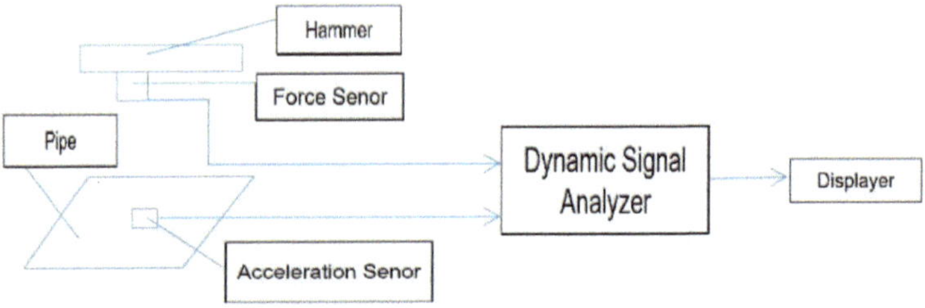

Fig. 2. Schematic of the experiment equipment

3.3 Experiment Parameter

The analysis frequency range is set to 0 ~ 250 Hz, the sampling time is 2 s, triggered by the force of hammer, and force-index window are added for FFT analysis, and each excitation is 3 times on average.

3.4 Experiment Result

A total of 66 frequency response functions of 2 excitation points and 11 acceleration response measuring points (each measuring point contains 3 directions) were obtained. The modal parameters identification showed that the first three modal frequencies of the hydraulic pipeline were 71.21 Hz, 145.20 Hz and 228.12 Hz respectively. The corresponding modal shapes are shown in Figs. 3, 4 and 5.

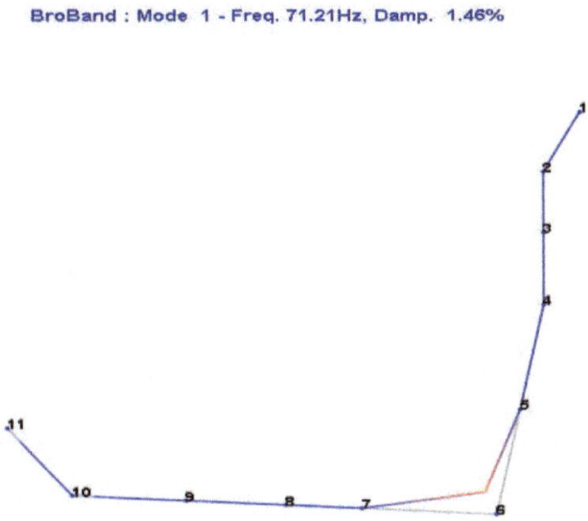

Fig. 3. First experiment modal

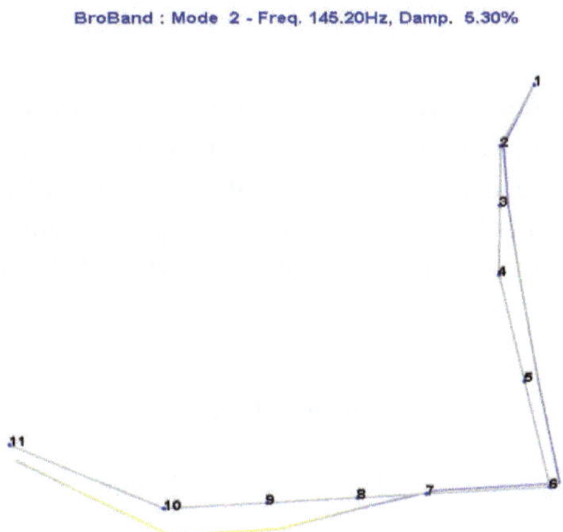

Fig. 4. Second experiment modal

Fig. 5. Third experiment modal

4 Modal Analysis

Based on a geometric model of hydraulic pipelines consistent with the experiment piece, the geometric model software tool is CATIA, the geometric cleaning software tool is HYPERMESH, the finite element modeling software tool is PATRAN, and the modal calculation software tool is NASTRAN. The unified unit system for numerical analysis is shown in Table 1 below. The finite element model is established through the following steps.

Table 1. Units

Names	Dimension	Units
Length	L	mm
Quality	M	ton
Frequency	T-1	Hz
Density	ML-3	ton/mm-3
Modulus	ML-3 T-2	Mpa

Import the geometric model, carry out geometric cleaning, extract the middle face of the hydraulic pipeline and fill the end face, extract the middle face of the support structure, and group different parts and structures of different thickness, including pipe joints, bracket, pipelines, etc.

Create finite element Meshing, Shell Element is used for pipelines and brackets, while Solid Element is used for pipe joints, clamps, and nuts.

Considering the quality of hydraulic oil, add the hydraulic oil quality to the hydraulic pipeline, and the calculation method is as follows:

$$\rho_{equ} = \frac{\pi\left(R_0^2 - R_i^2\right)\rho_{tubing} + \pi R_i^2 \rho_{fuel}}{\pi\left(R_0^2 - R_i^2\right)} \tag{7}$$

Where, ρ_{equ}, ρ_{tubing}, R_0, R_i is the equivalent density, is the pipe material density, is the outer radius, is the inner radius.

Add material properties and unit properties, and the material properties of the structure are shown in Table 2.

Table 2. Finite Element Material Properties

Group	material	Density	Modulus	Poisson ratio	Thickness
Pipeline	Titanium	8.3E-9	107000	0.3	0.5
Bracket 1, bracket 2	Aluminum	2.7E-9	70000	0.33	1.27
Clamp, pipe joint 1, pipe joint 2, nut 1, nut 2	steel	7.6E-9	120000	0.33	-

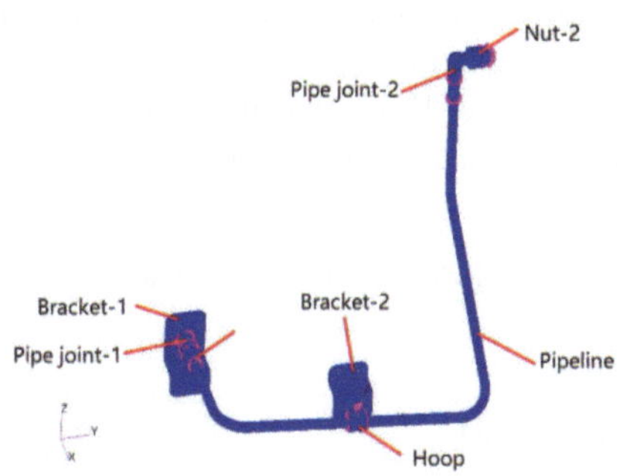

Fig. 6. Finite element model of structure

Create boundary conditions, the boundary conditions imposed on the structure of each part are shown in Table 3.

Get the finite element model, as shown in Fig. 6, including 1343793 elements, 313859 nodes.

Table 3. Boundary Conditions

Group	Constrain Type
Pipeline with pipe joint 1	MPC-RBE2
Pipeline with pipe joint 2	MPC-RBE2
Pipe joint 1 with nut 1	MPC-RBE2
Pipe joint 2 with nut 2	MPC-RBE2
Bracket 1	Clamp, constrain 6 DOF at the riveted hole of the support
Pipe joint 2	Clamp, constrain 6 DOF at the end face
Bracket 2	Clamp, constrain 6 DOF at the riveted hole of the support
Bracket 1 with pipe joint 1	Constrain 5 DOF, Release DOF for axial rotation
Bracket 2 with clamp	Beam Element
Clamp with pipeline	Spring

5 Calculation and Correction

Based on the above finite element model, a point unit is created at the measuring point No. 6 of the modal experiment, and the unit attribute of 13 g of the concentrated mass of the point unit is set up to establish a complete finite element model of the hydraulic pipeline containing the additional mass of the sensor. The modal parameters are obtained through modal analysis showing in Figs. 7, 8, 9 and 10 in which the second calculation modal is the local modal of bracket 1, which is irrelevant to the experiment results. The frequencies of the first three modals are shown in second row of Table 4.

Table 4. Modal Frequency Comparison Table

Group	First	Second	Third
Modal experiment (with added mass)	71.21	145.20	228.12
Modal analysis (with added mass)	73.22	150.28	236.64
Comparison between analysis (with added mass) and experiment (with added mass)	2.82%	3.50%	3.73%
Modal analysis (no added mass)	80.19	158.67	240.54
Modal analysis comparison between (without added mass) and (with added mass)	9.52%	5.58%	1.65%
Modal experiment - after correction	77.99	153.30	231.88

The frequency comparison method was used to evaluate the accuracy of the simulation analysis. The modal frequency error of the finite element analysis with concentrated mass and the experiment modal frequency of the hydraulic pipeline with sensor mass was less than 5% as shown in third row of Table 4. Moreover, according to the comparison between the simulation modal shapes and the experiment modal shapes, the maximum displacement was in a similar position, so the finite element model was considered accurate and effective.

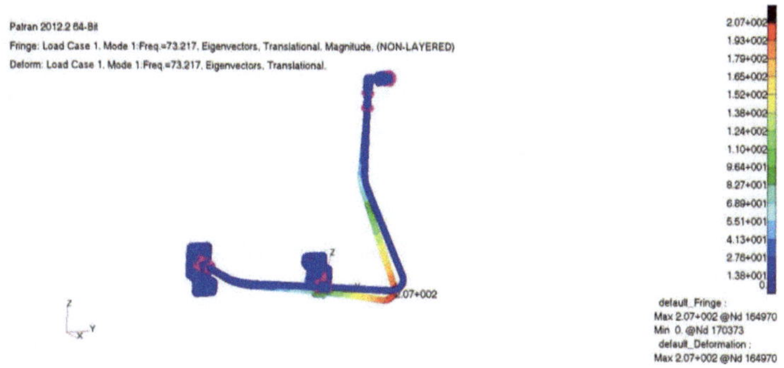

Fig. 7. First analysis modal

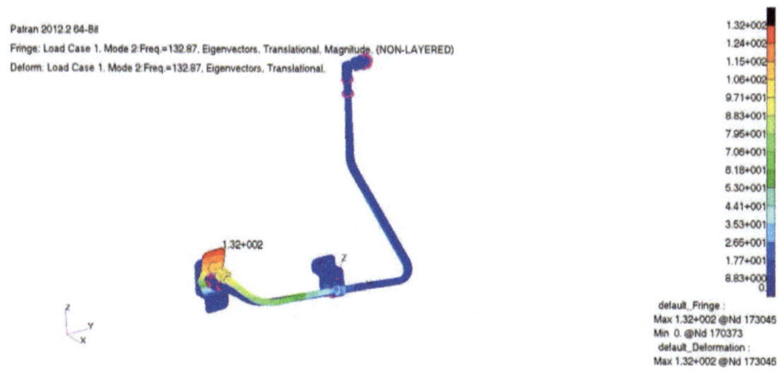

Fig. 8. Second analysis modal

Remove concentrated mass element from finite element model, Get the real finite element model that does not contain the additional mass of the sensor. The modal parameters are obtained from the modal analysis and the modal shapes is unchanged, the first three modal frequencies are increased by 9.52%, 5.58% and 1.65% respectively as shown in fourth and fifth row of Table 4. The coefficient is multiplied by the original modal experiment results to obtain the corrected results, that is, the real modal frequencies of the pipeline in this section are 77.99 Hz, 153.30 Hz and 231.88 Hz.

6 Conclusion

1. In this paper, by combining Shell Element and Solid Element, the quality of hydraulic oil is equivalent to the density of pipe structure, and a fine and complete finite element model including pipe, bracket, clamp, joint and nut is created through the equivalent constraint of the connection relationship between each part.

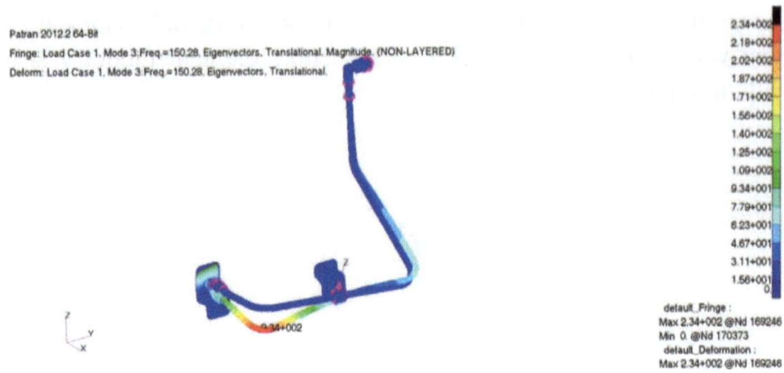

Fig. 9. Third analysis modal

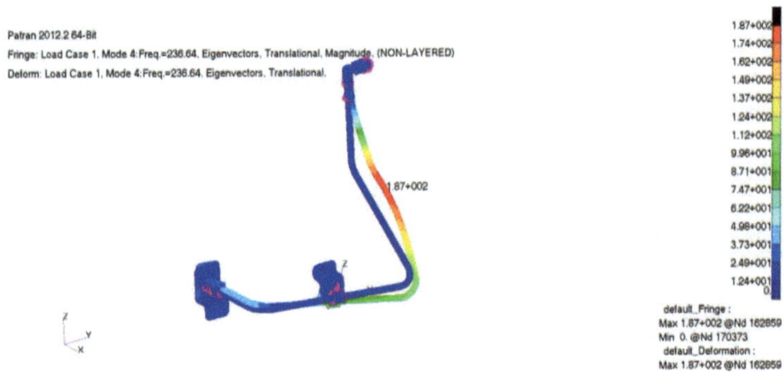

Fig. 10. Fourth analysis modal

2. According to the finite element modal analysis, the error of the modal frequency of the finite element analysis containing concentrated mass and the modal frequency of the hydraulic pipeline experiment with the sensor mass is less than 5%. The modal experiment result is corrected by removing the finite element concentrated mass, and the real modal parameters of the pipeline are obtained at 77.99 Hz, 153.30 Hz and 231.88 Hz.
3. The data show that due to the influence of the additional mass of the sensor, the pipeline modal frequency obtained by the pipeline modal experiment is obviously low, and the modal frequency of the hydraulic pipeline in this section can be lower by 9.52%.
4. This method can provide a reference for the finite element analysis of hydraulic pipeline, and the modal experiment results correcting method based on the finite element analysis can be applied to the modal experiment of lightweight structures such as thin-walled circular pipes.

References

1. Brahney, J.H.: Hydraulic pumps: the key to power generation. Aerosp. Eng. **12**, 9–13 (1991)

2. yong, C.X.: Analysis and experimental verification to effects of pipe initial installation stress on pipe's natural frequencies. China Mech. Eng. **4**(26), 512–517 (2015)
3. Zhenhua, Y.: Fault analysis and improvement methods of pipe crack for aero-hydraulic system. Hydraul. Pneum. Seals. (2023)
4. Marloff, R.H.: Resonant fatigue experimenting of riveted joints. Exp. Mech. **20**(2), 37–43 (1980)
5. Aggarwal, R.K., et al.: Development and qualification of alternative solutions for improved fatigue performance of deep water steel catenary risers. In: International Conference on Offshore Mechanics and Arctic Engineering. ASME (2007)
6. Yanhua, L., et al.: Analysis of vibration characteristics of typical pipe segment considering fluid-solid coupling. J. Vib. Shock. **29**(6), 50–53 (2010)
7. Peng, F., et al.: Analysis and fault diagnosis the flow impact response of a certain aircraft hydraulic pipeline. Hydraul. Pneum. Seals. **07**, 62–69 (2019)
8. Shengfa, Z., et al.: Dynamic characterisics of aviation hydraulic piping system. J. Dyn. Control. **6**, 58–63 (2022)
9. Qinghua, Z.: Elimination of effects of transducer mass on modal parameters in moving-point measurements. J. Nanjing Univ. Aeronaut. Astronaut. **5**(26), 692–695 (1994)
10. Qiuhai, L., Debao, L.: Engineering Measurement Experiment Analysis. Tsinghua University Press, Beijing (2015)

Metal Processing and Additive Manufacturing

Study on the Mechanism of Controlling Springback of Advance High-Strength Steel Bending Tube Component by Tube Hydro-Forging

Caiyuan Lin(✉), Chuan Xu, Wei Zhang, Keming Yan, and Sule Feng

China Aerospace Science and Technology Corporation, Shanghai Spaceflight Precision Machinery Institute, Shanghai 200000, China
lcyhiter@163.com

Abstract. In order to overcome the problem of advance high-strength steels (AHSS) bending tube component with serious springback defect, this paper proposes a new control method - Tube hydro-forging to eliminate the springback of such parts. Its mechanical principle is to apply a certain degree of compression deformation along the circumferential direction of the bending tube through the radial clamping of the upper die (perpendicular to the bending axis), so that the axial stress is all transformed into compressive stress, that can eliminate the axial stress difference between compressive stress on inner layers and tensile stress on outer layers which caused by bending, so reaching the purpose of basically controlling springback. That is exactly the opposite of the mechanical principle of the stretch bending. This paper adopted two methods containing finite element analysis and experimental verification to explain the eliminate springback mechanism for the RHFB. It is found that the springback value decreases with the increase of compression, and there is a certain minimum punch stroke. For AHSS with strength of 800 MPa, the springback angle could be reduced by more than 95%.

Keywords: Springback defect · Hydro-forging · Bending tube · Advance high-strength steels · Mechanical analysis

1 Introduction

In recent years, the bending tube component can not only adapt to the development trend of lightweight, but also make full use of its potention of bending/torsion resistance, which has been widely used in automobile manufacturing. However, springback is the most difficult problem existing in the manufacture process of such tube component with the bending axis. In particular, with the successful development of advance high-strength steels (AHSS) (such as dual-phase (DP), quench-and-partition (QP) and transformation-induced plasticity (TRIP) steels) after 2010, the material strength can be increased to 1000 MPa, causing the springback problem to be more prominent. Meanwhile, due to the residual austenite transformation and the dislocation accumulation, the material

Y. Zhao (Ed.): ICAMM 2025, SPM 96, pp. 105–112, 2025.
https://doi.org/10.1007/978-3-032-09862-7_10

properties of AHSS will change with the deformation, further increasing the difficulty of controlling springback [1–3]. Therefore, the springback analysis and corresponding control method of AHSS bending tube component has become the hot research topic.

The existing springback control technology can be divided into two categories from the mechanics principle: the springback compensation and the stretch bending. (1) Springback compensation This method has been the most common springback control in the past few centuries, however, because of the large error between calculation value and compensation value based on theory, there needs to modify process parameters and bending dies constantly. Especially, this problem is more prominent when applied to AHSS, thus, leading to the problems of high scrap-rate, low efficiency and high cost [4]. (2) Stretch bending Through applying a certain tensile deformation along the bending axis (parallel to the bending axis) to reduce the axial stress difference in the bending deformation area, the springback can be controlled efficiently. However, when applying it to the tube component, it is very easy to produce defects such as fracture, cross-sectional distortion and wall thickness thinning, more importantly, it can not be applied to three-dimensional bending tube [5].

Through the above analysis, it can be found that the existing springback control technology has their inevitable shortcomings. Consequently, how to eliminate the springback of AHSS bending tube component and ensure its cross-section accuracy simultaneously has become an urgent problem to be solved. Chu pointed out that under the sufficient normal support, the circumferential direction of the tube can be compressed through the clamping of the upper die, and this process is called tube hydro-forging technology (THFG) [6]. Based on this conclusion and inspiring by the stretch bending method, this paper points out that THFG process also can be used to lighten the springback of AHSS bending tube component. By the radial punch stroke of the die, the tube produces circumferential compression to reduce the springback. In this work, this new technology will be explained theoretically and verified by simulation and experiment.

2 Mechanical Principle of Eliminating Springback by THFG Process

The traditional springback calculation theory is as follows:

$$\Delta\theta = \theta - \frac{M}{EI} \tag{1}$$

Here, $\Delta\theta$ is the springback value, θ is the initial bending angle, M is the bending moment, E is the elastic modulus, and I is the product of inertia. M can be obtained by integrating the moment of axial stress σhf z on the neutral layer along the thickness direction. Therefore, the moment can be reduced by eliminating the axial stress difference between the outer/inner layers in order to control the springback, which is the basic idea of THFG to eliminate springback.

Take Fig. 1 as an example to explain the mechanical principle of controlling springback by THFG, which is contrary to the stretch bending. Fig. 1a shows the axial stress σbd z distribution after bending process: the inner layer is the compressive stress and the inner layer is the tensile stress. Springback will inevitably occur after this stress

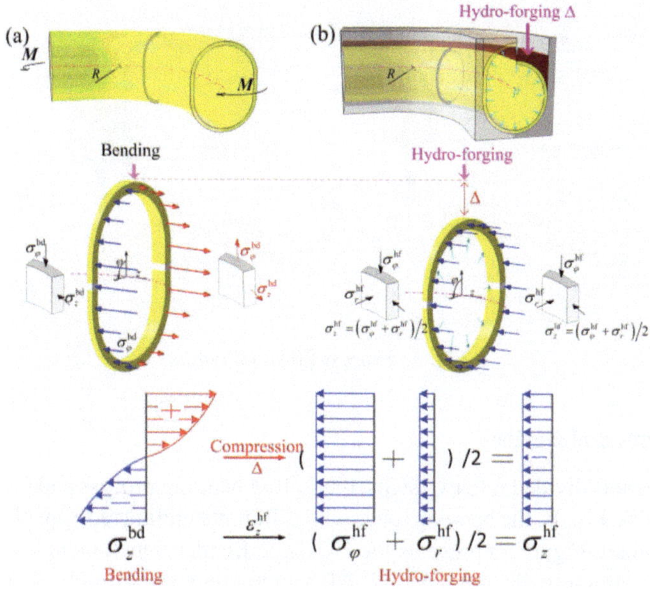

Fig. 1. Schematic diagram of THFG process mechanics.

distribution unloading. This stress characteristic can be removed by THFG process, as shown in Fig. 1b: Under the support of internal pressure *p*, a certain punch stroke Δ is continuously applied to the part along the radial direction, resulting in the circumferential compression and the subsequent yield, so that the axial stress is all changed to the compressive stress, eliminating the stress difference between inner/outer layer caused by bending. In addition, due to the plane strain state (εhf z = 0), the axial stress σhf z value is about 1/2 of the circumferential stress σhf φ. In other words, the axial stress value of each element is basically equal after THFG. So the mechanical basis of springback is eliminated and the springback also can be successfully controlled.

3 Experiment Verification

3.1 Process Design

In this work, the springback produced by the rolling bending will be eliminated by THFG. As shown in Fig. 2, the specific process flow is as follows: The bending tube obtained by rolling bending is placed in the lower die, and under the support of internal pressure *p*, the upper die closes Δ along the radial direction, that is, the circumferential perimeter of the tube is compressed by 2Δ. Fig. 2c shows the springback after bending; while Fig. 2d shows that the springback can be controlled after THFG.

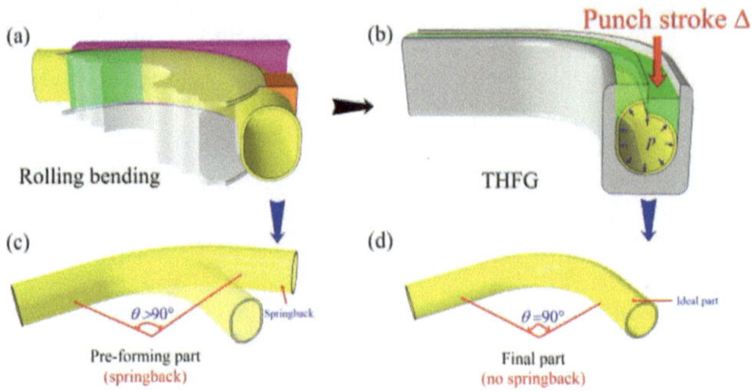

Fig. 2. Process flow diagram.

3.2 Experimental Device

The experiment is divided into two steps: the rolling bending process and the tube hydro-forging process. Firstly, the bending tubes with different bending radii are obtained by the NC Tube Bender. Figure 3a presents the NC Tube Bender which includes bending die, clamping die, pressure die and mandrel. After unloading, the springback with different bending radius can be obtained. Figure 3b is the experimental device for THFG process, which is mainly composed of the upper die and the lower die.

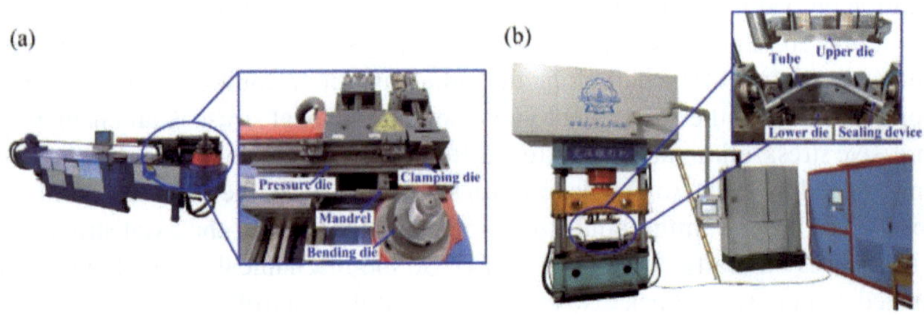

Fig. 3. Experimental device: (a) NC Tube Bender; (b) THFG device.

3.3 Analysis of Experimental Results

Relationship Between Springback Value and Punch Stroke As shown in Fig. 4, the experimental results were the DP800 tube under different punch stroke Δ when the bending radius is 150 mm and 200 mm respectively. It is obvious from the figure that with the increase of punch stroke, the deviation between the formed part and the die gradually decreases, that is, the bending springback can be controlled. For the bending radius of 150 and 200 mm, the required punch stroke Δ only needs 2 mm to control their springback basically. In addition, the cross-sectional accuracy obtained by THFG process is higher because of the combination of the internal pressure and the die.

As shown in Fig. 5, when the bending radius is 150 and 200 mm respectively, the change rule of springback value with punch stroke is shown. The maximum springback

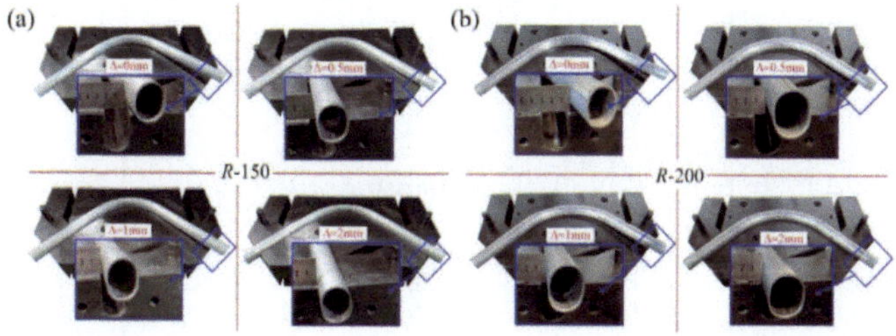

Fig. 4. Experimental results under different punch stroke: (a) *R*-150 mm; (b) *R*-200 mm.

value after bending can be reached 6° and 9° respectively. After THFG, the springback can be basically eliminated with the increase of punch stroke. Besides, it can be seen that the springback value will decrease rapidly firstly and then slowly with the punch stroke, that is, only a small punch stroke in the early stage is required in order to eliminate 70% ~ 80% of the springback, which shows the high efficiency of the process. This trend is independent of the bending radius.

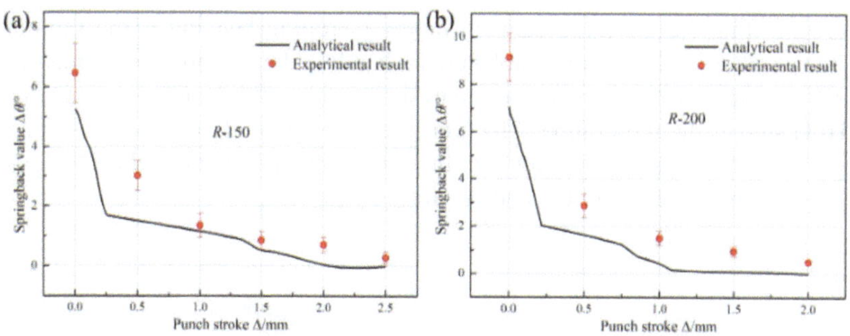

Fig. 5. Change rule of springback value with different punch stroke: (a) *R*-150 mm; (b) *R*-200 mm.

Minimum Punch Stroke Besides, we can found from Fig. 5 that when the punch stroke increases to a certain value, the springback gradually decrease to a constant value (nearly 0°). Therefore, this punch stroke can be defined as the minimum punch stroke to control the springback absolutely. As shown in Fig. 6, the theoretical and experimental results of minimum punch stroke under different bending radii are compared for DP600 and DP800 tube respectively. The minimum punch stroke required to basically control the springback is a key parameter in actual production. The minimum punch stroke decreases with the reduction of bending radius. The results show that there is minimum compression for each bending radius, which further confirms the universality of this technology.

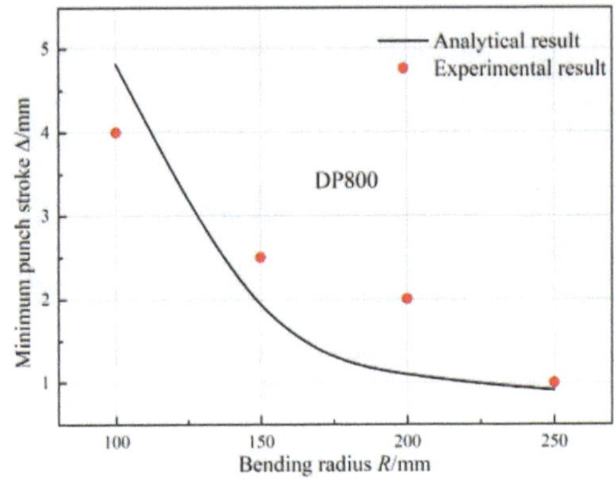

Fig. 6. Change rule of minimum punch stroke with different bending radius.

4 Finite Element Simulation

4.1 Minimum Punch Stroke

As shown in Fig. 7a, the finite element model of rolling bending is the same as the experiment, whose main components include: bending die, clamping die, pressure die, mandrel and tube. Through the restart analysis, the tube after bending is imported into the THFG step as the initial tube, which ensures the continuity of the stress and strain in the two simulation steps. Meanwhile, the finite element model of the THFG step is composed of upper die, lower die and pre-forming tube.

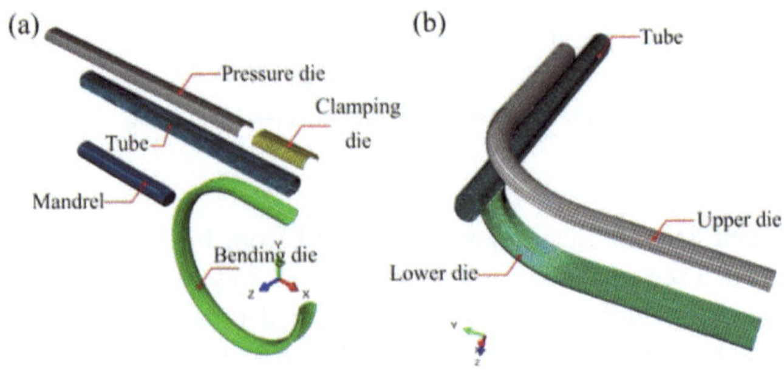

Fig. 7. Finite element model: (a) rolling bending; (b) THFG.

4.2 Mechanical Analysis

Transformation of Axial Stress. In this section, the finite element simulation is used to further reveal the mechanical principle of THFG technology to eliminate springback. When the bending radius is 150 mm, the transformation of axial stress and bending moment during THFG is shown in Fig. 8. When the punch stroke is 0.5 mm, the axial stress on the outer layer first undergoes sign transformation, from the initial tensile stress to the compressive stress. As the compression progresses, the axial stress on the inner layer begins to change, whose sign remains unchanged, but its value is reduced to half of the initial value. Finally, when the punch stroke reaches 2 mm, the axial stress of each element on the cross-section all change to the compressive stress, and according to the plane strain principle, their value is similar, which is approximately 1/2 of the circumferential stress.

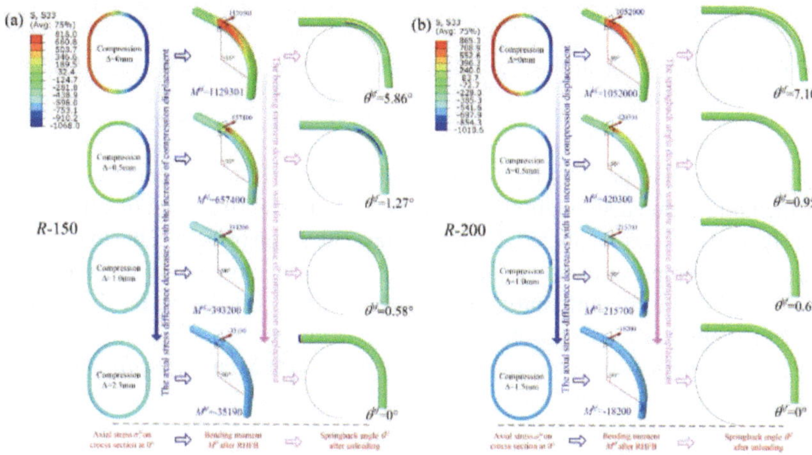

Fig. 8. Simulation results: (a) *R*-150 mm; (b) *R*-200 mm.

Transformation of Bending Moment. The transformation of bending moment on the 0° cross-section with the punch stroke in the THFG process can be shown in Fig. 9. It can be seen from the figure that the bending moment reaches 1.1 × 10^6 N*mm after the rolling bending, and its value decreases continuously with the growth of the punch stroke. When the punch stroke is 0.5 mm and 1 mm, its value is 6 × 10^5 N*mm and 4 × 10^5 N*mm respectively, until the compression reaches 2 mm, the bending moment is basically reduced to zero. Similarly, the bending moment decreases rapidly at first and then slowly with the increase of punch stroke, which verifies the change rule of the previous springback value with the punch stroke according to Eq. (1).

5 Conclusion

1. A new method to eliminate the springback of AHSS bending tube component by THFG is proposed.
2. Because the mechanical condition of springback is fundamentally eliminated, this method is less dependent on the material properties.
3. With the increase of punch stroke, the springback value first decreases rapidly and then decreases slowly.

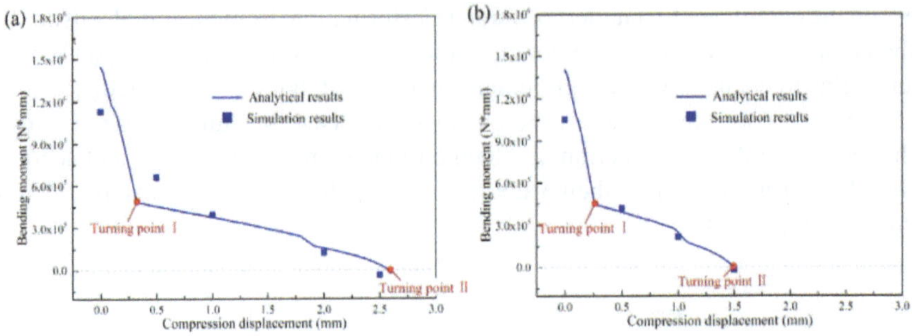

Fig. 9. Comparison of analytical and simulation results of bending moment: (a) *R*-150 mm; (b) *R*-200 mm.

4. With the growth of bending radius, the minimum punch stroke decreases gradually.
5. By the finite element simulation, the change rule of axial stress and bending moment in THFG process was analyzed. Based on the plane strain principle, the axial stress all change into compressive stress and their numerical value is also all uniform (approximately 1/2 of the circumferential stress), which further reveals the mechanism of the process to eliminate springback.
6. For advance high strength steels with strength of 800 MPa, the springback can be reduced by more than 95% after THFG.

References

1. Ghaei, A., Green, D.E., Aryanpour, A.: Springback simulation of advanced high strength steels considering nonlinear elastic unloading–reloading behavior. Mater. Design. **88**, 461–470 (2015)
2. Hassan, H., Maqbool, F., Güner, A., Hartmaier, A., Khalifa, N.B., Tekkaya, A.E.: Springback prediction and reduction in deep drawing under influence of unloading modulus degradation. Int. J. Mater. Form. **9**, 619–633 (2015)
3. Wagoner, R.H., Lim, H., Lee, M.G.: Advanced issues in springback. Int. J. Plast. **45**(45), 3–20 (2013)
4. Strano, M.: Automatic tooling design for rotary draw bending of tubes. Int. J. Adv. Manuf. Tech. **26**(7–8), 733–740 (2005)
5. Ueda, M., Ueno, K., Kobayashi, M.: A study of Springback in the stretch bending of channels. J. Mater. Process Tech. **5**(3–4), 163–179 (1981)
6. Chu, G.N., Chen, G., Lin, Y.L.: Tube hydro-forging – a method to manufacture hollow component with varied cross-section perimeters. J. Mater. Process Tech. **265**, 150–157 (2019)

Microstructure and Mechanical Behaviour Comparison between Conventional and AISI H11 Tool Steel Material (1.2343) Developed by Additive Manufacturing Wire Laser-Based DED Technology (DED-LB/M)

J. Ureña(✉), A. García-Díaz, and M. Alvarez-Leal

Advanced Materials Department, Technology Centre of Metal-Mechanical and Transport (CETEMET), Santana Science and Technology Park for Transport, Avda. 1° de Mayo s/n, 23700 Linares, Spain

j.urena@cetemet.es

Abstract. The main aim of this research was to study and compare the microstructure and mechanical properties of AISI H11 hot work tool steel samples produced by wire laser-based DED technology (DED-LB/M) with respect to the conventional material. Material properties of Material properties of metal Additive Manufacturing (AM) materials differ significantly from those produced by conventional methods due to the numerous AM phenomena, like thermal gradients, high cooling rates or complex parameters could successfully result in materials with outstanding properties. Dense samples were achieved by the process parameters developed with a material densification up to 99.89% and 99.75%, in the horizontal and vertical directions, respectively. Microstructural analysis and hardness measurements were also carried out. Horizontal and vertical samples were manufactured from the as-built H11 material obtained by DED wire to machine cylindrical tensile samples according to ISO 6892 standard. Tensile specimens were also machined from conventional AISI H11 material before and after performing the conventional heat treatment for AISI H11. Results for the conventional material indicated tensile strength of 718 MPa and 1435 MPa, before and after heat treatment, respectively. Whereas for the material developed by additive manufacturing in as-built condition, tensile strength values of 1778 MPa and 925 MPa in horizontal and vertical directions, respectively, were obtained. These promising properties could lead to an innovative method for processing the AISI H11 tool steel by wire DED-LB/M in one single step without the need for heat treatment and leading to properties even higher compared to the heat-treated conventional material. This could potentially save costs and time on the overall production of industrial components.

Keywords: Directed Energy Deposition (DED) · Wire Laser-based DED (DED-LB/M) · AISI H11 tool steel (1.2343) · process parameters · microstructure · mechanical properties

Y. Zhao (Ed.): ICAMM 2025, SPM 96, pp. 113–124, 2025.
https://doi.org/10.1007/978-3-032-09862-7_11

1 Introduction

AISI H11 steel, also designated as 1.2343 or X37CrMoV5-1, contains chromium (Cr) as the main element and uses molybdenum (Mo), silicon (Si), and vanadium (V) as its main alloying elements. Compared to other tool steels, it is characterized by relatively low carbon content and a high chromium percentage, which classifies it as a chromium hot-work steel [1, 2]. This chemical composition provides the steel with excellent toughness, high hardness, good ductility, as well as resistance to elevated temperatures and wear, high resistance to thermal shock and good thermal conductivity [2–4]. It is particularly suitable for high-temperature applications that involve extreme loads and highly stressful conditions [2], such as forging tools, shearing blades, extrusion tools, punches, screwdrivers, steel pins, shafts, and gears [1, 5, 6]. In addition, AISI H11 is also especially well-suited for applications in the plastic processing industry, such as the manufacturing of injection molds, where a combination of thermal and mechanical resistance is required [7].

In the context of injection molding, numerous studies have investigated AISI H11 steel to optimize its performance and adapt it to new manufacturing technologies. Another relevant approach evaluates the feasibility of processing AISI H11 using LPB-F without preheating the platform, with the aim of implementing it in injection molds [8]. Additionally, the effect of PVD coatings of CrN and CrAlN on AISI H11 cores has been studied, aiming to reduce ejection forces during the demolding of polypropylene parts [3]. Other investigations have focused on the development of inserts using multimaterial AM by metal extrusion (MEX/M), combining AISI H11 (selected for its high toughness) with AISI A2 and AISI 420 steels, known for their wear and corrosion resistance, respectively [7]. Various alternatives to the maraging steel 18Ni300, including AISI H11, have also been evaluated for mold manufacturing using the LPB-F 3D printing technology [9]. Finally, the use of metal-ceramic joints, particularly between zirconium oxide (ZrO_2) and AISI H11, has been explored to enhance the functional properties of plastic injection molds [6].

As demonstrated in previous research, Additive Manufacturing (AM) presents an effective alternative for both the production and repair of components, promoting cost reduction and increased process productivity. In this context, in recent years, AM has emerged as a promising alternative to traditional manufacturing methods, particularly to produce complex geometries, component repair, and material usage optimization. Among the various AM technologies, directed energy deposition using laser and metallic powder (DED-LB/M) has been established as an effective and well-established process for the repair and maintenance of injection molds or previously mentioned forming tools [10].

Several studies have explored the feasibility of processing AISI H11 steel using DED-LB/M. Hentschel et al. investigated the effect of adding tungsten carbide nanoparticles (WC-NP) at various concentrations (1, 2.5, and 5% by weight) to AISI H11 steel powder mixtures processed via DED-LB/M, aiming to analyze their influence on the processability, microstructure, and mechanical properties of the resulting samples [10]. In another study by the same authors, the DED-LB/M process of AISI H11 powder mixtures was investigated, modified by the addition of C-NP at concentrations up to 0.5% by weight, focusing on the influence of nanoscale carbon additives on hardness and mechanical properties [11]. Bold et al. compared the processability of HS DED-LB/M technology using a conventional AISI H11 alloy and a modified version (AISI H11m), to reduce the amount of austenite retained in the microstructure and minimize the need for subsequent heat treatments [12]. Meanwhile, Vetter et al. evaluated a coating strategy using DED of AISI H11 tool steel on a porous AISI H11/Cu preform, without the use of nickel-based interlayers [13].

As evidenced in the previous studies, the use of DED-LB/M technology for processing AISI H11 steel has been widely investigated. Nevertheless, most of these studies employ the material in powder form. In contrast, the application of AISI H11 in wire form is significantly less common, and research on this approach remains limited. A relevant study on this variant was conducted by Oliari et al., who carried out a parametric analysis of the additive manufacturing process by wire laser metal deposition (LMD-W) applied to AISI H11 hot work tool steel. The research focused on the optimization of the main deposition parameters, such as laser power, travel speed and wire feed speed, as well as on the layer-by-layer build strategy [5].

Intending to advance the research on the development of process parameters for AISI H11 hot tool steel using wire-based laser DED technology, the main objective of the present study was to investigate and compare the microstructure and mechanical properties of the produced samples with respect to the conventional material by using a low energy source of 1.2 kW to succesfully manufacture H11 parts by DED-wire with outstanding properties for the industry of injection molding.

2 Experimental Procedure

2.1 Material and Additive Manufacturing by Wire Laser-Based DED (DED-LB/M)

The AISI H11 conventional material, also named DIN (X 37 CrNiV 5-1); AFNOR (Z 38 CDV 5) or UNI (X 37 CrMoV 5-1 KU), was supplied by Meusburger Georg GmbH &Co KG (Austria) with Electro-Slag Remelted (ESR) quality suitable for high gloss polishing due to homogenous structure. Chemical composition (wt%): 0.38% C, 1.00% Si, 0.40% Mn, 5.30% Cr, 1.20% Mo, 0.40% V and Fe base.

The AISI H11 used for the additive manufacturing DED process was classified as EN14700 and supplied by Nippon Gases Welding & Cutting (Spain) in 1 mm diameter wire with the following chemical composition (wt%): 0.37% C, 1.00% Si, 0.40% Mn, 7.00% Cr, 2.20% Mo, 0.45% V, 0.30% Ti, $<$ 0.20% Ni and Fe base.

The Laser DED AM system, equipped with a 1.2 kW multilaser system composed of 6 individual fiber coupled diode lasers of 976 mm wavelength and 200 W each, was used to manufacture the samples. Argon with a purity higher than 99.996% was used as shielding gas the during the manufacturing process.. The effect of layer thickness wanted to be investigated in this research from the point of view of material densification and overhang processability. Therefore, the same process parameters combination was employed in this research, except the layer thickness parameter, which was 0.8 mm and 1 mm in process parameters set 1 (PP1) and set 2 (PP2), respectively, as summarized in Table 1.

Table 1. DED-wire process parameters for AISI H11 tool steel (1.2343).

Process param. ID	Print speed *(mm/min)*	Layer height *(mm)*	Laser power *(W)*	Track spacing *(mm)*	Gas flow *(L/min)*
PP1	300	0.8	1100	1	10
PP2	300	1	1100	1	10

2.2 Material Densification and Microstructure Analysis

The microstructural analysis was carried out by optical microscopy (Axiolab A1 equipped with an AxioCam ERc, Carl Zeiss ICS and ZeissCore software) after cross-sectioning and surface preparation by standard metallographic techniques to evaluate the density level of the different materials processed. The surface of the specimens was ground successively from 600 to 4000 grit with SiC papers, followed by polishing with 1 μm diamond suspension and finally with colloidal silica suspension. Material densification, porosity and defects of the AM samples were analyzed for samples processed under PP1 and PP2 process parameters configurations. Both horizontal and vertical directions were examined to assess the effect of the layer thickness parameter in the material anisotropy. Image analysis was carried out using ImageJ software to represent the average value after analyzing four images for each manufacturing plane. Then, the as-polished samples were chemically etched for a few seconds to reveal the microstructures by using a Vilella's etchant composed of picric acid, HCl and alcohol. Microstructure analysis in the XY (horizontal) and XZ (vertical) directions of the additive manufactured samples before and after heat treatment, as well as for the conventional material, was carried out.

2.3 Heat Treatment Process

Heat treatment of the conventional and DED processed AISI H11 tool steel material was carried out in NaberTherm LH 30/14 furnace. Conventional samples were first annealed to reduce hardness for easy machining, and then, optimized in terms of hardness and toughness by means of hardening and tempering (fully-heat-treatment). Specimens named Conventional-annealed were heated at temperature of 800 °C at a heating speed of 10 °C/min; temperature was held for 4 h and then, air cooling at room temperature. Other conventional samples followed the fully heat treatment recommended for tool steels. These samples were submitted to a hardening treatment after the annealing process. They were heated to the austenitization temperature selected of 1020 °C at a heating speed of 10 °C/min. Specimens were held at this temperature for 1 h and then, air cooling at room temperature. After that, specimens were subsequently double tempered at 550 °C for 1 h at each tempering stage. These specimens were named Conventional-fully-heat-treated.

Lastly, the additive manufactured as-built DED specimens directly were submitted to hardening plus double tempering. Temperature and holding times were the same as employed for the conventional samples.

2.4 Hardness Characterization

Hardness Rockwell C was measured for the additive manufactured and conventional material at the different conditions. 5 measurements were performed for each condition for reliability purposes. Results were represented as the average value with the standard deviation. Hardness measurements were performed by a Wilson Rockwell C hardness tester where a loading force of 150 Kg and 15 s indentation time were effective for achieving accurate measurements.

2.5 Tensile Testing

Tensile samples were machined from different AISI H11 tool steel conditions (conventional-annealed, conventional fully-heat-treated; and horizontal and vertical DED-as-built and DED-fully-heat-treated) for mechanical characterization. The mechanical anisotropy in the additively manufactured samples was considered (XY and XZ directions). Cylindrical specimens of 5 mm diameter x 32 mm length were tested in an INSTRON universal tester at room temperature according to ISO 6892-1:2020B.

3 Results and Discussion

3.1 Production of the AISI H11 Specimens by Additive Manufacturing (AM)

The multilaser head deposition DED-LB/M wire-based system used for processing H11 small cubes and overhang samples, with two different process parameters combinations, is shown in Fig. 1. The effect of manufacturing with 0.8 mm (PP1) or 1 mm (PP2) layer thickness was studied in terms of material properties and processability. From the point of view of material processability, both conditions were suitable for manufacturing simple geometries like small cubes but also overhang samples with different angles (10°, 15°, 20° and 25°) without any support structures. Results indicated that overhang angles up to 20° were successfully processed with high quality surface finishing. This was not observed for the 25° sample since the wire material came off.

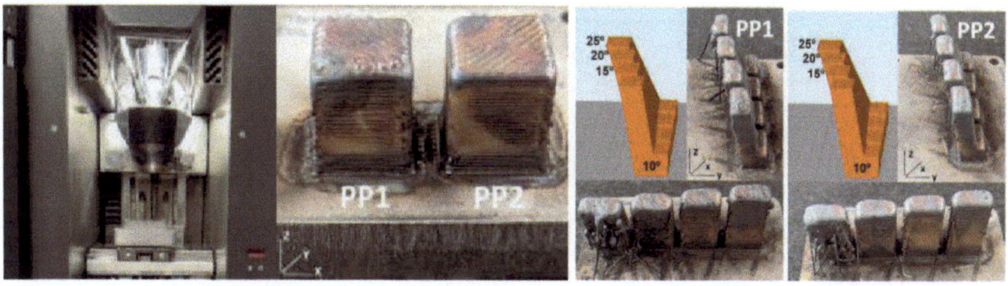

Fig. 1. Images showing from left to the right: Multilaser head deposition (M450) DED-wire technology located at CETEMET facilities (Spain), small cubes manufactured for material densification and microstructure characterization, and overhang samples with layer thickness of 0.8 mm and 1 mm labeled PP1 and PP2, respectively.

Material densification and porosity of the DED as-built specimens manufactured through PP1 and PP2 process parameters were analyzed in both directions, parallel and perpendicular to the material deposition layers, XY and XZ, respectively, as represented in Fig. 2. In general, PP1 (Fig. 2a–b) led to higher porosity compared to PP2 (Fig. 2c–d), and this was even more evident in the case of XZ direction, where the material densification achieved in PP1 and PP2 was 99.52% and 99.75%, respectively. Results were more similar for the XY direction in both samples, 99.69% (PP1) and 99.89% (PP2). Therefore, layers with 1 mm layer thickness (PP2 process parameters) resulted in better material densification with almost full density and no porosity, which means that microstructure, hardness and tensile behaviour were tested for samples processed with the PP2 configuration.

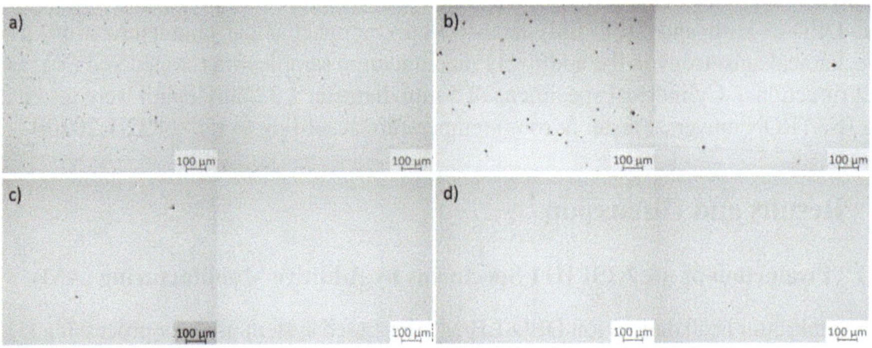

Fig. 2. Optical microscopy results showing the material densification level and porosity of the DED-wire manufactured AISI H11 as-built samples: a) PP1-xy, b) PP1-xz; c) PP2-xy and d) PP2-xz.

Results from the machining of tensile specimens from H11 DED-as-built in the horizontal direction, H11 DED-as-built in the vertical direction and H11 conventional-annealed were shown from left to right in Fig. 3. Six tensile samples were successfully machined for each condition with high quality result.

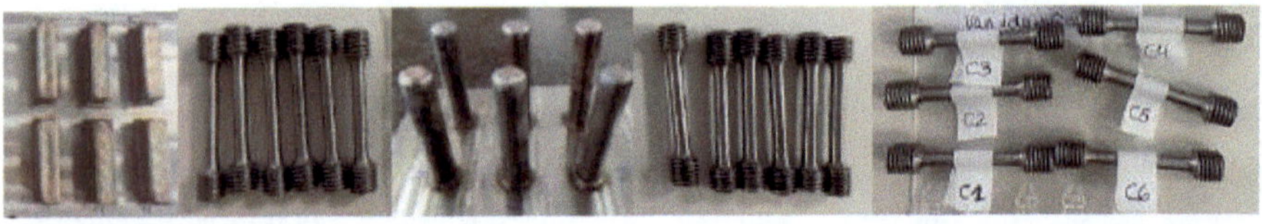

Fig. 3. Tensile specimens (from left to right) of H11 DED-as-built-xy, H11 DED-as-built-xz material and H11 conventional-annealed.

Six heat-treated tensile samples for each material condition (H11 DED-fully-heat-treated horizontal and vertical, and H11 conventional-fully-heat-treated) are presented, respectively, in the upper, middle and bottom at Fig. 4. All of them indicated good visual aspect without presence of defects.

Fig. 4. Tensile specimens of H11 DED-fully-heat-treated-xy, H11 DED-fully-heat-treated-xz and H11 conventionally-fully-heat-treated (upper, middle and bottom), respectively. Images at the different stages "as-built, heat treated and machined" (from left to the right).

3.2 Microstructure Results

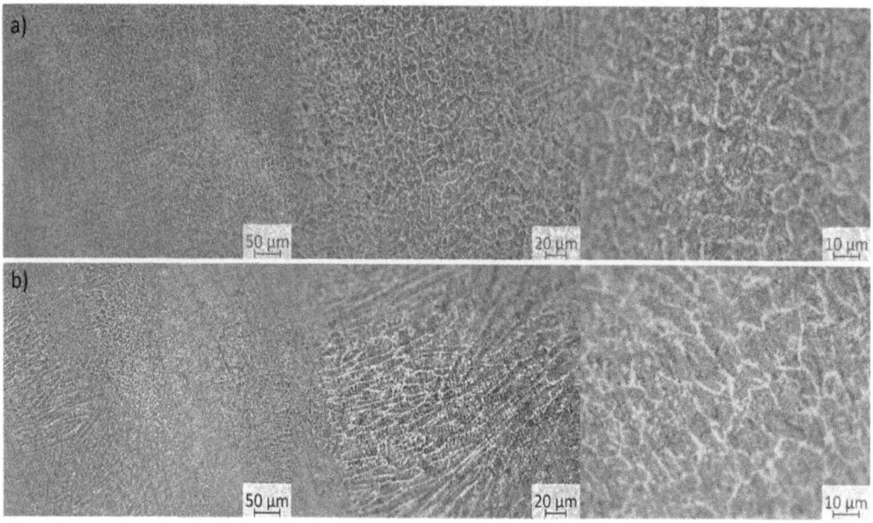

Fig. 5. Microstructure results of the DED-wire manufactured AISI H11 samples with highest level of material densification (PP2 process parameters) in as-built condition: a) xy-plane (horizontal direction) and b) xz-plane (vertical direction).

Microstructural characterization of the DED-wire manufactured AISI H11 tool steel processed with PP2 process parameters in as-built condition was represented at different magnifications in Fig. 5-a (xy-plane) and Fig. 5-b (xz-plane) with the aim to observe the material anisotropy. Mainly, very fine equiaxial grains with a grain size around 10 μm were found in the horizontal direction (Fig. 5-a) while some similar equiaxial but also very thin columnar grains were evidenced in the vertical direction as shown in Fig. 5-b. Results with very similar equiaxial grain sizes were reported in a similar study [5].

Tool steels are typically subjected to an initial annealing process to reduce their hardness and facilitate machining. Subsequently, they are optimized for both hardness and toughness through quenching and tempering, a process known as fully-heat-treatment. Microstructures of AISI H11 tool steel conventional-annealed and fully-heat-treated are shown in Fig. 6-a and Fig. 6-b, respectively. After performing the full heat-treatment, grain sizes were bigger and precipitation of carbide particles or inter-metallic phases were found, as expected [14].

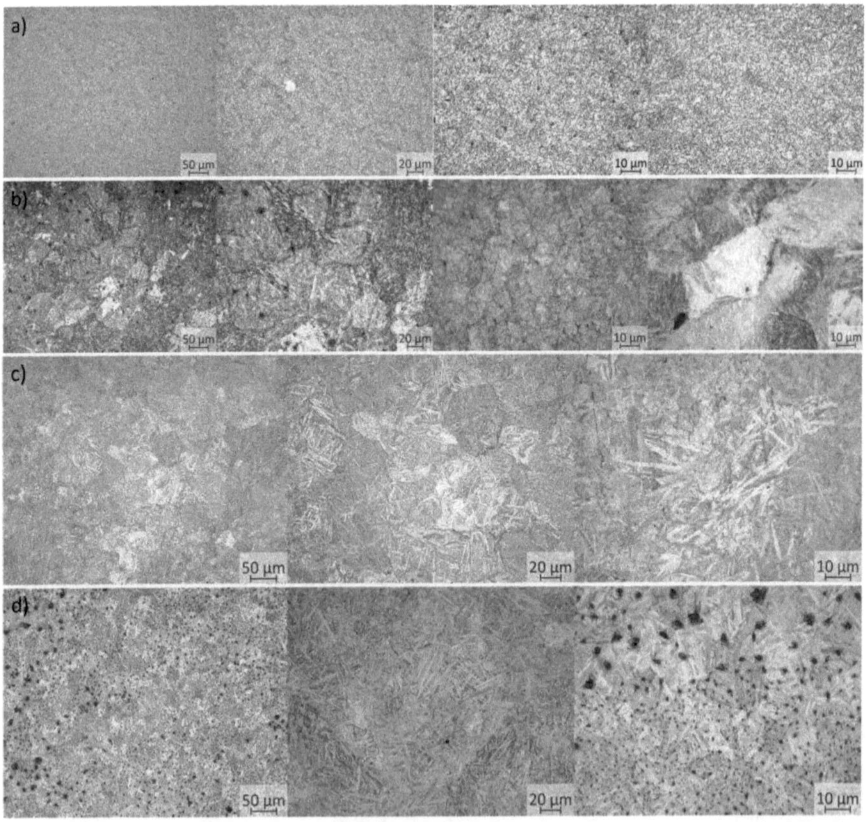

Fig. 6. Microstructure comparison results of the Conventional and DED-wire manufactured AISI H11 material after the different heat-treatments applied: a) conventional-annealed, b) conventional-fully-heat-treated, c) DED-fully HT-xy, and d) DED-fully HT-xz material.

In the case of the DED samples, microstructure in as-built condition was represented in Fig. 5-a and Fig. 5-b, which can be compared to microstructure after fully heat-treatment in Fig. 6-c and Fig. 6-d. Similar to the conventional AISI H11 tool steel, grain growth was observed after full-heat-treatment, which led to grain sizes around 30–50 μm. Martensitic laths were generated after the tempering process, a finding that is corroborated by other researchers [15]. However, carbide or intermetallic phases did not appear uniformly, since they were found only in the xz direction (Fig. 6-d). This could have a negative impact on mechanical performance in the vertical direction.

3.3 Hardness and Tensile Behaviour

Hardness values of the conventional and DED-processed hot-work tool steel at different heat treatment conditions were summarized in Table 2 to enable direct comparison. For the two conventional cases, hardness results showed similar behaviour to the as-built DED samples (40-46 HRC), which means that DED samples without heat treatment reached the same hardness as annealed and fully-heat-treated conventional H11 tool steel. Then, hardness increased in DED fully-heat-treated samples up to 56 HRC in both directions, XY and XZ. This behaviour shows good agreement with other studies where hardness varies between 51 HRC and 59 HRC depending on hardening and double tempering temperatures, and the highest value was reported for the highest hardening temperature at 1150 °C [16].

Table 2. Rockwell C hardness results of AISI H11 conventional-annealed, conventional-fully-heat-treated, DED-as-built-xy, DED-as-built-xz material, DED-fully HT-xy, DED-fully HT-xz material.

Rockwell C hardness of the AISI H11 material					
H11 Conv-annealed	*H11 Conv-fully HT*	*H11 DED-AB_XY*	*H11 DED-AB_XZ*	*H11 DED-fully HT_XY*	*H11 DED-fully HT_XZ*
46 ± 0,6	42 ± 0,2	46 ± 1,9	40 ± 0,8	56 ± 1,9	56 ± 0,7

Mechanical results of the conventional (annealed and fully-HT), and as-built DED H11 tool steel materials are presented in Table 3. On one hand, although DED-fully-heat-treated samples were analyzed, four of the six horizontal and vertical tensile specimens broke up before reaching RP0.2%. However, two horizontal-HT tensile specimens gave 2129 MPa of ultimate tensile strength, 1882 MPa yield strength and 18% elongation. Nevertheless, those results were not shown together with the rest of the successfully tested conditions since they were not reliable enough. This can be attributed to a non-homogeneous heat treatment, which probably led to an inhomogeneous microstructure as shown in Fig. 6c–d.

Table 3. Mechanical behaviour of the AISI H11 conventional-annealed, conventional-fully-heat-treated, DED-as-built-xy and DED-as-built-xz material.

Tensile behaviour of the AISI H11 material				
	H11 Conv-annealed	*H11 Conv-fully HT*	*H11 DED-AB_XY*	*H11 DED-AB_XZ*
Tensile strength (MPa)	718 ± 48	1435 ± 12	1778 ± 94	925 ± 68
Yield strength (MPa)	527 ± 36	1203 ± 16	1559 ± 64	888 ± 87
Elongation at break (%)	18 ± 2	10,5 ± 1	2,2 ± 1	1,4 ± 1

On the other hand, it was worth it to notice the positive results of the as-built DED horizontal and vertical specimens compared to the conventional-annealed material. Both materials (as-built DED and conventional-annealed) were not heated to austenitization temperature, however, ultimate tensile and yield strength were higher in the as-built DED material, both horizontal and vertical. Then, comparing H11 DED as-built_XY and H11 conventional-fully-HT, it can be pointed out that the highest result was shown by the as-built horizontal DED H11 (UTS = 1778 MPa and YS = 1559 MPa), being more than 300 MPa higher as compared to the conventional material after the full heat treatment of hardening and double tempering. Reduced elongation of just 2% was obtained in the DED as-built samples, similar results were reported in other studies [16]. However, this can be increased up to 12%, finding the right conditions for heat treatment. Therefore, the high tensile and yield results achieved even in as-built condition without any heat treatment, means that the H11 DED processed in horizontal direction would be suitable to be directly used without any heat treatment, giving excellent mechanical properties. This will reduce the number of stages of processing the AISI H11 tool steel, decreasing time and cost of manufacturing with even better mechanical behaviour than fully-heat-treated conventional material.

4 Conclusions

The present study aimed to evaluate and compare microstructure and mechanical properties between the developed AISI H11 hot work tool steel by Directed Energy Deposition (DED) wire-based technology and conventional material. The effect of a 0.8 mm or 1 mm layer thickness process parameters on material densification and overhang samples processability were studied. Higher material densification level, 99.89% (horizontal direction) and 99.75% (vertical), together with very low porosity results, were reached with 1 mm layer thickness. Thus, the process parameters were validated, and the microstructure, hardness and tensile study was conducted with the validated PP2 (1 mm layer thickness) set of process parameters.

The effect of heat treatment was studied on microstructure, hardness and tensile properties. Conventional samples were first submitted to just a single annealing process, and then, other conventional samples were also submitted to hardening and double tempering. These routes were selected since they are used in the tooling industry. Then, as-built DED specimens were processed in horizontal and vertical directions and were tested in as-built and after performing the same hardening and double tempering cycles.

Microstructure revealed very fine equiaxial grains in the as-built DED samples with some very thin columnar grains in the vertical direction. Carbide precipitation particles or intermetallic phases were found in the vertical DED samples after the fully heat treatment, however, they were not observed in the horizontal DED-HT samples. Martensitic laths were evidenced in both directions after fully heat treatment. The AISI H11 conventional material also showed grain growth and carbide or intermetallic phases formation after fully heat treatment.

Conventional conditions resulted in similar hardness as horizontal and vertical as-built DED samples (40-46 HRC). The highest hardness results were found in the horizontal and vertical DED samples (56 HRC).

In terms of mechanical properties, the highest result was shown by thc as-built horizontal DED H11 (UTS = 1778 MPa and YS = 1559 MPa), being more than 300 MPa higher as compared to the conventional material after the full heat treatment of hardening and double tempering. However, a reduced elongation was obtained in the as-built samples (2%), as expected, since elongation increases with the corresponding hardening and tempering treatments. This means that the H11 DED processed in the horizontal direction would be suitable to be directly used without any heat treatment, giving excellent mechanical properties. This will reduce the number of stages of processing the AISI H11 tool steel, decreasing time and cost of manufacturing with even better mechanical behaviour than fully-heat-treated conventional material.

References

1. Work, H., Steel, T.: Alta special steel 1.2581. Alta Steel, no. mm. No.86 road, huangshi, China
2. Mondal, A., Pilone, D., Brotzu, A., Felli, F.: Effect of heat treatment on mechanical properties of FeMnAlC alloys. Procedia Struct. Integr. **33**, 237–244 (2021). https://doi.org/10.1016/j.prostr.2021.10.029
3. Tillmann, W., et al.: Reduction of ejection forces in injection molding by applying mechanically post-treated CrN and CrAln PVD films. J. Manuf. Mater. Process. **3**(4), 88 (2019). https://doi.org/10.3390/jmmp3040088
4. Šebek, M., Falat, L., Kováč, F., Petryshynets, I., Horňak, P., Girman, V.: The effects of laser surface hardening on microstructural characteristics and wear resistance of AISI H11 hot work tool steel. Arch. Metall. Mater. **62**(3), 1721–1726 (2017). https://doi.org/10.1515/amm-2017-0262
5. Oliari, S.H., D'Oliveira, A.S.C.M., Schulz, M.: Additive manufacturing of H11 with wire-based laser metal deposition. Soldag. Inspeção. **22**(4), 466–479 (2017). https://doi.org/10.1590/0104-9224/si2204.06
6. Tillmann, W., Anar, N.B., Manka, M., Wojarski, L., Lehmert, B.: Development and construction of AISI H11/ZrO2 joints for injection molding tools. Weld. World. **63**(6), 1861–1871 (2019). https://doi.org/10.1007/s40194-019-00800-6
7. Vetter, J., Beneder, S., Schmidt, M.: Material-tailored tool inserts for injection moulding using material extrusion multi-material additive manufacturing. Procedia CIRP. **124**, 231–234 (2024). https://doi.org/10.1016/j.procir.2024.08.106

8. Fryzowicz, K., Dziurka, R., Bardo, R., Marciszko-Wiąckowska, M., Bała, P.: Point- by-point laser exposure for crack susceptibility reduction in Powder Bed Fusion processing of H11 tool steel. J. Mater. Process. Technol. **316**, 2023 (2022). https://doi.org/10.1016/j.jmatprotec.2023.117946
9. Saby, Q., et al.: Laser powder bed fusion printability of cobalt-free steel powders for manufacturing injection molds. Addit. Manuf. **44**, 2021 (2020). https://doi.org/10.1016/j.addma.2021.102031
10. Hentschel, O., et al.: Experimental investigations in the processing of AISI H11 powder blends enriched with tungsten carbide nanoparticles for the additive manufacturing of tailored hot working tools in the directed energy deposition (DED-LB/M)— impact of tungsten carbide nanoparticles on microstructural and mechanical characteristics. Metals. **14**(2), 188 (2024). https://doi.org/10.3390/met14020188
11. Hentschel, O., Kohlstruck, J., Krakhmalev, P., Nikas, D., Schmidt, M.: Processing of carbon nanoparticle-enriched AISI H11 tool Steel powder mixtures in DED-LB/M for the AM of forging tools with tailored properties (part II): influence of nanoscale carbon additives on microstructure and mechanical properties. Alloys. **2**(4), 288 (2023). https://doi.org/10.3390/alloys2040020
12. Bold, M.N., Raffeis, I., Adjei-Kyeremeh, F., Schleifenbaum, J.H., Bührig-Polaczek, A.: Rapid alloy development using Calphad simulation and powder blends in direct energy deposition. Metals (Basel). **14**(1), 1–15 (2024). https://doi.org/10.3390/met14010079
13. Article, J., Vetter, J., Hentschel, O., Chechik, L., Schmidt, M.: Interlayer-free laser coating of AISI H11 tool steel for H11 / Cu composite material general information, 9–12 (2024). https://doi.org/10.2351/7.0001533
14. Leskovšek, V., Šuštaršič, B., Jutriša, G.: The influence of austenitizing and tempering temperature on the hardness and fracture toughness of hot-worked H11 tool steel. J. Mater. Process. Technol. **178**(1–3), 328–334 (2006). https://doi.org/10.1016/j.jmatprotec.2006.04.016
15. Zhang, Z., Delagnes, D., Bernhart, G.: Microstructure evolution of hot-work tool steels during tempering and definition of a kinetic law based on hardness measurements. Mater. Sci. Eng. A. **380**(1–2), 222–230 (2004). https://doi.org/10.1016/j.msea.2004.03.067
16. Skela, B., Sedlaček, M., Kafexhiu, F., Podgornik, B.: Influence of microstructure and mechanical properties of hot-work tool steel on wear resistance subjected to high-stress wear conditions. Tribol. Lett. **68**(2), 1–12 (2020). https://doi.org/10.1007/s11249-020-01300-1

Effect of Laser Remelting on Microstructure, Porosity and Surface Roughness of Laser Direct Energy Deposited Inconel 718

Manik A. Patil[1(✉)], Dhananjay M. Kulkarni[1], Biswajit Das[1], and Digvijay G. Bhosale[2]

[1] Department of Mechanical Engineering, Birla Institute of Science and Technology Pilani, K. K. Birla Goa Campus, Goa 403 726, India
p20210062@goa.bits-pilani.ac.in
[2] Department of Mechanical Engineering, Dr. D. Y. Patil Institute of Technology, Pune, Maharashtra 411 018, India

Abstract. In laser-based additive manufacturing, surface modification to enhance corrosion resistance, wear performance, and hardness has become a prominent topic of discussion. However, these properties depend upon the microstructure and precipitation evolved during the solidification of laser-directed energy deposited Inconel 718 (L-DEDed). The cooling rate affects the microstructure and phase transformation during laser remelting. Thermal history was captured using Infrared camera during laser remelting to determine the cooling rate. Equiaxed grains and a lower percentage of Laves phase were observed for an accelerated cooling rate. Furthermore, surface roughness for various laser parameters was investigated. The result shows a decrease in surface roughness with reduced scanning speed at constant laser power. Also, porosity in the L-DEDed samples was examined utilizing a scanning electron microscope (SEM). It is observed that porosity in the L-DEDed samples was reduced after laser remelting. This work emphasizes the significance of laser remelting, which enhances the clad quality and refines the microstructure to strengthen the L-DEDed Inconel 718 samples.

Keywords: laser direct energy deposition · Inconel 718 · laser remelting · microstructure · surface roughness · porosity

1 Introduction

A nickel-based heat-resistant precipitation hardenable alloy, Inconel 718 finds extensive use in complex structural applications, including the aerospace, automotive, and marine industries (Fig. 1). Inconel 718 is strengthened through precipitation hardening via γ' (Ni3Al/Ti) and γ'' (Ni3Nb) phases. The alloy additionally comprises Laves phase (Ni2Nb), δ phase (Ni3Nb), and secondary carbides (MC) [1, 2].

The predominant techniques for laser additive manufacturing of metals are Electron Beam Melting (EBM) and Laser Metal Deposition (LMD). LMD is known as Direct Metal Deposition (DMD), Laser Engineered Net Shaping (LENS), Laser Cladding [3,

Y. Zhao (Ed.): ICAMM 2025, SPM 96, pp. 125–133, 2025.
https://doi.org/10.1007/978-3-032-09862-7_12

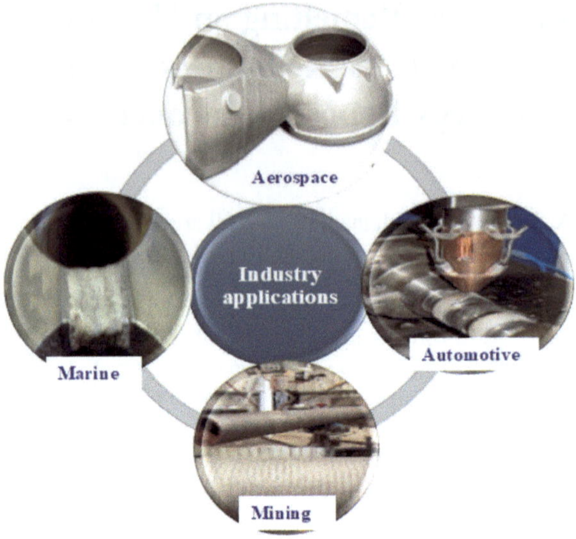

Fig. 1. Applications of Inconel 718

4]. The comprehensive classification of additive manufacturing is illustrated in Fig. 2. L-DED is a extensively used additive manufacturing technique for fabricating and repairing high-performance metal components, especially in aircraft [5, 6]. However, L-DED process components have defects like surface roughness [7], porosity [8, 9], and residual stress [10], which affect their mechanical performance.

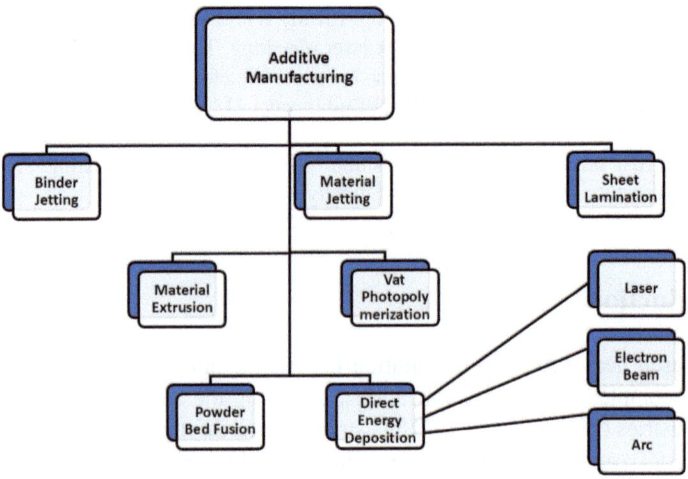

Fig. 2. Classification of additive manufacturing [3]

L-DED results in high surface roughness owing to uncontrolled solidification and layer-by-layer deposition. Laser remelting smoothens the irregularities in the deposited structure and improves the surface finish. This helps to enhance the functional performance of L-DED manufactured structures, especially in aerospace applications [10]. Also, the components obtained from L-DED sometimes have porosity in the laser-cladded part. This can reduce the strength and may result in crack initiation later. Laser remelting eliminates the pores by removing trapped gases and improving density. This leads to a fully densified structure, which leads to better mechanical strength [11].

Laser remelting also promotes grain refinement, improving toughness and performance. Laser remelting helps with stress relaxation, which reduces the probability of cracks and distortion. Shao et al. [12] examined the influence of laser remelting on the microstructure of IN718 produced through LPBF. The Laser remelting works as a grain modifier, removing stray grains and facilitating $\langle 001 \rangle$ BD columnar growth. Optimized parameters yield a nearly directional solidification structure with enhanced texture regulation. Jardon et al. [13] assess single-track roughness through diverse measurement methods and determine critical process parameters influencing roughness. Laser remelting is suggested as a remedy, attaining a roughness reduction of up to 86% with optimal laser power configurations.

Hodgir et al. [14] emphasize the efficacy of laser remelting in enhancing the performance of L-DED components. Laser remelting diminishes porosity by 72%, augments micro-hardness by 13%, and noticeably enhances surface finish by 68%.

2 Methodology

2.1 Laser Direct Energy Deposition

The Inconel 718 samples were fabricated using a L-DED setup with a 2 kW fiber laser and coaxial nozzle. The laser power was 1500 W, and the scan rate was 960 mm/min. The laser spot diameter is 3 mm, and the unidirectional scan pattern is used during the L-DED process. A handheld pyrometer (Make: Micro Epsilon) was used to monitor the molten pool temperature. The Inconel 718 substrate of $50 \times 50 \times 5$ mm size and Inconel 718 powder of size 45–90 μm was used.

2.2 Material Characterization

L-DED samples were cross-sectioned using a wire-EDM machine (ELECTRONICA ULTIMA OS), polished with SiC papers, and subsequently with diamond paste to attain a mirror finish. Afterwards, polished coupons were etched using Kalling's No. 2 etchant to examine the grain structure. The microstructure was examined using a SEM (Quanta-FEG-250, ELECMI-Spain).

2.3 Laser Remelting

The L-DEDed samples were examined for porosity, surface roughness, and microstructure. These samples were laser remelted (Fig. 3) using a 1.1 kW fiber laser (YLR-1000, IPG, Bangalore) with a D50 collimator, F300 focusing lens, and Gemorian chiller. The laser remelting parameters are mentioned in Table 1.

Table 1. Laser Remelting Parameters

Parameter	Values
Laser Power (W)	700,900,1100
Scanning Speed (mm/min)	600,720,840,960
Laser Beam Diameter (mm)	1 mm

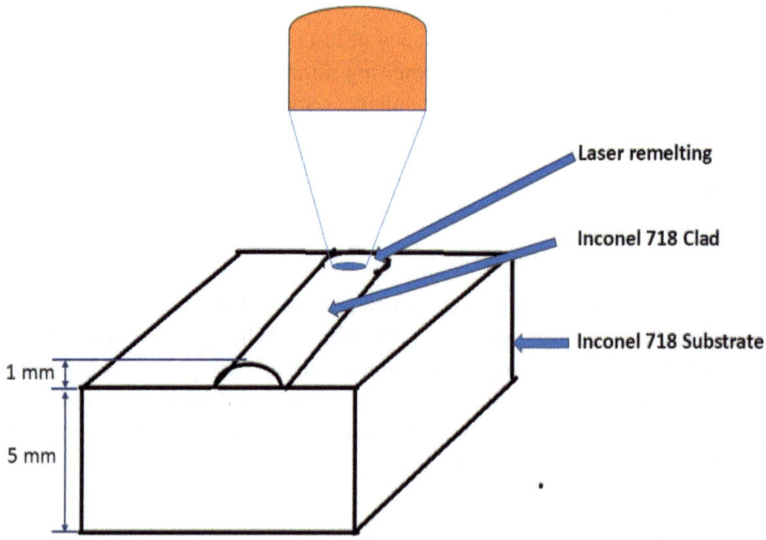

Fig. 3. Schematic diagram of laser remelting

An infrared pyrometer was used to capture thermocycles in laser remelting of L-DEDed Inconel 718 parameters, as shown in Fig. 4. The peak temperature was observed to vary with changing laser remelting parameters. This variation in the temperature brings the cooling rate into play, which has an influence on the microstructure and strength in the L-DED samples.

At a laser power of 900 W and 720 mm/min scan rate, the melt pool peak temperature was recorded value of 1790 °C. As the scan rate is decreased to 600 mm/min, the melt pool temperature increases to 1825 °C. For 700 W laser power and 480 mm/min, the temperature decreased to 1710 °C with a decrease in laser intensity. This trend continues as the

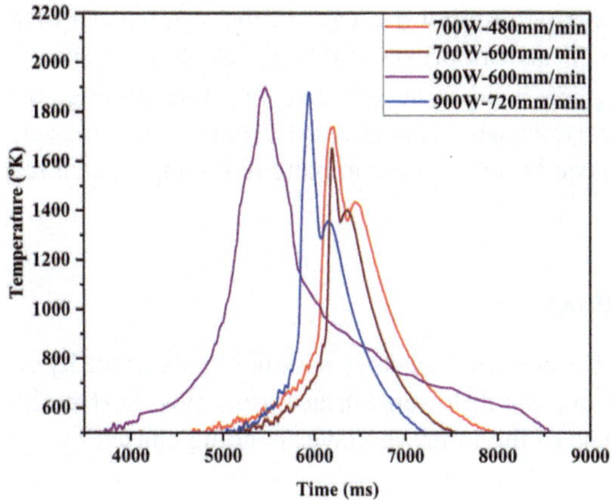

Fig. 4. Thermal history captured for different laser remelting parameters

molten pool temperature decreases to 1620 °C, and the scan rate reduces to 600 mm/min. This variation in the molten pool temperature changes the overall microstructure and clad quality of the L-DEDed Inconel 718 samples.

3 Results and Discussion

3.1 Microstructure

The directional solidification of DED-processed Inconel 718 usually results in elongated columnar grains. Equiaxed grains may form in remelted areas owing to a higher cooling rate, as shown in Fig. 5, enhancing the isotropic mechanical properties. Because of their rapid cooling, primary DED layers frequently exhibit element segregation (e.g., Nb, Mo, and Ti) [12, 15, 16].

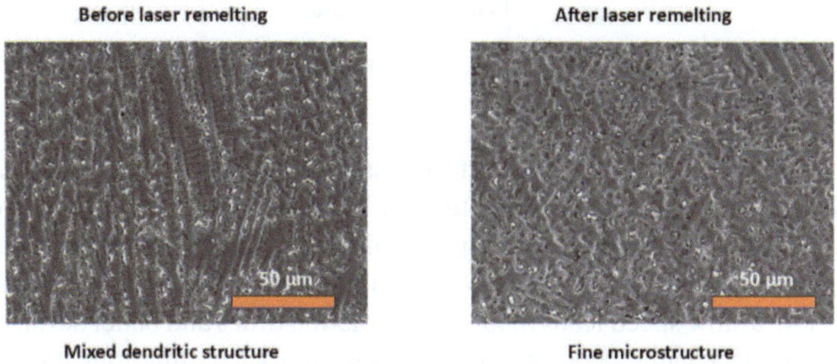

Fig. 5. Improvement in the laser remelted L-DED samples.

A uniform microstructure and decrease in the segregation of Nb are obtained by laser remelting, which also improves diffusion and homogenization. Deeper remelting may result in grain coarsening but also promote homogenization. It results in extended exposure periods, which enhance diffusion, but it increases heat accumulation. A slower scanning speed promotes incomplete remelting, which restricts the effects of homogenization.

3.2 Surface Roughness

In Fig. 6(a) and (b), Laser remelting creates a smoother surface by encouraging material flow and redistribution, which lowers surface asperities. Surface irregularities are less likely to form because of the improved layer bonding caused by localized melting and resolidification.

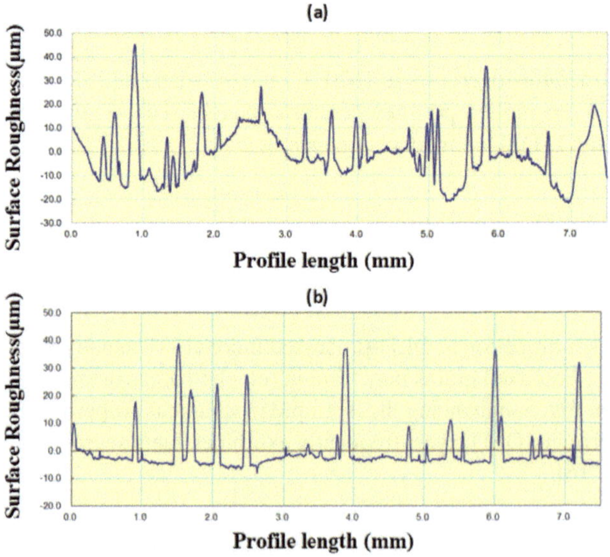

Fig. 6. Surface roughness (a) before (b) after laser remelting

The powder particles that cause surface roughness in DED because of incomplete melting can be entirely fused by remelting. A larger molten pool results from higher laser power, which improves smoothing but can also result in excessive material flow and the creation of new irregularities. While a high scanning speed could cause uneven resolidification, a slow speed leads to longer interaction times and better surface leveling. A surface roughness value decreased from 11.09 μm to 7.55 μm after laser remelting [10, 17].

Figure 7(a) shows a graph of surface roughness against laser power with a fixed scanning speed of 600 mm/min. The surface roughness decreases from 11.10 μm to 7.55 μm with an increase in laser power from 700 W to 1100 W.

Moreover, an increase in scan rate from 600 mm/min to 960 mm/min, surface roughness increases from 7.55 μm to 12.10 μm. The result shows that lower laser power increases heat accumulation, which promotes a good surface finish. On the other hand, an increase in scanning rate increases the surface roughness due to uneven solidification.

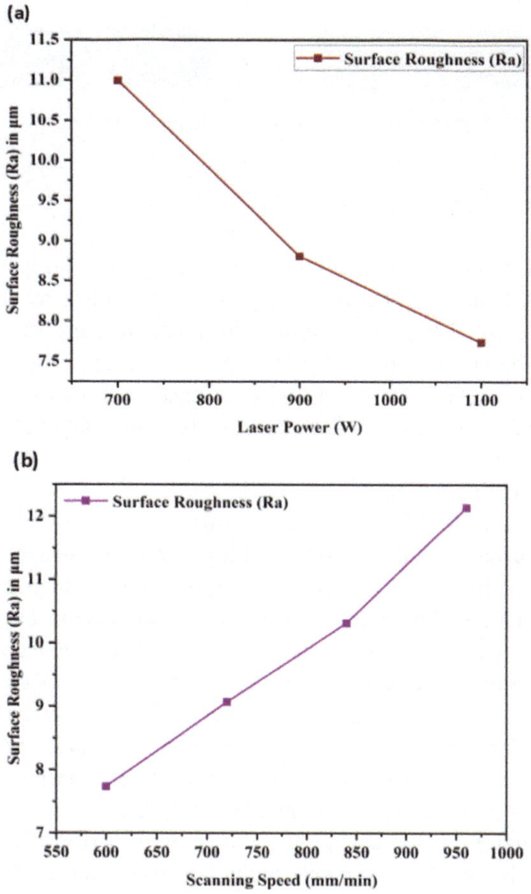

Fig. 7. Correlation of surface roughness with (a) laser power (scan rate–600 mm/min) and (b) laser scanning speed (laser power - 900 W)

3.3 Porosity

The deposited material is remelted by laser remelting, which lengthens the liquid-phase period and improves flow. In doing so, porosity is decreased by allowing trapped gas pockets or lack-of-fusion defects to escape. Although higher power guarantees better defect healing and deeper remelting, too much of it can result in keyhole porosity. Porosity healing takes more time and has slower scan speeds [18] (Fig. 8).

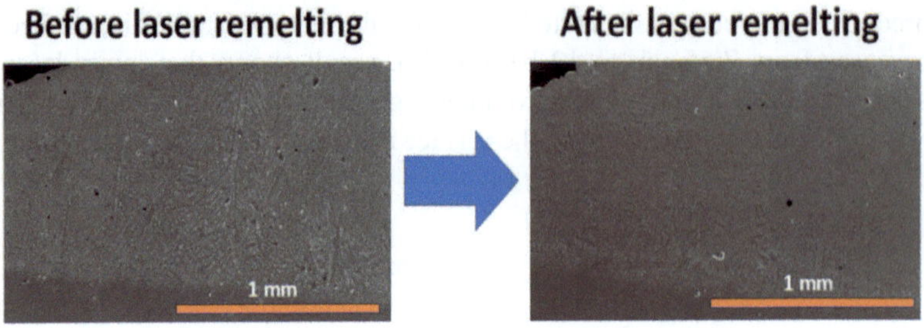

Fig. 8. Effect of laser remelting on the porosity

4 Conclusion

Laser remelting (LR), a critical post-processing technique, significantly enhances the quality of L-DEDed Inconel 718 by increasing surface finish, reducing porosity, and refining microstructure. LR process refines the microstructure of deposited samples and redistributes the surface material, effectively reducing surface irregularities and asperities. An optimal set of laser power and scan speed. ensures a superior surface finish while preventing excessive material flow and oxidation. In addition, LR becomes crucial because of benefits like eliminating the lack of fusion defects, promoting interlayer bonding, and minimizing porosity.

Laser remelting significantly alters the microstructure of DED Inconel 718, alongside improvements in surface quality and porosity. It generates refined, equiaxed grains that enhance mechanical properties by disrupting the typical columnar grain growth observed in as-deposited materials. Furthermore, the process facilitates the dissolution of Nb-rich Laves phases, thereby reducing microsegregation and enhancing overall material homogeneity. Post-processing heat treatments are essential to restore mechanical integrity, as excessive remelting may lead to residual stresses and grain coarsening. Despite these challenges, laser remelting is an efficacious technique for enhancing the mechanical performance and surface quality of L-DED Inconel 718. To maximize benefits and mitigate potential drawbacks, process parameters must be meticulously refined. Further insights into optimizing LR for aerospace and other high-performance applications can be obtained through additional research, including experimental characterization and thermomechanical simulations.

References

1. Hosseini, E., Popovich, V.A.: A review of mechanical properties of additively manufactured Inconel 718. Addit. Manuf. **30**, 100877 (2019)
2. Godec, M., Bala, M., Klob, D., Pambaguian, L., Conradi, M., Kocijan, A.: Hybrid additive manufacturing of Inconel 718 for future space applications. Mater. Charact. **172**, 110842 (2021)
3. Bennett, J.L., Wolff, S.J., Hyatt, G., Ehmann, K., Cao, J.: Thermal effect on clad dimension for laser deposited Inconel 718. J. Manuf. Process. **28**, 550 (2017)
4. Svetlizky, D., et al.: A laser-based directed energy deposition (DED-LB) of advanced materials. Mater. Sci. Eng. A. **840**, 142967 (2022)

5. Blakey-Milner, B., et al.: Metal additive manuf. in aerospace: a review. Mater. Des. **209**, 110008 (2021)
6. Patil, M.A., Ghara, T., Das, B., Kulkarni, D.M.: Influence of cooling rate on microstructure, dislocation density, and associated hardness of laser direct energy deposited Inconel 718. Surf. Coatings Technol. **495**, 131575 (2025)
7. Misal, N.D., Saraf, A.R., Sadaiah, M.: Experimental investigation of surface topography in photochemical machining of Inconel 718. Mater. Manuf. Process. **32**, 1756 (2017)
8. Ning, J., Wang, W., Zamorano, B., Liang, S.Y.: Analytical modeling of lack-of-fusion porosity in metal additive manuf. Appl. Phys. A Mater. Sci. Process. **125**, 1–11 (2019)
9. Slotwinski, J.A., Garboczi, E.J., Hebenstreit, K.M.: Porosity measurements and analysis for metal additive manufacturing process control. J. Res. Natl. Inst. Stand. Technol. **119**, 494 (2014)
10. Balbaa, M., Mekhiel, S., Elbestawi, M., McIsaac, J.: On selective laser melting of Inconel 718: densification, surface roughness, and residual stresses. Mater. Des. **193**, 108818 (2020)
11. de Formanoir, C., et al.: Healing of keyhole porosity by means of defocused laser beam remelting: operando observation by X-ray imaging and acoustic emission-based detection. Addit. Manuf. **79**, 103880 (2024)
12. Shao, W., He, B., Qiu, C., Li, Z.: Effect of hatch spacing and laser remelting on the formation of unique crystallographic texture of IN718 superalloy fabricated via laser powder bed fusion. Opt. Laser Technol. **156**, 108609 (2022)
13. Jardon, Z., Ertveldt, J., Lecluyse, R., Hinderdael, M., Pyl, L.: Directed energy deposition roughness mitigation through laser remelting. Procedia CIRP. **111**, 180–184 (2022)
14. Hodgir, R., Singh, R.K., Mujumdar, S.: Experimental investigation of laser remelting in directed energy deposition (DED) of CPM-9V. Manuf. Lett. **35**, 701–706 (2023)
15. Guan, J., Jiang, Y., Zhang, X., Chong, X.: Microstructural evolution and EBSD analysis of AlSi10Mg alloy fabricated by selective laser remelting. Mater. Charact. **161**, 110079 (2020)
16. Maruschak, P., Zakiev, I., Mocharsky, V., Nikiforov, Y.: Experimental study of the surface of steel 15kh13mf after the nanosecond laser shock processing. Solid State Phenom. **200**, 60–65 (2013)
17. Liu, B., Li, B.Q., Li, Z.: Selective laser remelting of an additive layer manufacturing process on AlSi10Mg. Results Phys. **12**, 982–988 (2019)
18. Yu, W., Sing, S.L., Chua, C.K., Tian, X.: Influence of remelting on surface roughness and porosity of AlSi10Mg parts fabricated by selective laser melting. J. Alloys Compd. **792**, 574–581 (2019)

Efficient Prediction of Additive Manufacturing Capabilities Using Hybrid Theoretical Method and FEA for Smart Manufacturing Systems

Chanawee Promaue(✉), Suchandrima Das, and Aydin Nassehi

School of Electrical, Electronic and Mechanical Engineering (EEME), University of Bristol, Bristol, UK BS8 1TR

th22301@bristol.ac.uk

Abstract. The development of resilient and adaptive manufacturing systems requires efficient methods for predicting the mechanical properties of additively manufactured (AM) components, particularly as these systems increasingly leverage digital technologies for rapid reconfiguration. This study presents a hybrid approach that integrates the theoretical method with finite element analysis (FEA) to accurately predict the tensile and torsional properties of AM parts while significantly reducing computational time. Two FEA models were evaluated: a standard model with detailed material and geometric representations, and a simplified model optimised for computational efficiency. The proposed methodology utilises G-code data from AM processes to generate a replicated CAD model, streamlining the analysis while maintaining predictive accuracy. The findings demonstrate that this approach can quickly establish manufacturing resource capabilities, thereby supporting the agile reconfiguration of production systems. By reducing reliance on extensive physical testing, the hybrid FEA-theoretical approach enhances sustainability through material and energy savings, contributing to the development of smart, resilient, and sustainable manufacturing systems aligned with the principles of Industry 4.0 and beyond.

Keywords: Additive Manufacturing (AM) · Finite Element Analysis (FEA) · Sustainable simulation tool

1 Introduction

The rapid evolution of manufacturing systems under Industry 4.0 and the emerging Industry 5.0 highlights the demand for adaptive, efficient, and resilient production methods. Additive Manufacturing (AM) [1], which fabricates complex components directly from digital models [2], has become integral to sectors such as aerospace, medical, and automotive. AM consolidates multiple production steps into a single process, enhancing efficiency and enabling rapid prototyping [3–5]. It constructs parts layer by layer via a heated extruder [6], and recent advancements have extended its material range to include metals, composites, and ceramics [7]. Its ability to produce lightweight, intricate geometries offers performance advantages over traditional methods [5, 8, 9]. However, predicting the mechanical properties of AM parts remains challenging due to the

Y. Zhao (Ed.): ICAMM 2025, SPM 96, pp. 134–142, 2025.
https://doi.org/10.1007/978-3-032-09862-7_13

complex interplay of process parameters, material behavior, and geometries [10–12]. Accurate predictions are critical for ensuring part reliability and promoting sustainable manufacturing by minimizing waste and energy use [7, 13, 14].

To address these challenges, integrating theoretical methods with finite element analysis (FEA) presents a promising solution. This study proposes a hybrid theoretical–FEA framework that reduces computational time while maintaining prediction accuracy for tensile and torsional properties of AM parts. By leveraging G-code data and optimizing the FEA model, the method supports resilient manufacturing and smart factory principles, offering rapid evaluation of production capabilities in agile, reconfigurable environments. Previous efforts to predict AM part behavior include Classical Laminate Theory (CLT) [15] and FEA techniques using stiffness matrices [16–18]. While effective, these often rely on experimental input. More recent approaches—such as the Replicated CAD Model Method—eliminate the need for parameter-specific material adjustments [19, 20], but challenges like modelling plasticity and computational cost remain.

Building on prior research [21], this study enhances the Replicated CAD Model Technique to better simulate elastic and plastic behavior using G-code path data. The new framework improves computational efficiency while maintaining moderate accuracy, addressing a key limitation of earlier models. Torsional behavior is also explored through a theoretical extension of the method. Overall, this work contributes to developing practical, reliable, and efficient simulation tools for AM component analysis, reducing reliance on physical testing and enabling smarter production system design. The structure of the paper is as follows: Sect. 2 presents the methodology; Sect. 3 discusses experimental and FEA results; and Sect. 4 concludes the study.

2 Methodology

The mechanical properties of AM parts are significantly influenced by process parameters such as infill pattern, density, and layer thickness. Ćwikła et al. [22] highlight how these variables affect strength, stiffness, and durability. Infill pattern influences stress distribution, infill density impacts porosity and weight, and layer thickness governs surface quality and interlayer bonding. Optimizing these parameters enhances mechanical performance and improves alignment between FEA and experimental results. To investigate this, four specimen batches were printed with varying settings (Table 1) using an Ultimaker S3 and 2.85 mm PLA filament at 200 °C, following manufacturer guidelines.

Table 1. Printing parameters for testing specimens

Condition	Layer thickness (mm)	Infill density (%)	Print pattern
1	0.2	70	Triangle
2	0.2	35	Grid
3	0.3	35	Triangle
4	0.3	70	Grid

Dog bone-shaped specimens were prepared according to ASTM D638 for tensile testing of plastics (Fig. 1) [23]. Both tensile and torsion tests were conducted. Tensile tests used an Instron 8872 machine at a 1 mm/min pull rate until failure, while torsion tests employed an Instron Electropuls E1000, twisting specimens at 5°/min to failure. Each condition was tested on five specimens to ensure accuracy and repeatability.

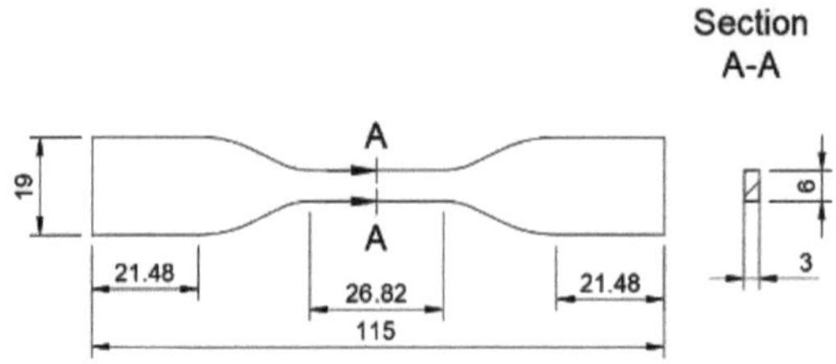

Fig. 1. Dimension of the test specimen

To replicate the microstructure of AM specimens, a CAD model generation process was developed using G-code extraction and profile extrusion, transforming extruder paths into a 3D CAD model that closely matches the printed geometry. The process began with parsing G-code line by line to extract key commands: "G0" for non-extrusion travel and "G1" for material deposition, using the corresponding XYZ coordinates to define the extruder path. This path was then processed with the SolidPython CAD library to generate 3D profiles resembling the microstructure seen in CT-scan images (Fig. 2), with fillet curves added at edges and junctions for realism. The profiles were extruded along the path to form a solid model representing the AM layer-by-layer deposition. Transformations such as rotation, translation, and merging were applied to ensure alignment, structural continuity, and accuracy, while layer thickness adjustments further refined the model. Building on the approach from prior work [21], refinements were introduced to improve both model accuracy and computational efficiency by optimizing path representation, transformation operations, and merging techniques. The meshing strategy was also revised to reduce computational load without compromising structural fidelity. Once finalized, the CAD model was exported as a SCAD file, processed in OpenSCAD to generate an STL, and imported into Abaqus for finite element analysis. These improvements enabled accurate simulation of mechanical behavior under various loading conditions, enhancing the model's reliability and computational performance. The overall process is illustrated in Fig. 3.

To enhance computational efficiency, the FEA model focused on a 5 mm segment within the gauge length, identified as the specimen's weakest and most critical region. This localized approach preserved mechanical accuracy while reducing resource use. Material properties, assumed isotropic based on tensile tests, were derived from the stress–strain curve of a solid AM specimen printed at 0° (Fig. 4). Boundary conditions included a fully fixed (Encastre) end and a displacement-controlled load on the opposite end, applied via a reference point coupled to the surface. Loading continued to 5% strain (Fig. 5a). Perfect layer adhesion was assumed to reflect realistic structural

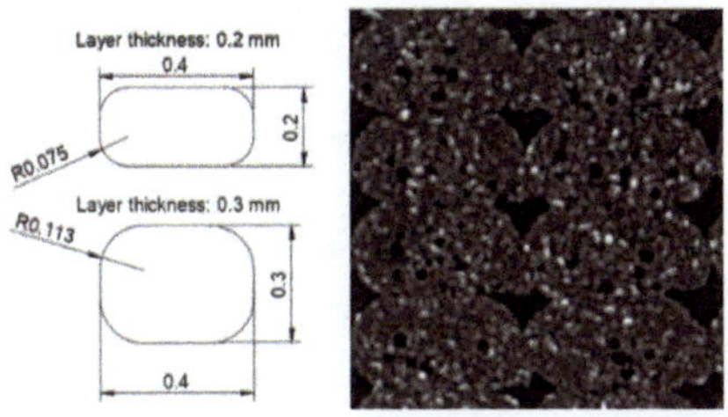

Fig. 2. Extruded profile dimension and CT scan image of microstructure study of AM parts

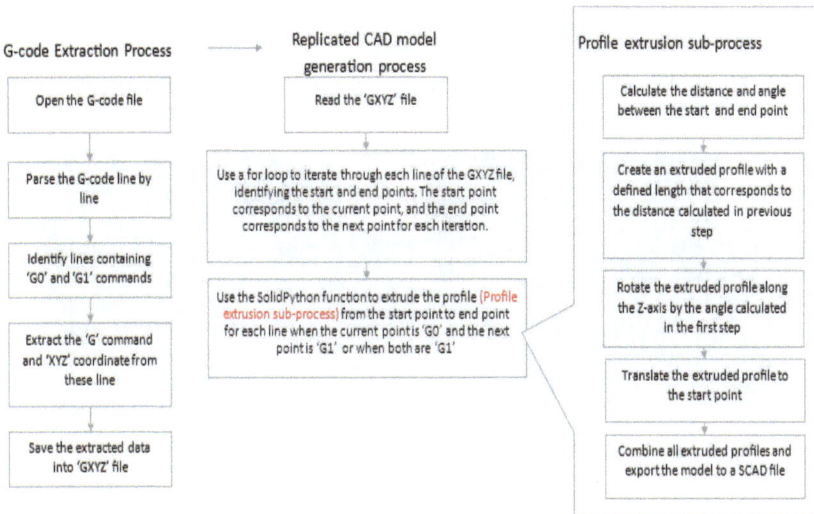

Fig. 3. Replicated CAD model generation process

integrity. A 0.1 mm tetrahedral mesh ensured high-resolution stress capture. The resulting load–displacement data were used to generate a stress–strain curve for comparison with experimental results, validating the model's accuracy.

A simplified FEA model was developed using the same conditions as the standard model but employed a course 0.25 mm mesh to improve computational efficiency and reduce hardware demands. To further streamline the geometry, the original extruded profile (Fig. 2) was replaced with a rectangular profile (Fig. 5b), aiming to lower complexity while preserving acceptable accuracy. Although this simplification may slightly alter local stress distributions, it is expected to have minimal effect on the overall mechanical response. Validation was performed by comparing stress–strain curves from both models with experimental data, focusing on tensile and torsional behavior. If results align closely, the simplified model offers a practical, efficient alternative for resource-intensive simulations.

The mechanical properties obtained from the FEA models, including longitudinal and transverse displacements, reaction force, cross-sectional area, initial specimen length, and cross-sectional dimensions, were used to compute the torque–angle relationship.

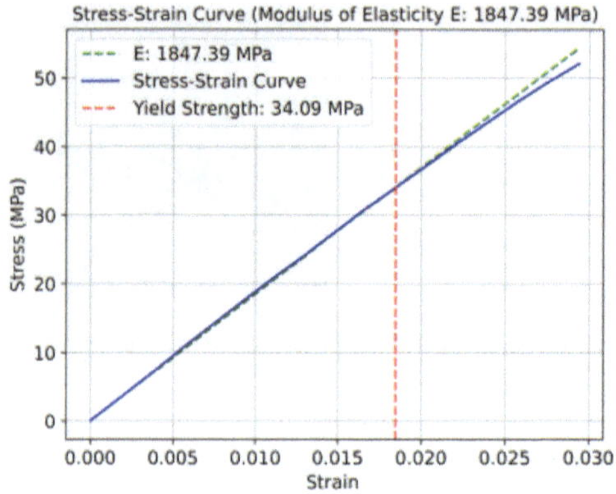

Fig. 4. Material properties of AM specimens applied in the FEA model

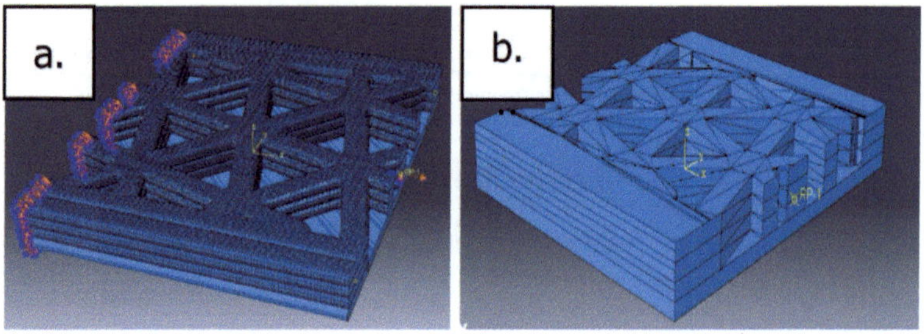

Fig. 5. **a.** FEA model of AM parts, **b.** Simplified FEA model of AM parts

The process began with calculating the elastic modulus (E_1) using the ratio of stress (σ_1) to strain (ε_1), where stress was derived from the applied force (F_1) divided by the cross-sectional area (A), and strain was the longitudinal displacement (U_1) over the initial length (L), as expressed in Eq. (1): $E_1 = \frac{\sigma_1}{\varepsilon_1}$, with $\sigma_1 = \frac{F_1}{A}$ and $\varepsilon_1 = \frac{U_1}{L}$. Poisson's ratio (ν) was then calculated as the transverse strain (ε_2) over longitudinal strain (ε_2), where $\varepsilon_2 = \frac{U_2}{L_2}$ and U_2 is transverse displacement, as shown in Eq. (2): $\nu = \frac{\varepsilon_2}{\varepsilon_1}$. Using these, the shear modulus (G) was determined through Eq. (3): $G = \frac{E_1}{2(1+\nu)}$. Torque (T) at varying twist angles (θ) was then calculated using Eq. (4): $T = \frac{\theta JG}{L_1}$, where J is the polar moment of inertia. For a rectangular cross-section, J was calculated using Eq. (5): $J = \frac{ab^3}{16} \times \left[\frac{16}{3} - 3.36\left(\frac{b}{a}\right)\left(1 - \frac{b^4}{12a^4}\right)\right]$, where a and b are cross-sectional dimensions. The resulting torque–angle relationship was compared with experimental torsion data to evaluate model accuracy.

3 Result and Discussion

Figure 6a to d present stress–strain curves for AM specimens under four conditions, comparing standard and simplified FEA models. The simplified model, which uses a rectangular cross-section (Fig. 5b), generally predicts higher stress values due to its slightly larger cross-sectional area. Both models overestimate stress compared to experimental data, with the standard model predicting elastic modulus values 11–28% higher and the simplified model 17–34% higher. For instance, in Condition 2 (Fig. 6b), the standard model estimates 2.3 MPa at 0.02 strain versus 1.8 MPa experimentally, while the simplified model predicts 2.4 MPa. These discrepancies arise from unmodeled experimental factors (e.g., clamping forces, strain rate variations), geometric simplifications (real filament shapes differ from rectangular assumptions), and the assumption of perfect interlayer bonding, which does not reflect actual variability in print quality. Despite these limitations, both models capture similar stress–strain trends, indicating their usefulness for estimating the mechanical behavior of AM parts.

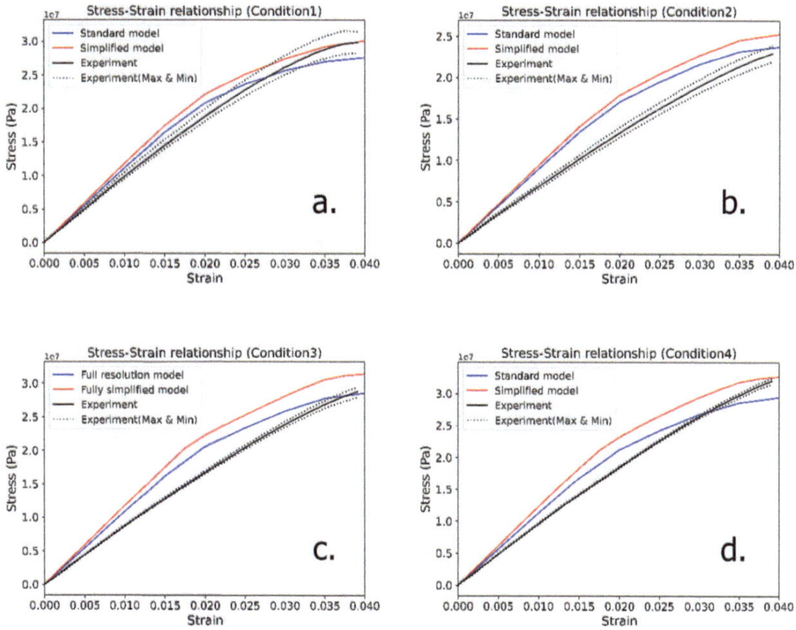

Fig. 6a–d. Stress-Strain relationship (Condition1–4)

Figure 7a to d illustrate the torque–angle responses of AM specimens under different conditions, revealing that the simplified FEA model consistently predicts higher torque values than both the standard model and experimental results, echoing trends observed in the tensile analysis. Notably, specimens with 0.3 mm layer thickness show greater deviation, with experimental torque values falling below simulation predictions, while 0.2 mm specimens demonstrate closer agreement. This suggests that layer thickness significantly influences torsional performance, although several factors contribute to these discrepancies. First, layer thickness affects interlayer bonding due to differences in material deposition, cooling rates, and print conditions, all of which influence shear resistance under torsional loads; FEA models assume perfect bonding and thus cannot fully account for such variations. Second, internal stress distribution is impacted by layer thickness, with thicker layers potentially causing uneven stress concentrations and thinner layers offering more uniform load transfer—another aspect that simplified simulations may not capture accurately. Third, experimental factors such as alignment, gripping force, and environmental variability can affect torque measurements, introducing inconsistencies between simulated and actual results. To improve understanding, further research is recommended in three areas: microstructural analysis (e.g., CT scanning) to examine bonding differences between layer thicknesses; thermal analysis to investigate cooling rate effects; and additional mechanical testing under controlled process variations to validate current findings. These efforts will help clarify the relationship between layer thickness and torsional behavior in AM components and enhance the predictive accuracy of FEA models.

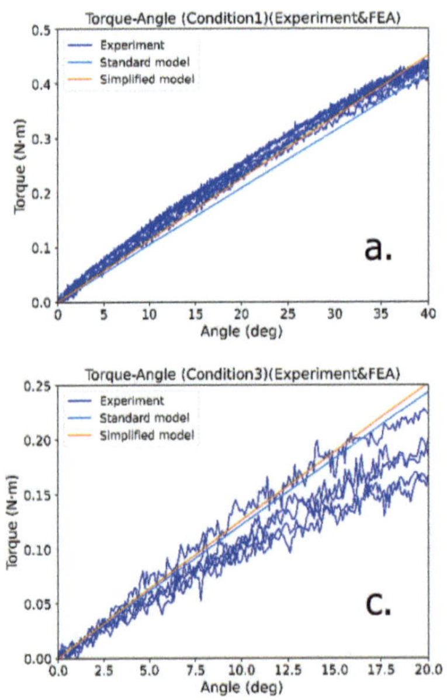

Fig. 7a–d. Torque-Angle relationship (Condition 4)

A major advantage of the simplified FEA model is its significantly reduced computational complexity compared to the standard model, which typically requires 1.5 to 2 million finely meshed elements. In contrast, the simplified model uses fewer than 30,000 elements, greatly lowering computational load and making it more practical for applications that require a balance between accuracy and efficiency. This reduction enables faster simulations and less hardware demand, making it ideal for preliminary design evaluations and iterative analysis in additive manufacturing. While both the standard and simplified models effectively predict the mechanical behavior of AM components, the simplified version tends to overestimate stress due to its use of a rectangular cross-section. Nevertheless, it captures the overall stress–strain trend. In torsional analysis, predictions for specimens with 0.2 mm layer thickness closely match experimental data, whereas 0.3 mm samples show higher torque values—likely influenced by unmodeled variations in interlayer bonding. Despite some discrepancies, the simplified model reduces the need for extensive physical testing, supporting sustainable manufacturing by minimizing material use, energy consumption, and production waste. Its ability to streamline the design validation process makes it valuable in reducing trial-and-error experimentation. Future enhancements, including the integration of interlayer bonding effects and material plasticity variations, could further improve predictive accuracy and strengthen the model's role in efficient, sustainable production workflows.

4 Conclusion

This study successfully develops a predictive tool using FEA to assess the mechanical properties of additively manufactured parts, focusing on both tensile and torsional behavior. By systematically comparing FEA simulations with experimental results across different printing conditions, the study highlights the capability of FEA in providing reliable mechanical property predictions. The simplified model, despite minor accuracy limitations, demonstrates significant computational efficiency, making it suitable for rapid assessments and early-stage design optimizations. Additionally, this approach reduces reliance on extensive physical testing, minimizing material waste and energy consumption, thus promoting sustainability in AM processes. Future research should focus on refining FEA models to better capture real-world interlayer bonding effects and material plasticity variations. Overall, this method enhances the understanding of AM part behavior while contributing to the advancement of sustainable and efficient manufacturing practices.

References

1. Somireddy, M., Singh, C.V., Czekanski, A.: Analysis of the material behavior of 3D printed laminates via FFF. Exp. Mech. **59**(6), 871–881 (2019). https://doi.org/10.1007/s11340-019-00511-5
2. Somireddy, M., de Moraes, D.A., Czekanski, A.: Flexural behavior of FDM parts: experimental, analytical and numerical study. In: International Solid Freeform Fabrication Symposium (2017)
3. Frazier, W.E.: Metal additive manufacturing: a review. J. Mater. Eng. Perform. **23**(6), 1917–1928 (2014). https://doi.org/10.1007/s11665-014-0958-z
4. Mohammed, R., et al.: Advancements and challenges in additive manufacturing: future directions and implications for sustainable engineering. Adv. Sustain. Sci. Eng. Technol. **7**(1), 0250108 (2025). https://doi.org/10.26877/asset.v7i1.1079
5. Mantalas, E.-M., Sagias, V.D., Zacharia, P., Stergiou, C.I.: Neuro-fuzzy model evaluation for enhanced prediction of mechanical properties in AM specimens. Appl. Sci. **15**(1), 7 (2024). https://doi.org/10.3390/app15010007
6. Wong, K.V., Hernandez, A.: A review of additive manufacturing. ISRN Mech. Eng. **2012**(1), 1–10 (2012). https://doi.org/10.5402/2012/208760
7. Gibson, D., Rosen, B.S., Khorasani, M.: Additive Manufacturing Technologies. Springer Nature, Switzerland (2020)
8. Martín-Sosa, E., Távara, L., Ojeda, J., Estefani, A.: Anisotropic and hyperelastic mechanical response of 3D printed TPU parts. Prog. Addit. Manuf. **10**, 5697 (2025). https://doi.org/10.1007/s40964-024-00937-x

9. Yao, Y., Wang, K., Chen, H., Le, H.: Characterisation and mathematical modelling of nonlinear mechanical behaviour of 3D printed short carbon fibre reinforced composites. Compos. Part C Open Access. **14**, 100455–100455 (2024). https://doi.org/10.1016/j.jcomc.2024.100455
10. Casavola, C., Cazzato, A., Moramarco, V., Pappalettere, C.: Orthotropic mechanical properties of fused deposition modelling parts described by classical laminate theory. Mater. Des. **90**, 453–458 (2016). https://doi.org/10.1016/j.matdes.2015.11.009
11. Manoj Prabhakar, M., Saravanan, A.K., Haiter Lenin, A., leno, I.J., Mayandi, K., Ramalingam, P.S.: A short review on 3D printing methods, process parameters and materials. Mater Today Proc. **45** (2020). https://doi.org/10.1016/j.matpr.2020.10.225
12. Cano-Vicent, et al.: Fused deposition modelling: current status, methodology, applications and future prospects. Addit. Manuf. **47**, 102378 (2021). https://doi.org/10.1016/j.addma.2021.102378
13. Somireddy, M., Czekanski, A.: Characterization of material behavior of the fused deposition modeling processed parts. In: International Manufacturing Science and Engineering Conference. American Society of Mechanical Engineers (2017)
14. Webbe Kerekes, T., Lim, H., Joe, W.Y., Yun, G.J.: Characterization of process–deformation/damage property relationship of fused deposition modeling (FDM) 3D-printed specimens. Addit. Manuf. **25**, 532–544 (2019). https://doi.org/10.1016/j.addma.2018.11.008
15. Singh, R., Garg, H.K.: Fused deposition Modeling – a state of art review and future applications. In: Reference Module in Materials Science and Materials Engineering, pp. 1–20 (2016)
16. Ruzy Haryati, H., Celik, H.K., Smith, P., Rennie, A., Ucar, M.: Effect of build orientation on FDM parts: a case study for validation of deformation behaviour by FEA. In: International Conference on Design and Concurrent Engineering (2010) Available: https://www.researchgate.net/publication/328305076_Effect_of_Build_Orientation_on_FDM_Parts_A_Case_Study_for_Validation_of_Deformation_Behaviour_by_FEA. Accessed:23 Oct 2024
17. Martínez, J., Diéguez, J.L., Ares, E., Pereira, A., Hernández, P., Pérez, J.A.: Comparative between FEM models for FDM parts and their approach to a real mechanical behaviour. Procedia Eng. **63**, 878–884 (2013). https://doi.org/10.1016/j.proeng.2013.08.230
18. Fernández, V., García, A., Camacho, A.M., Claver, J., Rodríguez, A., Sebastián, M.A.: Methodology to estimate the modulus of elasticity of parts manufactured by FFF/FDM combining finite element simulations and experimental tests. IOP Conf. Series: Mater. Sci. Eng. **1193**(1), 012103 (2021). https://doi.org/10.1088/1757-899x/1193/1/012103
19. Lalegani Dezaki, M., Mohd Ariffin, M.K.A.: The effects of combined infill patterns on mechanical properties in FDM process. Polymers. **12**(12), 2792 (2020). https://doi.org/10.3390/polym12122792
20. Rupal, B.S., Mostafa, K.G., Wang, Y., Qureshi, A.J.: A reverse CAD approach for estimating geometric and mechanical behavior of FDM printed parts. Procedia Manuf. **34**, 535–544 (2019). https://doi.org/10.1016/j.promfg.2019.06.217
21. Promaue, C., Das, S., Nassehi, A.: Finite element based mechanical properties prediction for material extrusion additive manufacturing to enable rapid production system design. Procedia CIRP. **130**, 59–64 (2024). https://doi.org/10.1016/j.procir.2024.10.056
22. Ćwikła, G., Grabowik, C., Kalinowski, K., Paprocka, I., Ociepka, P.: The influence of printing parameters on selected mechanical properties of FDM/FFF 3D-printed parts. IOP Conf. Series: Mater. Sci. Eng. **227**, 012033 (2017). https://doi.org/10.1088/1757-899x/227/1/012033
23. American Society for Testing and Materials (ASTM): Standard Test Method for Tensile Properties of Plastics. American Society for Testing and Materials (ASTM), Pennsylvania, United States (2014) Available: https://borgoltz.aoe.vt.edu/aoe3054/manual/expt5/D638.38935.pdf. Accessed: 19 May 2023

Structural Mechanics Analysis and Test Study of Additive Manufacturing Cabin Door Hinge Lug

Mingliang Song[1,2], Lei Li[1,2](✉), Mingde Chang[1,2], Dongsen Ye[1,2], and Hongbiao Ding[1,2]

[1] State Key Laboratory of Airliner Integration Technology and Flight Simulation, Shanghai, China
songmingliang@comac.cc
[2] Shanghai Aircraft Design and Research Institute, Shanghai, China

Abstract. Due to the advantages of high specific strength and low density, the titanium alloy additive manufacturing structures are widely used in civil airliners and other fields. This paper takes the additive manufacturing cabin door hinge lug as the research object, and carries out structural mechanical tests on test pieces of the three configurations of the axial loading, oblique loading, and transverse loading to get the carrying capacity and failure modes. The following conclusions can be drawn from the tests: the average failure load under axial loading is 28.78 kN, the average failure load under oblique loading is 29.45 kN and the average failure load under transverse loading is 24.96 kN, and the results show that the carrying capacity of axial loading and oblique loading is higher than the carrying capacity of the transverse loading. The failure modes of the three configurations of the axial loading, oblique loading, and transverse loading are all shear failure, with the configuration of the oblique loading showing the tensile-shear composite failure. From the load-displacement curves, it can be seen that when the failure modes of the three configurations reach the failure load, the load is dropped dramatically, while the axial loading test appears more regular yielding stage. This paper provides the reference basis for the airworthiness verification and installing application on the additive manufacturing cabin door hinge of the service aircraft and the subsequent aircraft.

Keywords: Additive Manufacturing · Cabin Door · Lug · Mechanical Analysis · Experimental Study

1 Introduction

Additive manufacturing (AM) adopts on the principles of discretization and accumulation. It uses three-dimensional software to create solid models and employs high-energy beams as the input heat source to melt or sinter metallic materials (such as powders, wires, or blocks) to manufacture solid parts through the layer-by-layer printing process [1]. Compared to traditional machining methods, additive manufacturing significantly reduces material utilization from the typical 90% down to 7–20% [2–4].

Y. Zhao (Ed.): ICAMM 2025, SPM 96, pp. 143–151, 2025.
https://doi.org/10.1007/978-3-032-09862-7_14

Titanium alloys, known for their low density, high specific strength, corrosion resistance, and high-temperature resistance, are extensively used in aerospace, biomedical, precision machinery, and instrument manufacturing industries. In particular, in the aerospace sector, titanium alloys are highly valued and widely used due to their lightweight and excellent comprehensive mechanical and corrosion resistance properties [5, 6]·Currently, the traditional forming methods for titanium alloys mainly include precision casting, isothermal forging, machining, and powder metallurgy. However, these conventional processing methods are complex, long production cycles, low material utilization, and high costs. Moreover, with the increasing demand for more complex and higher-performance components in the aerospace industry, there is an urgent need for more advanced and efficient titanium alloy forming technologies [7–9]. As a result, the application of additive manufacturing technology in titanium alloys has garnered significant attention from researchers worldwide. Depending on the heat source used for material heating and the powder supply method, metal additive manufacturing technologies are mainly divided into Laser Deposition Manufacturing (LDM), Selective Laser Melting (SLM), Electron Beam Melting (EBM), and Wire and Arc Additive Manufacturing (WAAM) [10, 11]·

This paper takes the cabin door hinge lug formed by the SLM additive manufacturing process as the research object, and carries out structural mechanical tests to get the carrying capacity and failure modes, which provides a certain reference basis for the estimation and judgment of the load-bearing capacity of titanium alloy components for civil aircraft.

2 Test Objects

The test object of this paper is the additive manufacturing cabin door hinge lug structure of the civil aircraft, the configuration of test piece as shown in Fig. 1. The width of test piece is W ($W = 15$ mm), the hole diameter of test piece is D ($D = 10$ mm), the thickness of test piece is t($t = 6$ mm), and the length of test piece is L($L = 40$ mm or $L = 100$ mm). The material of the test specimen is Ti6Al4V, with a tensile strength of 970 MPa.

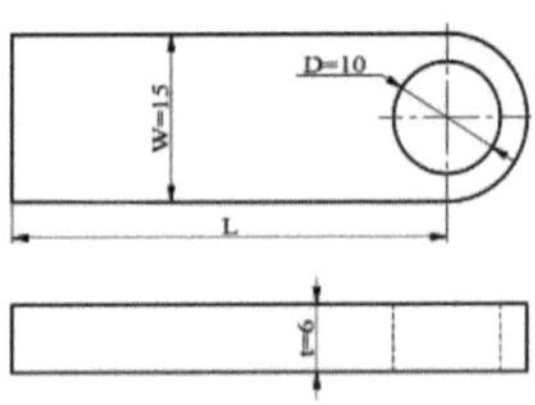

Fig. 1. The diagram of the test piece

The configurations of test loading include axial loading (0°), oblique loading (45°), and transverse loading (90°), as shown in the following Fig. 2.

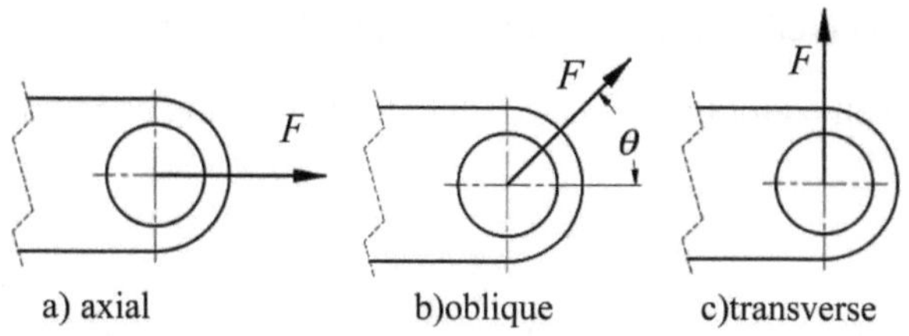

Fig. 2. Force diagram of the lug structure

According to the test loading configurations, the length L of the specimen for axial loading is 100 mm, while the length of the specimens for oblique and transverse loading is 40 mm. Three specimens were prepared for each test configuration, and their identification numbers are listed in Table 1.

Table 1. The configuration of test piece

The configuration of test piece	Environment	Number
Axial loading(0°)	room temperature	L001-L003
Oblique loading (45°)	room temperature	M001-M003
Transverse loading (90°)	room temperature	S001-S003

3 Structural Mechanics Analysis and Calculation

3.1 Strength Calculation under Axial Load ($\theta = 0°$)

(a) Tensile failure of lug

The estimated failure load for the lug under tensile loading can be calculated using the method shown in Eq. (1):

$$P_{tu} = K_t \times \sigma_{tu} \times A_t \quad (1)$$

Where: P_{tu}—lug tensile failure load; K_t—lug tensile efficiency coefficient; σ_{tu}—ultimate tensile stress of the lug, $\sigma_{tu} = 970$ MPa; A_t—lug tensile area, $A_t = (W\text{-}D) \times t$; W—width of the lug, as shown in Fig. 1; D—diameter of the hole in the lug, as shown in Fig. 1; t—thickness of the lug.

(b) Lug shear tear and compression failure

The calculation method for the failure load due to shear tear and compression in the lug is given by Eq. (2):

$$P_{bru} = K_{br} \times \sigma_{tu} \times A_{br} \quad (2)$$

Where: P_{bru}—shear tear and compression failure load; K_{br}—lug compression efficiency coefficient, $K_{br} = 0.51$; A_{br}—lug compression area, $A_{br} = \text{D} \times \text{t}$.

3.2 Strength Calculation Under Transverse Load ($\theta = 90°$)

The calculation method for the ultimate failure load of the lug under transverse loading is given by Eq. (3):

$$P_{tru} = K_{tru} \times \sigma_{tu} \times A_{br} \tag{3}$$

Where: P_{tru}—lug ultimate failure load under transverse loading; K_{tru}—lug transverse ultimate load efficiency coefficient, $K_{tru} = 0.41$; A_{br}—lug compression area, $A_{br} = D \times t$.

3.3 Strength Calculation Under Oblique Load ($\theta = 45°$)

For the lug under oblique loading, the ultimate load can be converted using relevant equations. The calculation method is given by Eqs. (4), (5) and (6):

$$R_a^{1.6} + R_{tr}^{1.6} = 1 \tag{4}$$

$$R_a = \frac{P \sin\theta}{\min(P_{bru}, P_{tu})} \tag{5}$$

$$R_{tr} = \frac{P \cos\theta}{P_{tru}} \tag{6}$$

Where: P—ultimate load; R_a—ratio of the axial component of the ultimate load P to the minimum value between shear tear and compression failure load P_{bru} or tensile failure load P_{tu}; R_{tr}—ratio of the transverse component of the ultimate load P to the transverse ultimate failure load P_{tru}; θ—angle of oblique loading.

4 Test Process

In order to test the mechanical properties of the lug structure, the axial loading tests, oblique loading tests, and transverse loading tests were carried out on the electronic universal material static testing machine. Schematic diagrams of the loading processes for the three tests are shown in Fig. 3a~b.

Before the formal test, 30% estimated load pre-test should be carried out for three times to eliminate the tooling gap, and then the test pieces, tooling and test loading equipment should be checked for abnormalities to ensure that there are no abnormalities before the formal test. The formal tests were carried out at room temperature, and the specimens were loaded until failure. The testing process was conducted without any anomalies, and the test data were deemed valid.

a) Axial loading process b) Oblique loading process c) Transverse loading process

Fig. 3. The diagram of the loading process

5 Test Results

5.1 Axial Loading Test ($\theta = 0°$)

The average failure load for the lug under axial loading at room temperature was 28.78 kN. The failure mode and failure load of the test specimens is shown in Fig. 4 and Table 2, and the failure mode of all specimens is consistent. The failure mode for specimens L001, L002, and L003 was the axial shear failure.

The load-displacement curves of L001, L002 and L003 for the axial loading test are shown in Fig. 5.The load-displacement curves show that the load-displacement of test pieces L001, L002 and L003 overlap well. According to the shear tear and compression failure formula for the lug, the estimated failure load is 29.68 kN. The actual failure load obtained from the test differs from the theoretical calculated load by only 0.9 kN, indicating a high degree of overlap in the load values.

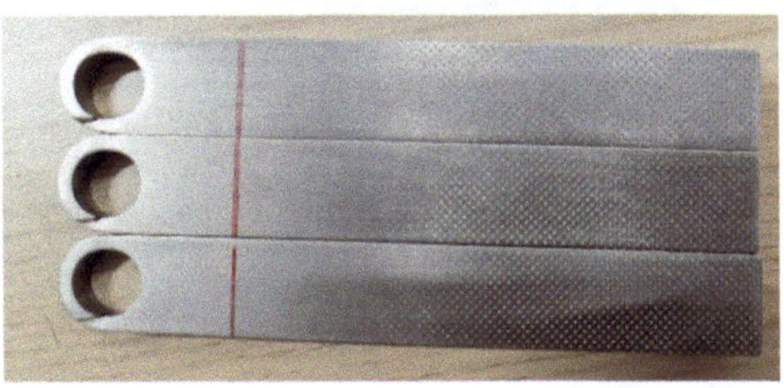

Fig. 4. Failure modes under axial loading

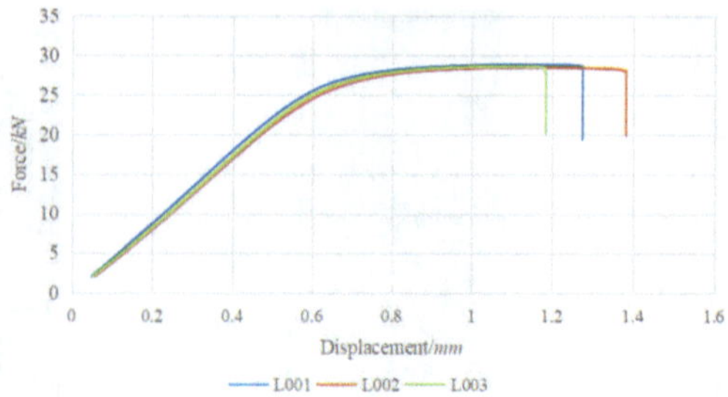

Fig. 5. Load-displacement Curve for the axial tensile test

5.2 Oblique Loading Test (θ = 45°)

The average failure load for the lug under oblique loading at room temperature was 29.45 kN. The failure mode and failure load of the test specimens is shown in Fig. 6 and Table 2, and the failure mode of all specimens is consistent. The failure mode for specimens M001, M002, and M003 was the tensile-shear composite failure.

The load-displacement curves of test specimens M001, M002, and M003 for the oblique loading are shown in Fig. 7. The curves for these specimens overlap moderately. According to the oblique loading formula for the lug, the estimated failure load is 24.19 kN. The actual failure load obtained from the test differs from the theoretical calculated load by 5.26 kN, indicating a significant difference.

Fig. 6. Failure modes under oblique loading

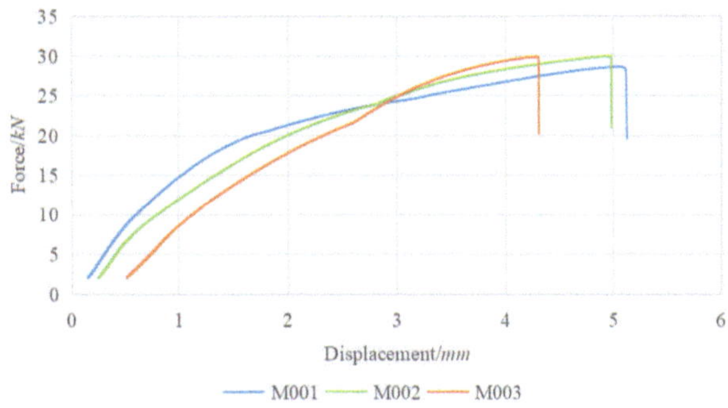

Fig. 7. Load-displacement curve for the oblique tensile test

5.3 Transverse Loading Test ($\theta = 90^\circ$)

The average failure load for the lug under transverse loading at room temperature was 24.96 kN. The failure mode and failure load of the test specimens is shown in Fig. 8 and Table 2, and the failure mode of all specimens is consistent. The failure mode for specimens S001, S002, and S003 was transverse shear failure.

The load-displacement curves for the transverse loading test specimens S001, S002, and S003 are shown in Fig. 9. The curves for these specimens overlap moderately. According to the oblique loading formula for the lug, the estimated failure load is 23.86 kN. The actual failure load obtained from the test differs from the theoretical calculated load by only 1.1 kN, indicating a high degree of overlap in the load values.

Fig. 8. Failure modes under transverse loading

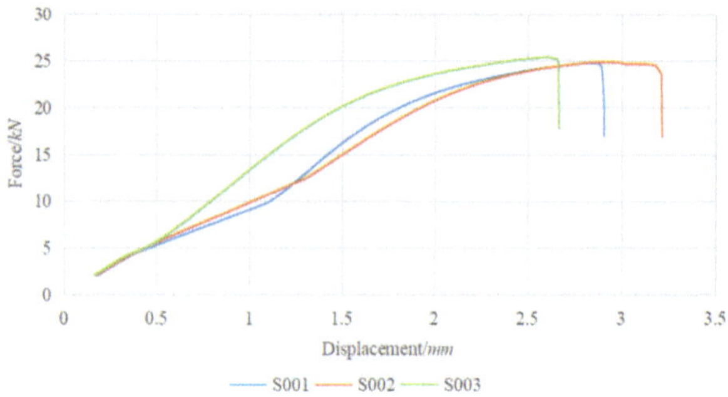

Fig. 9. The Load-displacement curve for the transverse tensile test

Table 2. Failure modes and failure loads

Number	Load type	Failure mode	Failure load/kN
L001	axial	axial shear failure	28.86
L002	axial	axial shear failure	28.85
L003	axial	axial shear failure	28.62
M001	oblique	tensile shear failure	28.63
M002	oblique	tensile shear failure	29.89
M003	oblique	tensile shear failure	29.82
S001	transverse	transverse shear failure	24.74
S002	transverse	transverse shear failure	24.82
S003	transverse	transverse shear failure	25.33

6 Conclusion

In this study, axial loading tests ($\theta = 0°$), oblique loading tests ($\theta = 45°$), and transverse loading tests ($\theta = 90°$) were conducted on additive manufactured lugs, with all key dimensions kept consistent. Use the static testing machine to test the bearing capacity of each test piece, and summarize the failure mode of each test piece. This paper provides the reference basis for the airworthiness verification and installing application of the additive manufacturing door hinge lugs in civil aircraft. The following conclusions can be drawn from the tests:

1) The following conclusions can be drawn from the tests: the average failure load under axial loading is 28.78 kN, the average failure load under oblique loading is 29.45 kN and the average failure load under transverse loading is 24.96 kN, and the results show that the carrying capacity of axial loading and oblique loading is higher than the carrying capacity of the transverse loading.

2) The failure modes of the three configurations of the axial loading, oblique loading, and transverse loading are all shear failure, with the configuration of the oblique loading showing the tensile-shear composite failure. And from the load-displacement curves, it can be seen that when the failure modes of the three configurations reach the failure load, the load is dropped dramatically, while the axial loading test appears more regular yielding stage .
3) From the test results of three configurations, it can be seen that the error range between the actual load value and theoretical load value under axial loading is 3%, the error range between the actual load value and theoretical load value under oblique loading is 17.9% and the error range between the actual load value and theoretical load value under transverse loading is 4.4%. The prediction calculation methods for the ultimate failure load of the lug under axial loading and transverse loading have good compliance and applicability.
4) This test can provide certain experimental basis for the assessment of load-bearing capacity and the installation application of the ramp area of the additive manufacturing cabin door hinge.

References

1. Westerweel, B., Basten, R.J.I., Van, H.G.J.: Traditional or additive manufacturing ? Assessing component design options through life cycle cost analysis. Eur. J. Oper. Res. **270**(2), 570–585 (2018)
2. Rosen, D.W.: A review of synthesis methods for additive manufacturing. Virtual Phys. Prototyp. **11**(4), 305–317 (2016)
3. Williams, J.C., Boyer, R.R.: Opportunities and issues in the application of titanium alloys for aerospace components. Metals. **10**(6), 705 (2020)
4. Najmon, J.C., Raeisi, S., Tovar, A.: Review of additive manufacturing technologies and applications in the aerospace industry. In: Additive Manufacturing for the Aerospace Industry, pp. 7–31 (2019)
5. Kumar, L.J., Nair, C.G.K.: Current Trends of Additive Manufacturing in the Aerospace Industry, pp. 39–54. Springer, Singapore (2017)
6. Liu, R., Wang, Z., Sparks, T., et al.: Aerospace Applications of Laser Additive Manufacturing, pp. 351–371. Woodhead Publishing (2017)
7. Kranz, J., Herzog, D., Emmelmann, C.: Design guidelines for laser additive manufacturing of lightweight structures in TiAl6V4. J. Laser Appl. **27**(S1), S14001 (2015)
8. Zhang, M.K., Yang, Y.Q., Wang, D., et al.: Effect of heat treatment on the microstructure and mechanical properties of Ti6Al4V gradient structures manufactured by selective laser melting. Mater. Sci. Eng. A. **736**, 288–297 (2018)
9. Arcieri, V.E., Baragetti, S., Božić, Ž.: Failure analysis of Ti6Al4V titanium alloy under fatigue loading: an experimental and numerical study. Eng. Fail. Anal. **164**, 108715 (2024)
10. Manjunath, K., Tewary, S., Khatri, N., et al.: Simulation-based investigation on ultra-precision machining of additively manufactured Ti-6Al-4V ELI alloy and the associated experimental study. Proc. Inst. Mech. Eng. B J. Eng. Manuf. **238**(10), 1554–1567 (2024)
11. Song, X., Chen, L., San, L., et al.: Design and realization of static verification test for additively manufactured civilian aircraft components. J. Harbin Inst. Technol. **27**(05), 70–78 (2022)

Multiscale Characterization and Extreme Environment Impact Response of Composite Structures

Characterizing Origami-Inspired Piezoresistive Folds for Foldable Composite Structures

Kamran A. Khan(✉), Israr Ud Din, and Rajasekar Ramachandran

Department of Aerospace Engineering, Khalifa, University of Science and Technology, PO Box 127788, Abu Dhabi, United Arab Emirates
kamran.khan@ku.ac.ae

Abstract. Origami principles can be efficiently exploited in the development of foldable structures when composite materials are used. In this study, glass fiber reinforced polymer composites (FRPCs) with dual-matrix system (epoxy-elastomer) was successfully used to manufacture folding hinge with four different sizes such as 3, 6, 9 and 12 mm. In addition, the glass fabric was coated with rGO to serve also as a piezoresistive sensor to monitor the folding and unfolding process. A cost-effective hot compression molding process was used to fabricate multiple samples in one batch and processing of laminate was monitored using in-situ measurement of resistance change. Then, the folding hinges were tested in single and 4 cycles tests. Moment-curvature and fractional changes in resistance (FCR) were simultaneously recorded in these tests. The results demonstrated that the folding region size in such structures plays a significant role in the electromechanical performance. Smaller folding region sizes of 3 mm and 6 mm demonstrated a higher value of FCR, and moment as compared to the 9 mm and 12 mm. It is concluded that these sizes shall be carefully selected in the design and development of origami-inspired deployable composite structures.

Keywords: Piezoresistivity · rGO · Deployable structures · Origami · Foldable structures · Composites

1 Introduction

The unique properties and wide-ranging applications of foldable composite structures have driven increased attention in recent years. These structures are designed to be flexible and capable of being folded or unfolded without compromising their structural integrity [1, 2]. The development of foldable composite structures has been driven by advancements in materials science, engineering, and manufacturing techniques [3]. One area of research in foldable composite structures focuses on stretchable and foldable conductors. Carbon-based nanomaterials, for example carbon nanotubes and graphene, have been widely studied for their stretchable and foldable properties. For example, Won et al. demonstrated the potential application of carbon nanotubes (CNTs) and silver flakes helical ribbon-structured composites in stretchable conductors [4].

Foldable and deployable flexible composite thin-walled structures have been developed with characteristics such as light weight, excellent mechanical properties, and

Y. Zhao (Ed.): ICAMM 2025, SPM 96, pp. 155–165, 2025.
https://doi.org/10.1007/978-3-032-09862-7_15

large deformation ability exploiting fiber reinforced polymer composites (FRPCs) [2, 3]. These structures have the potential to be used in space missions due to small volumes prior to deployment in full-scale space structures. The development of simplified analytical models and algorithms has facilitated the prediction of the neutral cross sectional locations during tensile deformation for lenticular deployable composite booms [5]. Bio-inspired nanocomposites have also been investigated for their potential in creating foldable electrodes. These nanocomposites mimic the microstructures of natural materials, such as cocoon silk, to achieve self-adaptive stress dispersion and enhance foldability. The use of biomimetic structures in composite materials enables the development of lightweight and flexible electrodes with excellent mechanical properties [6]. Origami-inspired foldable structures have also been investigated for their potential in creating reconfigurable materials [1, 7]. By folding and unfolding origami patterns, it is possible to achieve shape reconfiguration and tune the stiffness of the materials [8–11]. These foldable structures have applications in lightweight deployable materials, actuators, packaging, and lockable robotic systems.

The coating of graphene oxide (GO) on the glass fabric enables the fabric to become conductive, allowing for the detection of strain, pressure, and other mechanical stimuli. To construct the origami-inspired FRPCs hinge with dual-matrix system, we utilized rGO coated glass fabric and rGO embedded elastomeric lamina as the substrate. The substrate was stiffened with glass fiber prepreg utilizing a cost-effective hot compression molding process. The folding action was enabled by elastic hinges made of either rGO coated glass fabric or rGO embedded elastomeric lamina. The choice of an extremely flexible and ultra-thin strong fiber substrate allowed the structure to be folded and packaged efficiently, while also facilitated low-energy deployment by utilizing the stored strain energy. To characterize the piezoresistive behavior of the deployable composite structures, the width of folding region in the deployable structure was varied from 3 mm to 12 mm with 3 mm intervals. Moreover, piezoresistivity was also monitored during the hot compression molding process. The light-weight folding hinges with sensing capability developed in this study are manufactured using a facile cost-effective approach from the dual-matrix FRPCs which have not been investigated in literature focusing origami-inspired composite structure.

2 Materials

For the manufacturing of the origami-inspired folding hinge, E-glass plain weave fabric with trade name of RE210D™ of aerial density of 202 g/m^2 was used. This fabric was coated with graphene oxide (GO) purchased from Abalonyx AS, Norway. Two similar size prepreg (2 × 2 twill weave glass/epoxy) known as Hexply® M26 of weight 300 g/m^2 defining the stiff regions were placed. These prepreg patches were placed such that to achieve the gap sizes of 3, 6, 9, and 12 mm in the folding region. These gap sizes are highly important in achieving an appropriate folding/unfolding operation. These sizes were identified in our previous study to facilitate a symmetric 180° fold without failure, thereby requiring a lower bending energy and faster recovery [3].

3 Sample Manufacture

3.1 Preparation of Graphene Oxide (GO) Coated Fabric

To make the piezo resistive sensor, initially, a GO solution was prepared by taking 6 g of GO paste and 300 ml of ionized water. Then, sonication was done on the GO solution by keeping it at the bath sonicator for an hour at 47 °C. After that, 210 × 130 mm plain-woven E-glass fabric were cut and placed in a clean glass tray. Solution-dip technique was used for the GO coating [12, 13]. The GO solution was poured into the tray. Finally, the GO-coated fabrics were thermally reduced at 170 °C for 24 h to partially reduced to GO fabric as shown in Fig. 1a. The partially rGO-coated E-glass fabric was then cut into sample sizes. To measure the conductivity of the rGO-coated glass fabric, wiring connections were made by using copper wires and copper tapes as depicted in Fig. 1b.

3.2 Fabrication of Deployable Composite Structures

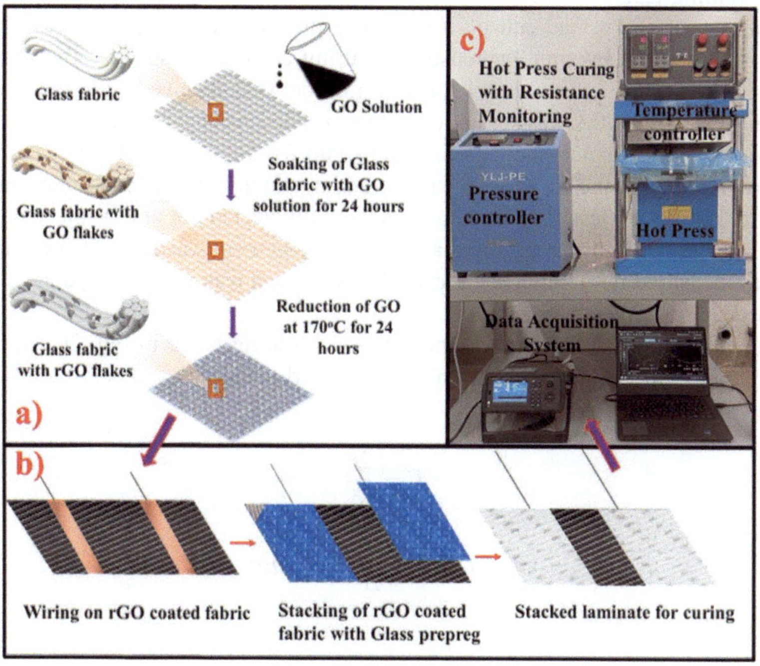

Fig. 1. Steps involved in the manufacturing of piezoresistive origami-inspired hinge structure (a) Graphene oxide (GO) coating and partial thermal reduction to rGO, (b) connections, (c) consolidation of prepreg patches to form the stiff region.

Glass fiber prepreg were laid up with rGO-coated E-glass fabric at the positions to make the stiffer regions of deployable composite structures as shown in Fig. 1b. The prepreg patches were cured keeping under the hot compression molding process at

1.8 bar and 125 °C for 90 min. The equipment used (MTI, YLJ-HP300) for this purpose has been given Fig. 1c.

The variation in resistance was also monitored during the curing in hot compression molding process. For this purpose, (DAQ) Keysight DAQ970A) was used as shown in Fig. 2c. DAQ970A displays the real time live variation in the resistance connected through a BenchVue® acquisition software.

4 Cyclic Folding and Unfolding Experiments

In the first set of tests, a single cycle experiment consisting of a folding and unfolding was carried out for the four types of folding hinge samples. After that, four cycles tests were completed for the four different types of samples classified on the basis of the size of folding region. These were selected as 3, 6, 9, and 12 mm, in order to investigate the effect of foldable region width on the folding mechanism of the deployable composite structure. The thickness of all the samples was around 0.45 mm whereas the width was kept as ~50 mm.

The folding tests were carried out on a SALT™ universal testing machine (UTM) installing an auxiliary load cell of 100 N. The piezoresistive response was obtained for all of these tests by simultaneously measuring electrical resistance from the rGO coated E-glass fabric sensors and load-displacement data received from the load cell. The Keysight DAQ970A was set up to capture piezoresistivity during the tests for this purpose. Figure 2a depicts the experimental setup for a typical deployable composite structure mounted in the UTM for a folding and cyclic test and to record the change in resistivity during the test. The folding tests were all carried out at a displacement rate having a value of 20 mm/min. The maximum displacement was set to 95 mm, and the load was measured throughout each test. The cyclic test was done on separate samples in each system with 4 cycles.

The angle (θ) as shown in the schematic of a folding hinge in Fig. 2b region was measured digitally during the test using a camera. Moment (M) and curvature (k) were calculated using the Eqs. (1), (2) and (3) [2].

$$k = \frac{2\theta}{b} \tag{1}$$

$$d = Lsin\theta + \frac{b}{2\theta}(1 - cos\theta) \tag{2}$$

$$M = F \times d \tag{3}$$

Piezoresistivity was measured using the fractional change in resistance (FCR) expressed as $(R - R_o)/R_o \times 100$. Where R_o is the reference resistance of rGO coated glass fabric at the beginning of the experiment when no stress or strain has been applied.

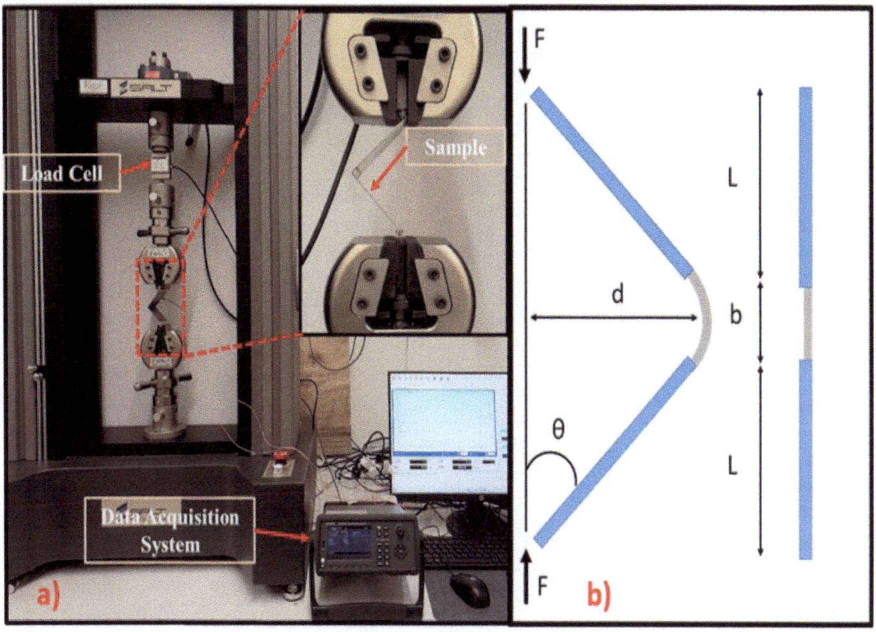

Fig. 2. (a) Electromechanical folding test setup, (b) schematic of deformed state of folding hinge and the geometrical parameters form which movement was calculated.

5 Results and Discussion

5.1 Piezoresistivity Measurement During Manufacturing

Prior to the hot compression molding process, the electrical resistance of each rGO coated fabric was measured to assess the uniformity and quality of the rGO coating. In addition, continuous monitoring of the electrical resistance change was done throughout the curing process for a total of 90 min. Figure 3 shows the FCR for all the samples for the initial duration of 30 min. The resistance of the folding hinge having a 3 mm folding region size demonstrated higher increase of 50% after the resin from prepreg started to flow through rGO coated fabric as compared to the remaining variants. The reason for the higher rise in the FCR is due to the greater amount of resin present since the size of the prepreg is greater for the gap of 3 mm. Samples with the folding hinge size greater than 3 mm exhibited a lower change in the piezoresistivity. They showed about a maximum FCR of ~20%. Once the resin started gelling, all the coupons responded in a same manner by decreasing the resistivity and the resistance stayed constant when the resin hardened.

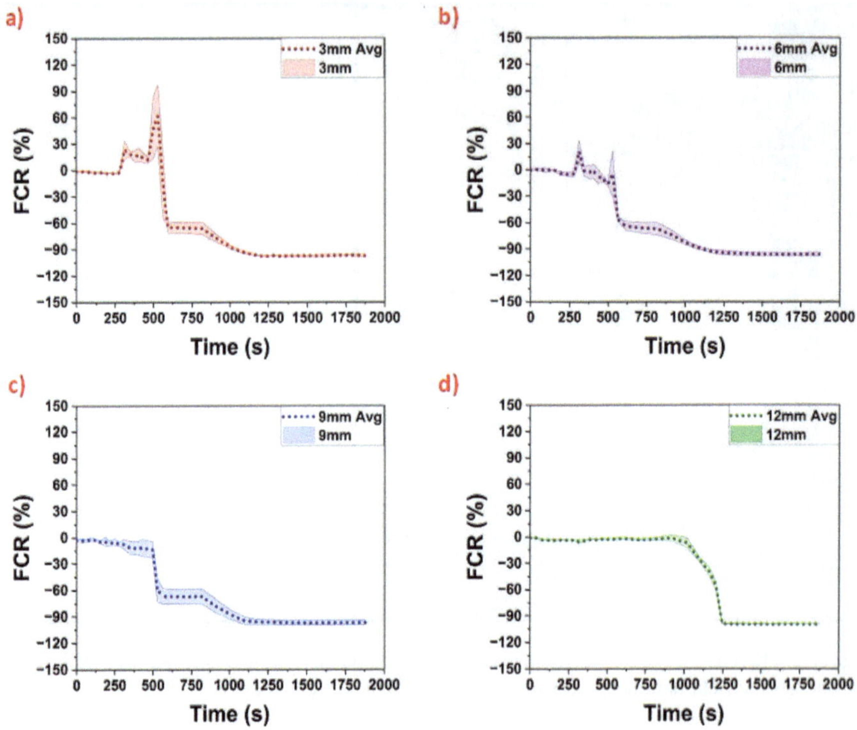

Fig. 3. Piezoresistivity measurement during the hot compression molding process, (a) folding region of 3 mm, (b) folding region of 6 mm, (c) folding region of 9 mm, (d) folding region of 12 mm.

5.2 Folding Experiments

Single Cycle

The moment curvature and piezoresistive response of all the four samples tested are shown in Fig. 4. These results are presented for a single cycle folding and unfolding cycle. Figure 4 shows that there is a considerable amount of increase in resistivity when the width of the folding region is reduced. Higher values of FCR can be observed in case of the 3 mm and 6 mm folding region width. On the other hand, origami-inspired folding samples having 9 mm and 12 mm sizes of the folding region demonstrated an initial spike in FCR when the compressive load applied to initiate folding operation. This phenomenon is due to the crushing or micro buckling of fibers in the foldable region. In the folding phase, the FCR decreased at the start and varied throughout the folding phase with a modest slope. This trend is due to the contact of rGO particles network when the coupon is folding. Following that, in the folding phase, a considerable rise in FCR was observed due to the breakage rGO network.

Multi-cycles Tests

The cyclic moment vs. curvature response for all the variants of the samples considered in this study are compared in Fig. 5. Moreover, Fig. 6 compares the piezoresistivity and

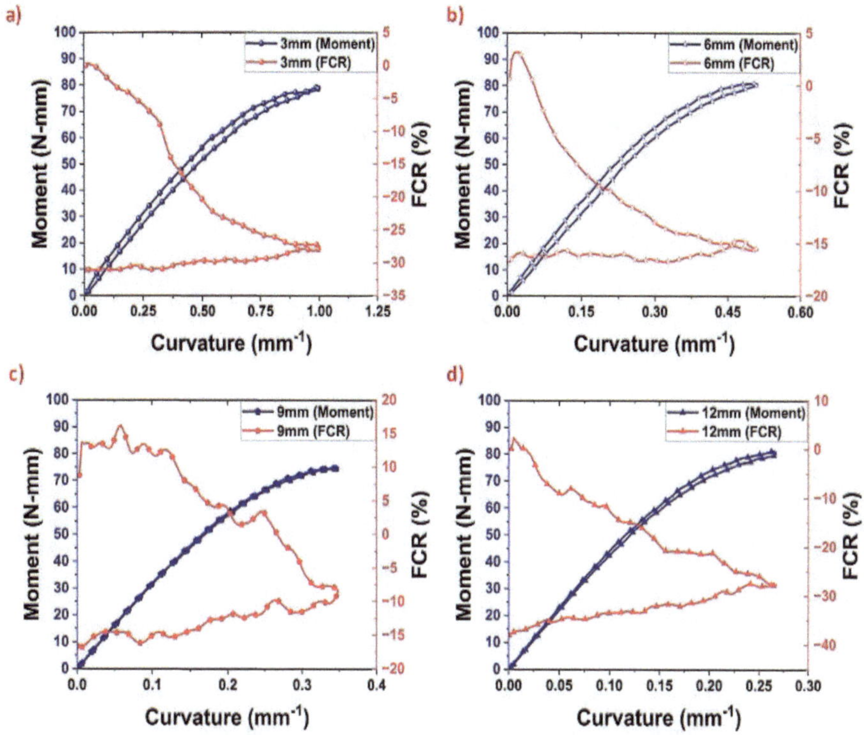

Fig. 4. Moment vs. curvature and piezoresistive response of single cycle folding tests for the four variants of samples, (a) folding region of 3 mm, (b) folding region of 6 mm, (c) folding region of 9 mm, (d) folding region of 12 mm.

moment vs. curvature tested for four cycles. The results demonstrate a typical behavior of a rise in electrical resistance followed by a drop based on the contact between rGO flakes in the foldable region. The variations in the contact between the rGO flakes and the deformation of the rGO coated glass fabric are responsible for the piezoresistivity.

Changes in resistance due to contact of rGO flakes where a significant component since the latter had a lower impact on the overall electrical resistance change compared to the former (Fig. 6a). It was observed that while cyclic load was applied to the samples, the foldable region of the sample was stretched on the outer surface and compressed in the inner surface. Thus, inner contraction resulted in dense rGO-based networks that reduced electrical resistance by reconfiguring the three-dimensional rGO-based networks. Consequently, there is negligible amount of change in resistance even in the maximum curvature state in the samples which have foldable region width higher than 3 mm.

Furthermore, the first cycle produced high stress when the fibers encountered frictional resistance within the folds and attempted to adjust themselves. The fibers modification permits compression and deployment more easily in successive cycles, as demonstrated from the second cycle. More specifically, first loading process weakens the structure, resulting in a loss in stiffness.

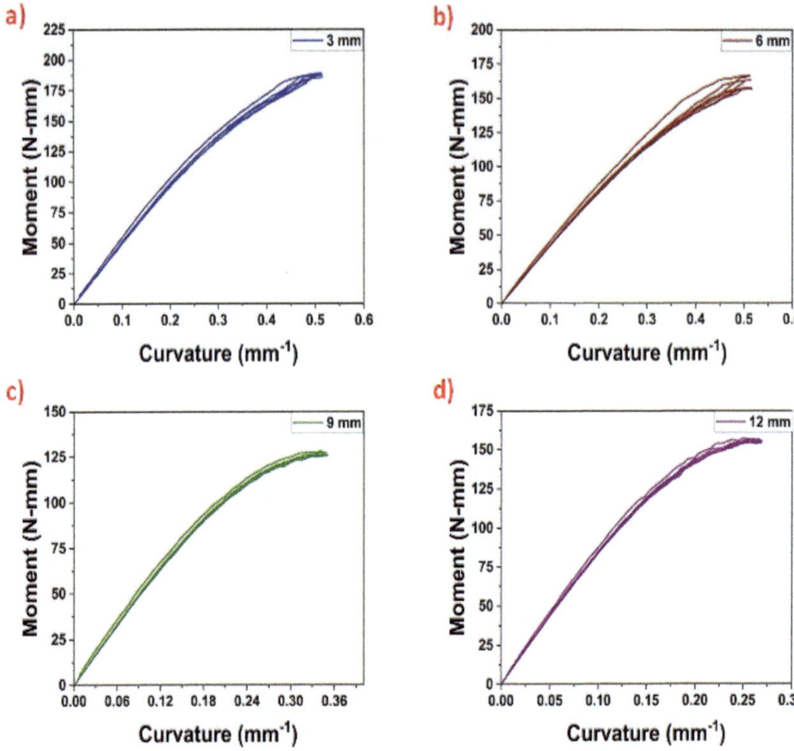

Fig. 5. Moment against curvature response during four cycles of folding and unfolding, (a) folding region of 3 mm, (b) folding region of 6 mm, (c) folding region of 9 mm, (d) folding region of 12 mm.

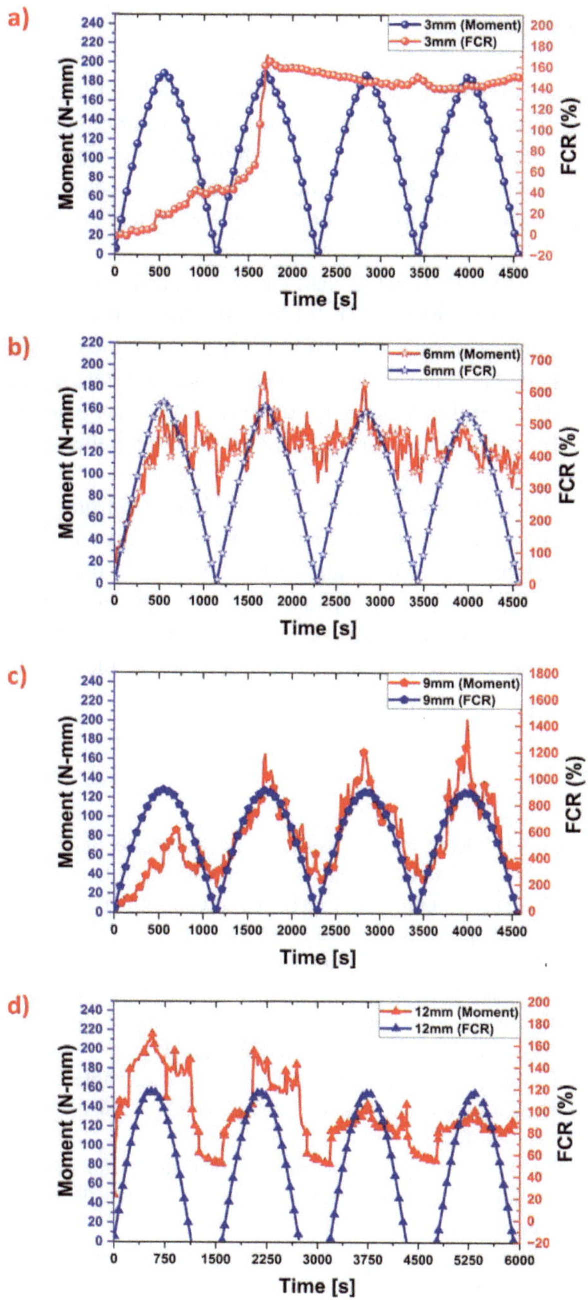

Fig. 6. Moment vs. time and FCR vs. time for four cycles of folding and unfolding, (a) folding region of 3 mm, (b) folding region 6 mm, (c) folding region of 9 mm, (d) folding region 12 mm.

6 Conclusions

Piezoresistive origami-inspired deployable composite folding hinge was developed and tested with four different design configurations. The folding hinge consists of stiff and folding regions. Single cycle and multicycles folding and unfolding tests were carried out in which the electromechanical behavior was recorded. rGO coated glass fabric was used as a piezoresistive sensor in the folding and cyclic loading test on deployable composite structure.

The findings show that rGO coated glass fabric demonstrated the sensing capabilities when employed in the origami-inspired folding structures. The folding test conducted on the deployable composite structures exhibited hysteresis due to the loss of energy by friction between fibers. The recovery from folded to unfolded state was achieved by using the strain energy stored in the deployable structure. The effect of folding region width on the folding performance and piezoresistivity of the deployable composite structures were evaluated.

It is concluded that the size of the folding region has strong effects on the electromechanical behavior in terms of FCR and moment curvature in the origami-inspired foldable composite structures. The folding region is highly important and should be carefully selected during the designing phase. Finite element analysis (FEA) is required to complement the experimental results presented which will be the focus of our future research work. Moreover, the electromechanical behavior of the folding hinges shall be evaluated under long-term fatigue testing.

References

1. Ahmed, A., Din, I.U., Kulkarni, S., Khan, K.A.: Multiscale finite element modeling of origami-inspired dual matrix deployable composite with visco-hyperelastic hinge. Compos. Struct. **343**, 118301 (2024). https://doi.org/10.1016/j.compstruct.2024.118301
2. López Jiménez, F., Pellegrino, S.: Folding of fiber composites with a hyperelastic matrix. Int. J. Solids Struct. **49**, 395–407 (2012). https://doi.org/10.1016/j.ijsolstr.2011.09.010
3. Al-Mansoori, M., Khan, K.A., Cantwell, W.J.: Harnessing architected stiffeners to manufacture origami-inspired foldable composite structures. Compos. Sci. Technol. **200**, 108449 (2020). https://doi.org/10.1016/j.compscitech.2020.108449
4. Won, Y., Kim, A., Yang, W., Jeong, S., Moon, J.: A highly stretchable, helical copper nanowire conductor exhibiting a stretchability of 700%. NPG Asia Mater. **6**, e132–e132 (2014). https://doi.org/10.1038/am.2014.88
5. Wang, L.-W., Bai, J.-B., Shi, Y.: Simplified analytical model for predicting neutral cross-section position of lenticular deployable composite boom in tensile deformation. Materials. **14**, 7809 (2021). https://doi.org/10.3390/ma14247809
6. Zhu, P., Xie, X., Sun, X., Soto, M.A.: Distributed modular temperature-strain sensor based on optical fiber embedded in laminated composites. Compos. Part B. **168**, 267–273 (2019). https://doi.org/10.1016/j.compositesb.2018.12.078
7. Ud Din, I., Ahmed, A., Tarek, F., Cantwell, W., Khan, K.A.: Ultra-thin composites membrane for deployable structures: XCT driven characterization and FE modeling of folding structure. Compos. Sci. Technol. **245**, 110341 (2023). https://doi.org/10.1016/j.compscitech.2023.110341

8. Jamalimehr, A., Mirzajanzadeh, M., Akbarzadeh, A., Pasini, D.: Rigidly flat-foldable class of lockable origami-inspired metamaterials with topological stiff states. Nat. Commun. **13**, 1816 (2022). https://doi.org/10.1038/s41467-022-29484-1
9. Hu, W., et al.: Light-induced topological patterning toward 3D shape-reconfigurable origami. Small. **18**, 2107078 (2022). https://doi.org/10.1002/smll.202107078
10. Ma, J., Zang, S., Feng, H., Chen, Y., You, Z.: Theoretical characterization of a non-rigid-foldable square-twist origami for property programmability. Int. J. Mech. Sci. **189**, 105981 (2021). https://doi.org/10.1016/j.ijmecsci.2020.105981
11. Ahmed, A., Ud Din, I., Ramachandran, R., Sikandar Bathusha, M.S., Cantwell, W., Khan, K.A.: Micro-CT driven characterization and validation of constituents' properties used in origami-inspired foldable composites fabricated via different manufacturing processes. Compos. Struct. **322**, 117382 (2023). https://doi.org/10.1016/j.compstruct.2023.117382
12. Ud Din, I., Medhin, Y., Aslam, N., Bathusha, M.S.S., Umer, R., Khan, K.A.: Rate dependent piezoresistive characterization of smart aerospace sandwich structures embedded with reduced graphene oxide (rGO) coated fabric sensors. Compos. Commun. **36**, 101382 (2022). https://doi.org/10.1016/j.coco.2022.101382
13. Ud Din, I., Bathusha, M.S.S., Khan, K.A.: Effects of liquid rubber-modified epoxy on the fracture toughness of rGO-coated fabric piezoresistive composites. Compos. Commun. **56**, 102368 (2025). https://doi.org/10.1016/j.coco.2025.102368

Interface Characterization in 3D Printed Continuous PA6 Composites Using Nano-Indentation

Yasser H. Alattas, Majed A. Alharbi(✉), and Anas I. Alburayt

Space Technologies Institute, King Abdulaziz City for Science and Technology, Riyadh 11442, Saudi Arabia
{yalattas,mbharbi,abareet}@kacst.gov.sa

Abstract. High demand in the last years for products with specific characterization has led to the development and advancement of materials. Carbon fibre reinforced polymer (CFRP) is one of the advanced composite materials used. The development in materials has continued by using additive manufacturing of CFRP in different applications. Studying the characterization of CFRP manufactured in additive manufacturing is important to understand its impact. Therefore, this study focuses on the characterization of the fibre-matrix interface in 3D-printed continuous carbon fibre-reinforced thermoplastic composites (CFRTPCs) using Nano-indentation techniques. The performance of composite materials is highly dependent on the quality of the fibre-matrix interface. In 3D printed composites, defects such as voids and poor interfacial bonding, inherent to the layer-by-layer deposition process, can play a vital role in mechanical properties. The study employs Nano-indentation to characterize the micromechanical properties of the fiber-matrix interface in 3D-printed continuous carbon fiber-reinforced polymer composites. Results are compared with compression-moulded specimens to evaluate manufacturing-induced variations. Findings reveal that interface properties are strongly influenced by fibre volume fraction, voids, and processing parameters. Optimization of 3D printing parameters, coupled with compression moulding, is shown to enhance interfacial adhesion, thereby improving mechanical performance. The findings suggest that optimization of 3D printing parameters and the use of compression moulding can significantly improve interfacial adhesion, leading to enhanced mechanical performance of the composites.

Keywords: 3D printing · carbon fibre reinforced polymers · Nano-indentation · interface characterization · push-in test

1 Introduction

Carbon fibre-reinforced polymer (CFRP) is an advanced composite material that is made by embedding carbon fibre within a polymer matrix. Carbon fibre is a strength in the weight ratio, while the polymer matrix is low-cost and easy to prepare. CFRP commonly benefits from its lightweight, corrosion resistance and high-performance structures. It is

Y. Zhao (Ed.): ICAMM 2025, SPM 96, pp. 166–174, 2025.
https://doi.org/10.1007/978-3-032-09862-7_16

composites fabricated by 3D printers that are most commonly used across diverse industries, including aerospace, electronics, and medical engineering, owing to their exceptional mechanical and thermal characteristics [1]. These materials are distinguished by their high stiffness, elevated modulus, superior strength, and minimal thermal expansion, properties that render them indispensable in applications requiring lightweight yet robust solutions [2]. Structurally, CFRP composites comprise three fundamental constituents: the carbon fibres, which provide reinforcement; the polymer matrix, which binds the fibres; and the interface, the critical region mediating their interaction [3]. The fibre-matrix interface plays a pivotal role in determining the mechanical performance of composite materials. In 3D printed composites, the layer-by-layer deposition process often introduces defects such as voids and weak interfacial bonding, which compromise load transfer efficiency [1, 2]. These challenges are exacerbated in continuous fibre-reinforced composites, where interfacial integrity directly governs tensile, flexural, and fatigue properties [4]. Recent advancements in Nano-indentation techniques have enabled obtaining the micromechanical properties at the fibre-matrix interface to study the interfacial adhesion quality in 3D printed continuous carbon reinforced composites at filament/matrix [5]. While compression moulding produces composites with fewer void defects due to uniform pressure and temperature [6], 3D printing offers design flexibility but struggles with achieving comparable interface quality [7]. Recent studies have shown that optimizing printing parameters can significantly improve interfacial properties in 3D printed composites [7, 8]. Therefore, investigating the fiber-matrix interface offers a promising avenue for advancing the design of CFRP composites, enabling improvements in mechanical performance and durability [9]. The primary goal of this study is to outline a methodology to (1) characterize the micromechanical properties of 3D printed CFRTPCs by Nanoindentation, (2) compare results with compression-moulded specimens, and (3) identify key manufacturing parameters influencing interfacial adhesion.

2 Experimental Procedure

2.1 Materials and Preparation

Continuous carbon fibre-reinforced polyamide (PA6) composites were fabricated via FDM (Markforged Mark Two) [10]. The filament comprised 35 vol.% carbon fibres and embedded in a PA6 matrix. Two specimens have been prepared for this experiment, which are regular and compression-molded specimens. Figure 1 shows 3D-printed specimens of CFRTPCs where thirteen layers were printed with a layer thickness of 125 μm. The first post-printing specimen was then sectioned, epoxy-embedded, and polished to a mirror finish as presented in Fig. 2. However, the second post-printing specimen was compression moulded at 230 °C under a uniform pressure to minimize voids and ensure full consolidation.

2.2 Test Methods

Nano-indentation tests were conducted using a Nano indenter (XP, MTS Nano Instruments) and equipped with a conical diamond indenter tip of radius 0.7 μm as in Fig. 3.

Fig. 1. 3D printed of CFRTPCs.

Fig. 2. Epoxy resin

Dynamic Nano-indentation measurements were performed under a continuous stiffness measurement (CSM) mode at a frequency of 70 Hz. Prior to each test series, a calibration procedure was performed to ensure instrument accuracy and minimize thermal drift. Indentations were executed on polished cross-sectional surfaces under displacement control. Five linear arrays of 20 indents each were generated, with adjacent indents spaced 0.8 μm apart to avoid inter-indentation stress field interactions. A trapezoidal loading profile was applied. The tip was loaded to a maximum depth of 100 nm at a constant strain rate of 0.05 s^{-1}. This is followed by a 60 s hold period at peak load to account for time-dependent material behaviour. Subsequently, the sample was unloaded at the same rate as 10% of the maximum load, with a final load plateau to correct the thermal drift effects.

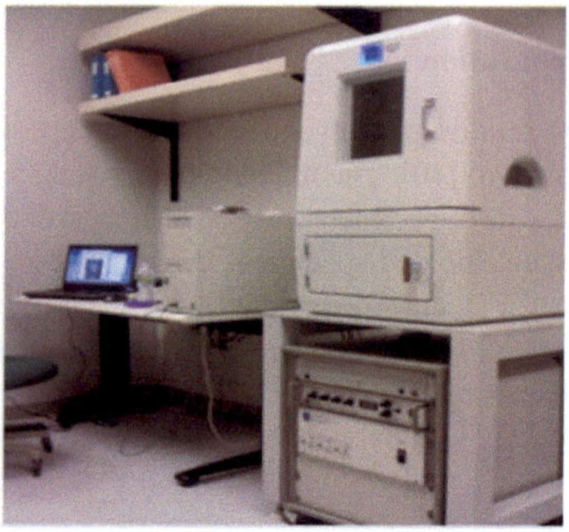

Fig. 3. MTS Nano instruments

3 Results and Discussion

3.1 Microstructure Analysis of CFRP6

Microscopic image analysis is employed to evaluate the interfacial bonding quality and defect distribution within the 3D-printed CFRTPCs. Representative micrographs of cross-sectional specimens, as presented in Figs. 4, 5 and 6, revealed distinct layer-wise structures, interlayer voids, and irregular fiber-matrix distribution. Figure 4 illustrates a cross-section with 0° fibre orientation, highlighting the alignment of thirteen printed layers (125 μm thickness each) and localized voids between adjacent filaments. The average filament width measured is approximately 910 μm, consistent with the printing parameters. The fibre distribution exhibits significant spatial heterogeneity, attributed to inherent irregularities in the CF/PA6 feedstock filament as presented in Fig. 5. A columnar void pattern is observed, correlated with variations in the printing path and inconsistent overlap between deposited filaments as shown in Fig. 6. These defects contrast sharply with the idealized microstructure underscoring the challenges posed by the layer-by-layer FDM process. Such voids and fibre misalignments are known to degrade interfacial load transfer and mechanical performance.

Fig. 4. Microscopic image of CF/PA6.

3.2 Nano-Indentation Analysis of 3D Printed Specimens

Nano-indentation tests were conducted to evaluate the hardness and Young's modulus of the fiber-matrix interface in 3D-printed CFRTPCs. The indentation was performed on the fibre within the matrix to capture the interfacial properties. The hardness values are stabilized at indentation depths greater than 40 nm, prompting the selection of a maximum depth of 100 nm for modulus determination as shown in Table 1.

Figure 7 presents the hardness values as a function of the number of indentations. It is clear that the hardness values reveal significant heterogeneity in the mechanical properties of the 3D-printed specimens. The hardness values for carbon fibre typically range between 0.2 and 0.3 GPa, but the measured values vary widely from 0.01 to 1.5 GPa. Similarly, Young's modulus values are expected to fall within 40–80 GPa, while the measured values are in the range of 2.5–50 GPa as shown in Fig. 8. This high variability

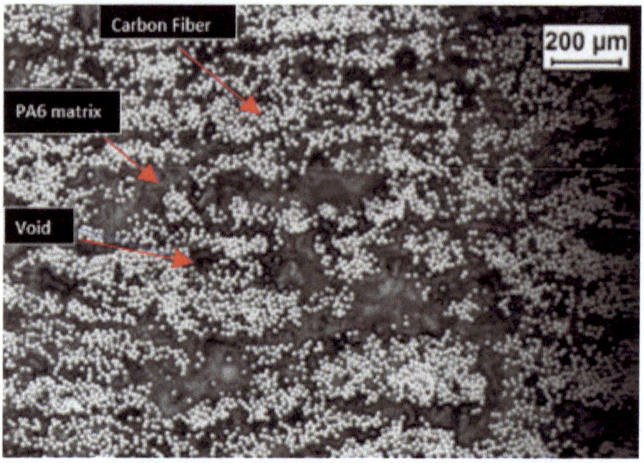

Fig. 5. Irregular distribution of continuous fibres.

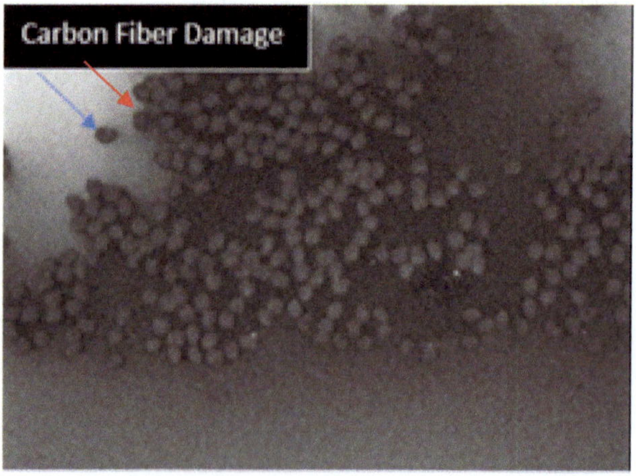

Fig. 6. Indentation debonding failure of CF/PA6 filament

Table 1. Results of the CFRP6 hardness and Young's modulus in different loading forces.

Force (nN)	Hardness H (GPa)	Young's modulus Er (GPa)
100	1.1038	19.1435
50	3.136	28.8972
20	6.9415	39.8018

may be attributed to sample preparation artefacts or intrinsic defects, such as voids, inherent to the 3D-printing process. For comparison, compression-moulded specimens with minimal void content will be evaluated to isolate the influence of process-induced defects.

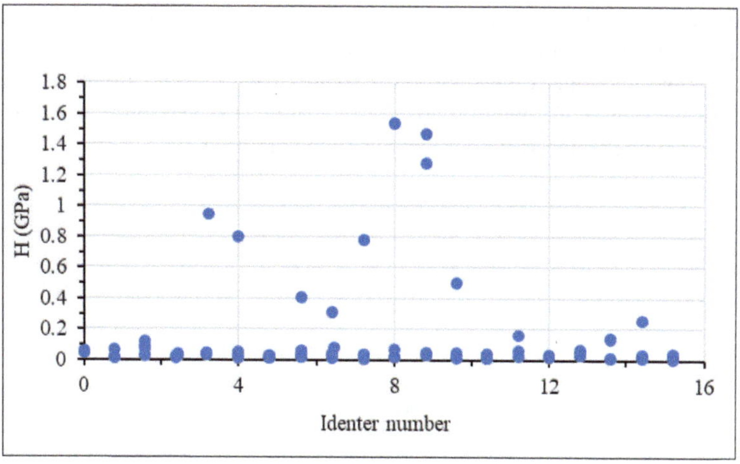

Fig. 7. Nano indentation hardness results curve.

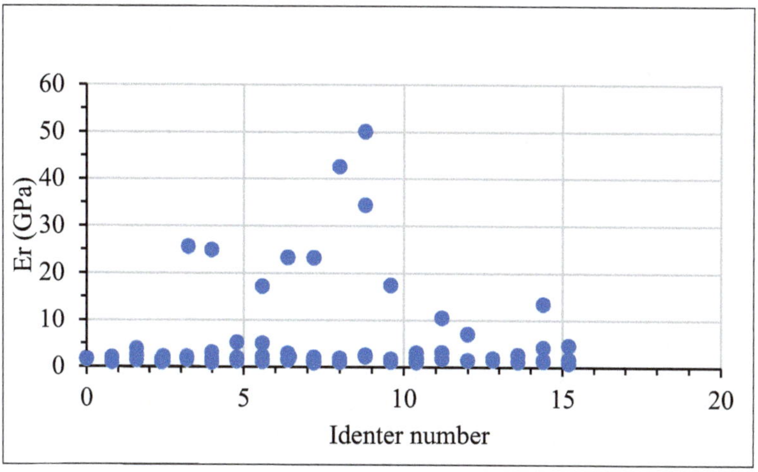

Fig. 8. Nano indentation hardness results curve.

3.3 Comparison of Compression-Molded and 3D Printed Specimens

Carbon fibre-reinforced polymer composites were fabricated using both fused deposition modelling (FDM) and compression moulding (CM) to assess the variability in Nano-indentation results. Compression-moulded specimens exhibit superior interfacial uniformity, attributed to void-free consolidation and the presence of transcrystalline structures at the fibre-matrix interface. In contrast, 3D-printed specimens display a higher prevalence of voids, as illustrated in Figs. 9 and 10, which contribute to localised stress concentrations and reduce load-bearing capacity. Nano-indentation tests on compression-moulded specimens yield hardness values ranging from approximately 0.8 GPa for the fibre to less than 0.2 GPa for the matrix as shown in Fig. 9. Similarly, as presented in Fig. 10, Young's modulus values are in the range of 9.0 GPa for the fibre to less than 4.0 GPa for the matrix. Although these results exhibit less variability compared to the 3D-printed specimens, the variation remains relatively high.

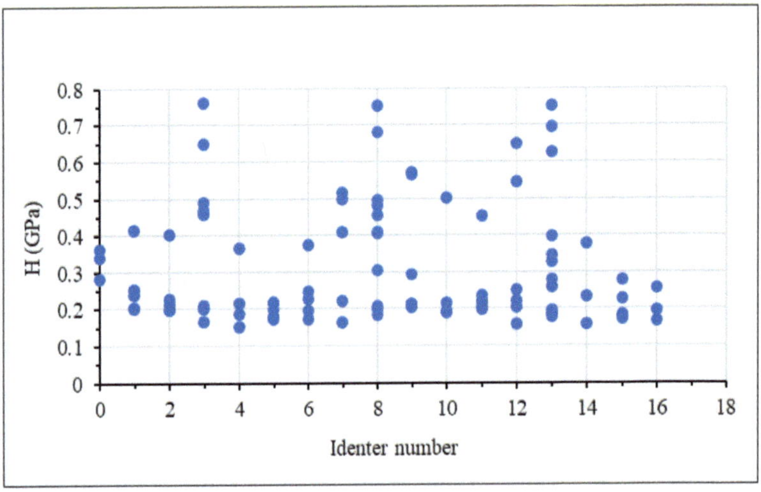

Fig. 9. Nano indentation hardness results curve (CM).

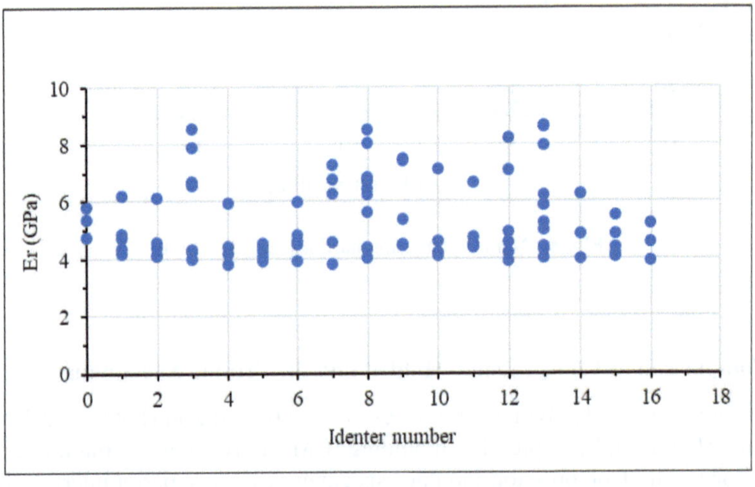

Fig. 10. Nano indentation young modules results curve (CM).

4 Conclusion

The study found that Nano-indentation as a robust technique for characterizing the interfacial properties of (CFRPA6) was ineffective. The key findings are summarized as follows. Firstly, for interfacial Variability and Void Content, the 3D-printed composites exhibited higher void content and higher variability values in interfacial properties compared to compression-moulded specimens. This heterogeneity is attributed to process-induced defects and irregular fibre-matrix distribution inherent to the fused deposition modelling (FDM) process. Secondly, for the case of Compression Molding Advantages, Compression-molded specimens displayed less variation as shown previously, achieved through void-free consolidation and the formation of trans crystalline structures at the fibre-matrix interface. However, the variation remained relatively high. This might also resulted in subpar sample preparation that adversely affected the results. for future work, printing parameters such as layer thickness, nozzle temperature, and printing speed can be optimized to reduce void content and enhance interfacial properties in 3D-printed composites. It can underscore the importance of process optimization to achieve mechanical properties comparable to those of traditionally manufactured composites [11]. Also, future work can be applied on advanced characterization techniques (e.g., fibre push-out tests) and hybrid manufacturing strategies to tailor interfacial properties for high-performance applications [3]. Moreover, another 3D printing method such as SLS (Selective Laser Sintering) can be employed for comparison with FDM method in terms of mechanical properties and bonding strength especially for short CFRPA6.

Acknowledgement. The authors would like to thank King Abdulaziz City for Science and Technology (KACST) for supporting the research work.

References

1. Wang, X., Jiang, M., Zhou, Z., Gou, J., Hui, D.: 3D printing of polymer matrix composites: a review and prospective. Compos. Part B. **110**, 442–458 (2017)
2. Brenken, B., Barocio, E., Favaloro, A., Kunc, V., Pipes, R.B.: Fused filament fabrication of fiber-reinforced polymers: a review. Addit. Manuf. **21**, 1–16 (2018)
3. Chen, K., Yu, L., Cui, Y., Jia, M., Pan, K.: Optimization of printing parameters of 3D-printed continuous glass Fiber reinforced polylactic acid composites. Thin-Walled Struct. **64**, 107717 (2021)
4. Man, Z., Wan, B., Wang, H., Li, Q., Chang, L.: Experimental and numerical study on scratch performance of additively manufactured continuous carbon Fiber reinforced polyamide 6 composites. Compos. Sci. Technol. **230**(2), 109314 (2022)
5. Chabaud, G., Castro, M., Denoual, C., Duigou, L.: Hygromechanical properties of 3D printed continuous carbon and glass fibre reinforced polyamide composite for outdoor structural applications. Addit. Manuf. **26**, 94–105 (2019)
6. Wang, Q., et al.: Experimental and numerical investigations on microstructures and mechanical properties of hybrid fiber reinforced thermoplastic polymer. Polym. Test. **70**, 215–225 (2018)
7. Tian, X., Liu, T., Yang, C., Wang, Q., Li, D.: Interface and performance of 3D printed continuous carbon Fiber reinforced PLA composites. Compos. A: Appl. Sci. Manuf. **88**, 198–205 (2016)

8. Downes, K.A., Thomason, J.L.: A method to measure the influence of humidity and temperature on the interfacial adhesion in polyamide composites. Compos. Interfaces. **22**(8), 757–766 (2015)
9. Liao, G., et al.: Properties of oriented carbon fiber/polyamide 12 composite parts fabricated by fused deposition modeling. Mater. Des. **139**(5), 283–292 (2018)
10. Markforged company. https://markforged.com/3d-printers/mark-two
11. Becker, C., et al.: Experimental quantification of the variability of mechanical properties in 3D printed continuous Fiber composites. Appl. Sci. **11**, 11315 (2021)

Study on Probability of Detection for Visual Inspection of Impact Damage in Large Composite Laminates

ZiYang Chen[1,2](✉)

[1] Shanghai Aircraft Design and Research Institute, Shanghai, China
chenziyang1@comac.cc
[2] State key Laboratory of Airliner Integration Technology and Flight Simulation, Shanghai 201210, China

Abstract. Composite aircraft fuselages often suffer damage from unexpected impacts during service, making the assessment of these damages crucial. This study investigates the factors affecting the probability of detecting impact damage in composite laminate panels in the aviation field. Based on visual inspection data, the probability curves for detecting pit damage under different inspection conditions were analyzed. The results show that, under the same lighting conditions, a larger visual inspection angle increases the detection probability, while white glossy surface treatments make damage detection more challenging.

Keywords: composite laminate panels · visual inspection · probability of detecting

1 Introduction

With the advancement of modern aviation technology, composite materials have become increasingly prevalent in aircraft. However, the anisotropic nature of composite materials results in significantly different impact damage compared to traditional materials [1]. During service, composite aircraft fuselages often sustain damage from unexpected impacts, making damage detection and assessment crucial. Visual inspection, one of the commonly used non-destructive testing methods, is widely applied across various fields [2].

Visual inspection includes ground walk-around (WA), general visual inspection (GVI), detailed visual inspection (DET), and specialized detailed inspection (SDI) [3]. By studying the impact damage dent depth of composite fuselage structures under varying lighting conditions and observation angles, this research aims to ensure that detection results meet the reliability assessment requirements of a 95% confidence level and 90% detection probability. The goal is to obtain probability of detection (POD) data for composite fuselage damage and to establish a visual inspection POD curve for impact damage dent depth in composite fuselage structures [4].

Y. Zhao (Ed.): ICAMM 2025, SPM 96, pp. 175–183, 2025.
https://doi.org/10.1007/978-3-032-09862-7_17

Using visual inspection methods, the study involves independent and repeated inspections of test specimens containing specific damage by relevant civil aviation maintenance personnel with varying experience, under prescribed inspection conditions [5]. By statistically analyzing the obtained detection results, POD curves for different inspection conditions are established.

2 Experimental Conditions

Under various lighting conditions and inspection angles, personnel conduct non-destructive testing on each test specimen according to the testing requirements.

2.1 Test Specimens

Based on a certain model of large domestic aircraft, the fuselage wall panels are chosen as the original configuration for the test specimens. The test specimens are composite fuselage curved panels, manufactured using automatic filament winding and high-temperature curing processes in an autoclave. There are three test specimens, labeled A1, A2, and A3, with projection dimensions of 3000 mm × 2000 mm. The surface treatment of the specimens involves painting with white gloss paint.

2.2 Test Conditions

The test specimen is placed on a specialized fixture that can adjust the inspection angle of the specimen. The inspection angles of the test specimen (the angle between the line of sight and the surface normal) are 15°, 45°, and 60°. During detection, the test specimen's position and orientation are adjusted according to the actual indoor light source to avoid reflections and ensure uniform, soft lighting on the specimen's surface. The indoor lighting intensity includes three conditions: 200 lux, 750 lux, and 1000 lux, which represent the lighting levels around the test specimen. Refer to Table 1 for details.

The personnel involved in routine inspection trials include both specialized inspectors and non-specialized inspectors, with general visual inspections and detailed visual inspections being performed by specialized inspectors only. Specialized inspectors are those who are current employees of airlines or MROs (Maintenance, Repair, and Overhaul organizations) and are engaged in visual inspections of aircraft structures. Non-specialized inspectors, on the other hand, are required to have no experience in composite material damage or visual inspection work [5].

The personnel involved in the project's WA are individuals without experience in visual inspection of composite material damage. The personnel participating in the GVI testing should meet one of the following three conditions: a) MH/T 3001/NAS410 nondestructive testing personnel; b) airline maintenance personnel; c) structural maintenance personnel. The above personnel will carry out impact damage dent detection tests using WA and GVI visual methods.

The test specimen and visual inspection platform are shown in Fig. 1.

Table 1. Visual Inspection Conditions for Test Specimens

Test Specimens	Color/Roughness	Test Method	Test Angle	Light intensity
A1	White/Bright Light	WA	15°	Outdoor (daytime)
	White/Bright Light	WA	45°	Outdoor (daytime)
	White/Bright Light	WA	60°	Outdoor (daytime)
A2	White/Bright Light	GVI	15°	indoor (750 lux)
	White/Bright Light	GVI	45°	indoor (750 lux)
	White/Bright Light	GVI	60°	indoor (750 lux)
A3	White/Bright Light	GVI	45°	indoor (20 0lux)
	White/Bright Light	GVI	45°	indoor (750 lux)
	White/Bright Light	GVI	45°	indoor(1000 lux)

Fig. 1. Test Specimen Visual Inspection Platform

2.3 Dent Damage Introduction

To facilitate the implementation of damage introduction into the test specimen, a specialized fixture has been developed, as shown in Fig. 2. The impact damage is introduced using a steel hemispherical (or conical) indenter. To introduce damages of varying sizes, the indenter diameters include 16 mm and 25.4 mm. The range of impact indentation depths includes from 0.1 mm to penetration (with a tolerance of ±0.1 mm for indentation depth). The number of impact points is 12, and their locations are required to be randomly distributed. Then Dent depths are measured and recorded.

Fig. 2. Damage introduction fixture.

3 Data Result Calculation and Processing

3.1 Data Result Calculation

Probability of Detecting Discrete Damage. Based on the detection test records of all participants, tally the number of times each type of damage was detected, obtain valid data pairs for damage size and test results, and calculate the detection probability of each type of damage on each test specimen under specific inspection conditions. Plot the detection probability points for discrete damage under each test condition, with the horizontal axis representing the depth of the dent and the vertical axis representing the damage detection probability.

Continuous Damage Detection Probability Function. Use the log-normal cumulative distribution model to fit the discrete damage detection probability data points [6]. As in (1), The POD function corresponding to this model is:

$$POD(a) = \Phi\left(\frac{Ina - u}{\sigma}\right) \tag{1}$$

In the formula, Φ is the standard normal cumulative distribution function.

Estimate the parameter values of the model using the maximum likelihood method [4, 7]. a_i represent the size of the damage i, and z_i represent the detection result. If the damage is detected, then $z_i = 1$; if not detected, $z_i = 0$. $POD(a_i)$ is the probability function for detecting damage of size a_i. Based on experimental data (a_i, z_i), its likelihood function is (2):

$$L(\theta) = \Pi[POD(a_i)]^{z_i}[1 - POD(a_i)]^{1-z_i} \tag{2}$$

In the formula, θ is the vector of unknown function parameters, which is determined by maximizing the likelihood function.

Detection Probability Function at 95% Confidence Level. Since the damage detection threshold required for airworthiness must have a detection probability with 95%

confidence, it is necessary to determine the damage detection probability function at this confidence level.

a_p represent the depth of $POD(a_p) = p$, and z_p represent the quantile p of the standard normal distribution. There is (3):

$$\frac{Ina_p - u}{\sigma} = z_p \tag{3}$$

Namely, $a_p = \exp(u + z_p\sigma)$.

Use $\hat{u}$ and $\hat{\sigma}$ to represent the maximum likelihood estimates of u and σ, respectively. Since u and σ follow an asymptoticnormal distribution, $X_p = \hat{u} + z_p\hat{\sigma}$ also follows an asymptotic normal distribution. Its mean, variance, and standard deviation are respectively, as in (4):

$$M(X_p) = u + z_p\sigma, D(X_p) = V_{11} + 2z_pV_{12} + z_p^2V_{22}, SD(X_p) = \sqrt{D(X_p)} \tag{4}$$

V_{11} represents the variance of $\hat{u}$, V_{22} represents the variance of $\hat{\sigma}$, and V_{12} represents the covariance between $\hat{u}$ and $\hat{\sigma}$, as in (5):

$$a_{p/q} = \exp[\hat{u} + z_p\hat{\sigma} + z_qSD(X_p)] \tag{5}$$

At a 95% confidence level, the mean and variance of the POD curve are respectively, as in (6, 7):

$$u_{95} = \ln(a_{50/95}) \tag{6}$$

$$\sigma_{95} = \frac{\ln(a_{99/95}) - \ln(a_{50/95})}{2.33} \tag{7}$$

The shape of the POD curve is shown in Fig. 3.

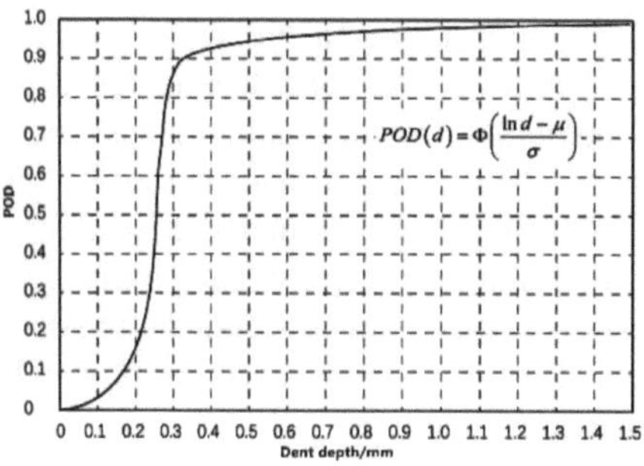

Fig. 3. Schematic diagram of the damage detection probability curve.

3.2 Data Processing

Conduct experiments according to the above requirements and estimate the detection probability of the damage based on the experimental results. Conduct impact damage dent detection testing using both WA (Wide Angle) and GVI (General Visual Inspection) methods in the order of the listed test specimen numbers. Each test specimen should be examined under various lighting intensities and inspection angles as specified. Each visual inspection test must involve at least 20 participants, who must meet the required criteria in 2.2.

4 Results and Analysis

Due to the large volume of data, efforts were made to maintain consistent surface cleanliness during the estimation process and to disregard the influence of professional inspectors. Only the total number of inspections and the number of detected instances under different lighting conditions and inspection angles for the three large panels at varying depths of pits were recorded.

Figures 4, 5 and 6 show the POD (Probability of Detection) estimation results under different inspection conditions. The solid lines represent the cumulative log-normal distribution curves, while the dashed lines represent the POD curves with 95% confidence limits. The POD curves align well with the experimental points. Figure 4 displays the pit depth detection probability curves for the test specimens under the same lighting intensity but different visual inspection angles using the general visual inspection (GVI) method: (a) visual angle of 15°, (b) visual angle of 45°, and (c) visual angle of 60°. Figure 4 shows the pit depth detection probability curves for the test specimens during outdoor daylight conditions using ground patrol (WA) with different visual inspection angles. Figure 6(a) shows a visual angle of 15°, (b) shows a visual angle of 45°and (c) visual angle of 60°. Comparing Figs. 4 and 6, it is evident that as the visual inspection angle increases, the visible damage (BVID) of the laminate decreases, while the detection rate at 95% confidence level increases. Compared to Fig. 4, the detectable damage for ground inspection (WA) is significantly greater than for general visual inspection (GVI). Figure 5 illustrates the probability curves for detecting dent depth under the same visual inspection angle of 45°, but with different lighting intensities. In Fig. 5(a), the lighting intensity is 200 lux; in Fig. 3(b), it is 750 lux; and in Fig. 5(c), it is 1000 lux. The BVID (Visible Damage Threshold) values under these three lighting conditions are 1.29 mm, 1.02 mm, and 1.69 mm, respectively. The proximity of these values indicates that the white glossy surface of the laminate significantly affects the damage detection probability. Further research could explore detection probabilities for laminates with different colors and surface treatments.

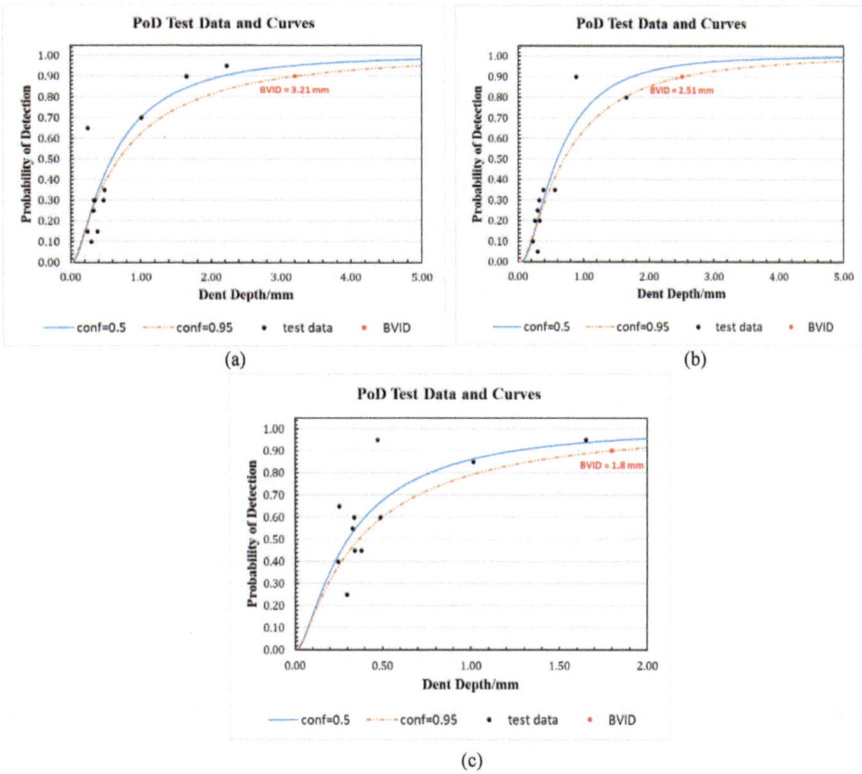

Fig. 4. The POD estimation results for specimens at a lighting intensity of 750 lux, with varying visual inspection angles

5 Conclusion

This study investigates the probability of detection (POD) curves for detecting pit depths in large composite panels under various visual inspection methods, angles, and lighting conditions. The POD curves were estimated and analyzed for each condition, leading to the following conclusions.

(1) Under different inspection conditions, the threshold value for BVID (Barely Visible Impact Damage) varies and is proportional; the greater the inspection angle, the easier it is to detect.
(2) The impact of different visual inspection methods on the probability of detecting pit damage is more significant.
(3) The color white has a considerable impact on the probability of detecting pit damage, whereas the intensity of lighting has a relatively minor effect. Future research could explore the effects of different panel colors.

By analyzing the obtained inspection results statistically, visual inspection POD curves for pit depth sizes of impact damage on aircraft composite fuselage structures

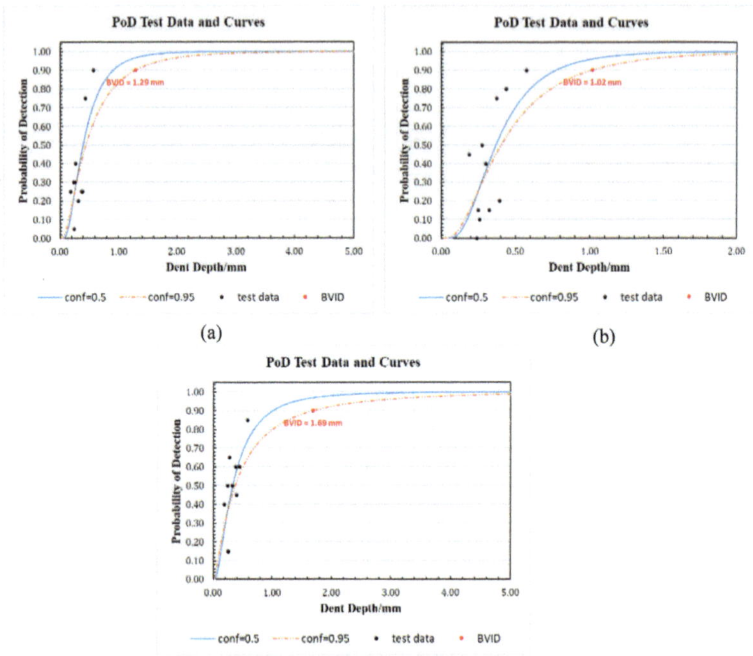

(c)

Fig. 5. The POD estimation results for specimens inspected at a 45°, under different lighting intensities.

were established. This provides foundational data support for non-destructive testing in damage tolerance assessment of aircraft composite fuselage structures.

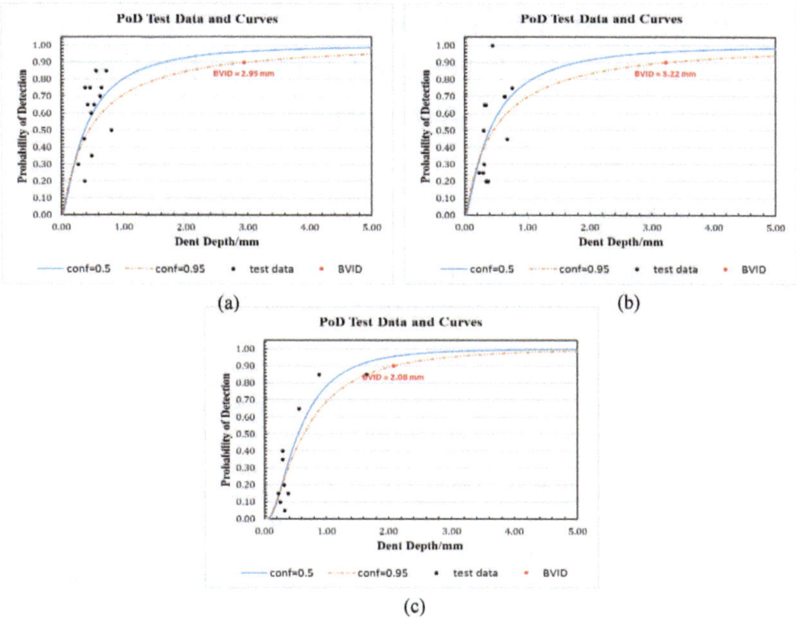

Fig. 6. The POD estimation results for specimens inspected at various visual angles during daytime outdoor conditions.

References

1. Ma, B., Zhou, Z.: Advances and trends in non-contact non-destructive testing technologies for aerospace composite structures. J. Aerosp. **35**(7), 1787–1803 (2014)
2. Xiao, S., Chen, P., Shen, W.: Experimental study and analysis of the probability of detecting impact damage in composite structures. J. Compos. Mater. **30**(4), 231–237 (2013)
3. Polymer Matrix Composites, Materials Usage, Design, and Analysis, MIL-HDBK-17-3F // Department of Defense Handbook. USA: Department of Defense, pp. 7-1 to 7-9 (2002)
4. Nondestructive Evaluation System Reliability Assessment, MIL-HDBK-1823 // Department of Defense Handbook. USA: Department of Defense, pp. 61–76 (1999)
5. Erhart, D., Ostrom, L.T., Wilhelmsen, C.A.: Visual detectability of dents on a composite aircraft inspection specimen: an initial study. Int. J. Appl. Aviat. Stud. **4**(2), 111–122 (2004)
6. Fan, J., Cheng, M., Chang, L.: Analysis of Data, pp. 126–131. Science Press, Beijing (2002)
7. Beren, A.P.: Probabilistic Detection (POD) Analysis for the Advanced Retirement for Cause (RFC)/Engine Structure Integrity Program (ENSIP) Nondestructive Evaluation (NDE) System – Volume 1: POD Analysis, AFRL-ML-WP-TR-2001-4011 [R]. Dayton, OH: 300 College Park, Dayton University, pp. 34-40 (2000)

Design and Analysis of the Crashworthiness of Composite Fuselage Segments for Civil Aircraft

Zhang Pengfei(✉), Li Xuhui, Zhu Zhengyi, and Shao Xingxiang

Shanghai Aircraft Design and Research Institute, Shanghai, China
{zhangpengfei1,lixuhui,zhuzhengyi,shaoxingxiang}@comac.cc

Abstract. A finite element model for the crashworthiness analysis of the composite fuselage cabin section was developed based on the structural design of the civil aircraft passenger fuselage, and calculations were performed by the explicit dynamic method. The influence of three cabin floor spar angles on the crashworthiness of the fuselage section was analyzed, and the proposed design angles based on crashworthiness were derived. Based on this, the impact of cargo floor spars and energy-absorbing structure of composite thin-walled tubes filled with aluminum foam on the crashworthiness of the fuselage section was analyzed by adding spars and energy-absorbing structure under the cargo floor. The results of the analysis show that the floor spar has a significant effect on the crashworthiness of the fuselage section, and the energy-absorbing structure improves the crashworthiness of the fuselage section.

Keywords: Composite Fuselage Segments · Crashworthiness · Floor Spars · Energy Absorbing Structures · Peak Acceleration

1 Introduction

Investigations of many aviation accidents have shown that no matter how advanced and reliable the aircraft design is, and how perfect the manufacturing quality is, because of a variety of reasons that cannot be accurately predicted, such as maintenance, meteorology, and human error, etc., aircraft crashes cannot be completely avoided. In civil aircraft accidents, more than 70% of them occur in the takeoff and landing stages. The flight altitude of these two phases is low, belonging to the category of survivable accidents. Detailed studies of a large number of flight accident investigation reports at home and abroad have shown that improving the structural design of an airplane can dramatically increase and improve the survivability of the crew as well as the passengers in the event of a crash [1]. In order to ensure that the people on board the airplane can survive in a survivable accident, the structural design must be considered so that the airplane has a crashworthiness characteristic, which is used to fully absorb the impact energy generated by the crash, and limit the load of the crash process to the human body's tolerable range. Therefore, in the field of civil aviation, more and more research organizations pay attention to the study of crashworthiness of civil aircraft [2]. The core issue of

Y. Zhao (Ed.): ICAMM 2025, SPM 96, pp. 184–191, 2025.
https://doi.org/10.1007/978-3-032-09862-7_18

crashworthiness research is to improve the crashworthiness of the aircraft structure so as to increase the probability of survival of the occupants. For the design of crashworthiness of aircraft, the United States and Europe and other countries have long regarded it as a key factor in the preliminary design stage of aircraft, as important as weight, load factor and fatigue life. Various airworthiness regulations have strict requirements for the crashworthiness of aircraft, and Part 25 of the Civil Aviation Regulations of China (CAAR) has clear provisions for "emergency landing situations", which require that every crew member be given every reasonable opportunity to avoid serious injury.

Nowadays, carbon fiber composites have been used in large quantities in the fuselage structure of advanced civil aircraft. The material damage form and energy absorption mechanism of the composite structure in the collision are different from those of the metal material commonly used in the past, and the difficulty of calculation and analysis has been improved. Therefore, the large number of applications of composite materials in the fuselage structure of civil airplanes also puts forward newer and higher requirements for the crashworthiness assessment and design of civil airplanes. In this paper, the crash impact of a composite fuselage is analyzed by a validated explicit dynamics computational model [3], and the structural design of the fuselage for crashworthiness is investigated.

2 Crashworthiness Analysis of Fuselage Segment with Three Cabin Floor Spacer Angles

The angle of cabin floor struts has an impact on the overall structural airworthiness of the aircraft, this paper establishes a finite element model of a typical fuselage section of a civil aircraft with three frames and two spans, and calculates the crash characteristics of the composite fuselage section with explicit dynamics, and comparatively analyzes the acceleration relative to the time of the course curves at the position of the passengers' seats and the damage form of the lower half of the fuselage section under the three kinds of cabin floor strut angles.

2.1 Finite Element Model of Composite Fuselage Segment

Referring to the metal fuselage segment crash model [4], the composite fuselage analysis model is established. The finite element model simplification principles are as follows:

1. Transverse beam, longitudinal beam and spacer frame are simplified as beam units;
2. Skin, floor and spars are simplified into shell units;
3. The ground is modeled as a rigid wall.

The crossbeam, longitudinal beam and spacer frame are made of 7075-T73511 aluminum alloy, and the material properties are shown in Table 1 the skin, long truss, floor and gusset are made of composite materials, and the T300/BMP-316 system is selected, and the material properties are shown in Table 2 the layup sequence is [45/−45/0/0/0/90/0/90/0/]s, and the thickness of each layer is 0.125 mm, and a total of 16 layers are laid.

The loads and constraints are as follows:

Table 1. Material Properties

E	μ	ρ	σ_Y
71.7GPa	0.33	2.81 g/cm3	507 MPa

Table 2. Material Properties

E_{11}	E_{22}	ν_{12}	G_{12}
128.8GPa	10.1GPa	0.355	4.1GPa
X_t	X_c	Y_t	Y_c
1446.51 MPa	1171.41 MPa	69.64 MPa	187.29 MPa
S			
95.939 MPa			

(1) The composite fuselage segment impacts the ground at a speed of 7 m/s and the solution time is 100 ms.
(2) The weight of each passenger is calculated as 100 kg with 75 kg passengers, 15 kg seats, and 10 kg hand baggage, and the high-density seating class is taken with 9 persons per row, 768 mm row spacing, and 512.25 mm frame spacing, which is loaded on the nodes in the form of discrete 0D mass elements. According to the geometric dimensions, material properties and constraints of the fuselage segments, a finite element model is established, and the model is a three-frame, two-span full-size fuselage model.

2.2 Influence of Cabin Floor Spar Angle on the Fallability of Fuselage Segments

The cabin floor struts affect the crashworthiness of fuselage segments. In the following section, finite element modeling of fuselage segments with three cabin floor strut angles (90°, 75° and 60°) and crashworthiness analysis with Ls-Dyna software are performed to obtain the damage modes of the composite fuselage segments and the acceleration history at the seats with time.

The composite fuselage segments with the three cabin floor strut angles were damaged to different degrees after crashing into the ground, and the typical damage form of the 90° spar is shown in Fig. 1.

Although the fuselage segments with the three cargo floor strut angles were damaged to different degrees during the crash, the damage of the passenger floor struts with different angles occurred differently during the same crash time, and thus had different effects on the damage of the corresponding fuselage segments. An important index for evaluating the crashworthiness of civil aircraft fuselage structure is the acceleration characteristics at the passenger seat position [5], so the reduction of the acceleration peak can be taken as the design goal in the structural design for crashworthiness. The acceleration versus time curves at the seat position for the fuselage segments corresponding to the three cargo compartment floor strut angles after crash analysis are shown in Fig. 2.

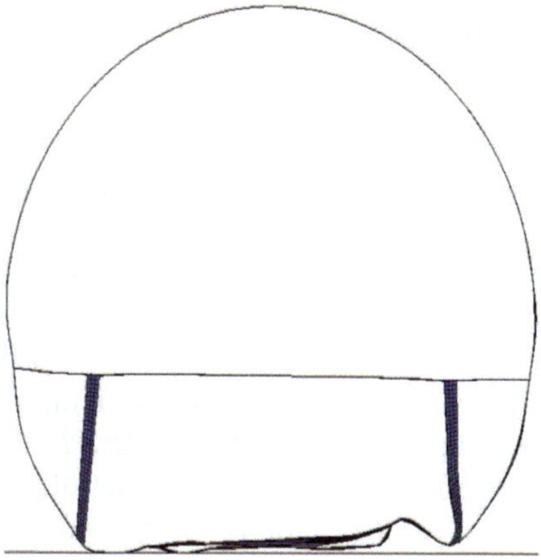

Fig. 1. Destruction of fuselage section at 90° cabin floor spar angle

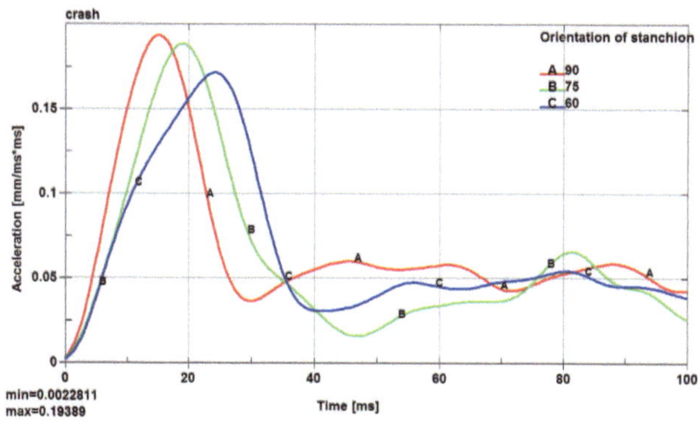

Fig. 2. Acceleration vs. time curve at passenger seat

From the curves in Fig. 2, it can be seen that different cabin floor strut angles have a large effect on the peak acceleration at the seat. From 90° to 60°, the peak acceleration is gradually delayed, and the peak value is also gradually reduced. Therefore, from the perspective of crashworthiness design alone, a smaller angle of cabin floor struts should be selected.

3 Crashworthiness Analysis of a Fuselage Section with Cargo Floor Struts and Energy-Absorbing Structures

The lower cargo floor structure has a significant effect on the impact overload and the damage pattern of fuselage frame segments [6]. The following section analyzes the crashworthiness of fuselage frame segments with additional stanchions and energy absorbing structures in the lower part of the cargo hold floor.

3.1 Finite Element Modeling of Cargo Floor Struts and Energy-absorbing Structures

In the above analysis, in order to emphasize the influence of the angle of the passenger compartment floor struts, the crashworthiness calculation of the cargo compartment underfloor structure was not included, but in fact the cargo compartment underfloor structure has a great influence on the crashworthiness of the fuselage section. According to the cargo floor structure of civil aircraft, the finite element model for crashworthiness analysis is modified by adding four cargo floor stanchions. The stanchions are simulated using shell units, and the material is T300/BMP-316 system, and the material properties are shown in Table 2. The layup sequence is $[45/-45/0/90/]_S$, and the thickness of each layer is 0.125 mm, 8 layers. The finite element model of the fuselage compartment section retrofitted with cargo floor stanchions was changed as shown in Fig. 3.

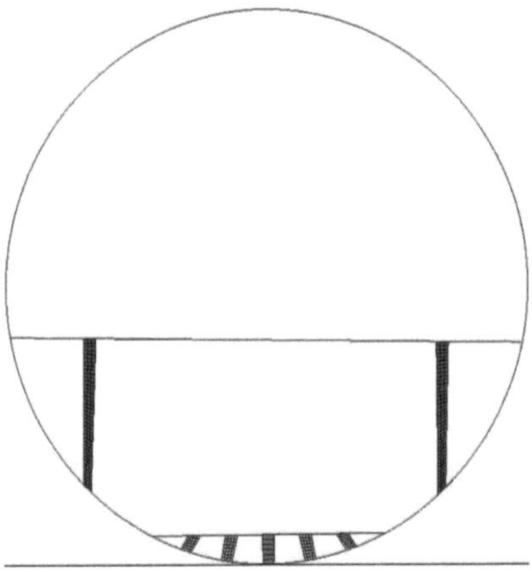

Fig. 3. Fuselage section with floor struts and energy-absorbing structures.

The energy-absorbing structure is a structure designed for crashworthiness, which aims to utilize the collapse of the structure after crash to absorb the impact energy, thus reducing the risk of passenger injury in a crash. Composite thin-walled tubes are an important load-bearing structure in their own right and can also be used as energy-absorbing members [7]. Aluminum foam is a new type of lightweight functional material, with a density of only 0.1 to 0.2 times that of solid metal aluminum, with high energy absorption efficiency and better energy absorption performance than other commonly used materials. In the automotive field, aluminum foam has been applied to fill steel or aluminum structural hollow parts to change the form of collapse of structural parts after impact, thus improving energy absorption and impact resistance [8]. In this paper, an energy-absorbing structure under the floor of the cargo hold is designed and added to the center of the lower side of the fuselage, as shown in Fig. 3. The details of the energy-absorbing structure are shown in Fig. 4, and the outer wall is a composite rectangular cross-section thin-walled tube, and the material is T300/BMP-316 system, and the material properties are shown in Table 2.

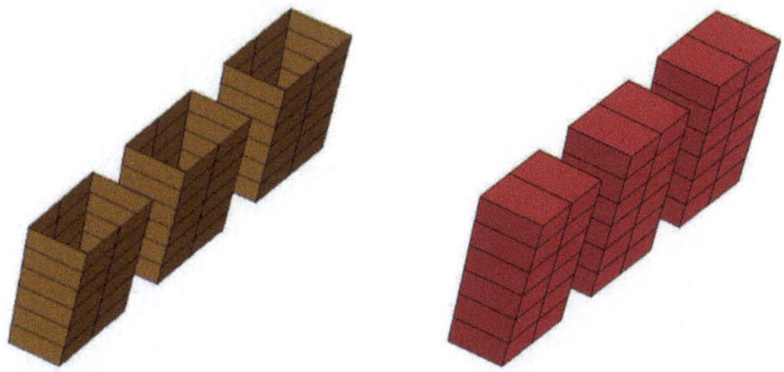

(a)Composite thin-walled tube (b)Filled aluminum foam

Fig. 4. Energy-absorbing structures under the cargo floor.

The layup sequence is [45/−45/0/90/]$_S$, and the thickness of each layer is 0.125 mm, with a total of 8 layers. It was filled with aluminum foam material, and the material properties are shown in Table 3.

Table 3. Material Properties

ρ	E	μ	α
250 kg/m3	590 MPa	0.0	2.02
γ	EPSD	α2	β
0.45 MPa	2.35	63.3GPa	8.29
SIGP	CFAIL		
1.8	0.025		

3.2 Effect of Cargo Floor Struts and Energy-absorbing Structures on the Crashworthiness of Fuselage Segments

As described above, three analytical models were developed, the first with only four floor stanchions, the second with a composite thin-walled tube structure, and the third with an aluminum foam filled in the thin-walled tube. The corresponding damage forms of the compartment segments obtained after analytical calculations are shown in Fig. 5.

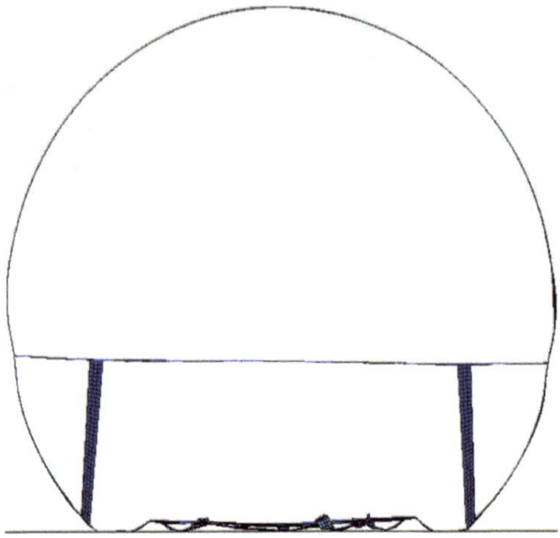

Fig. 5. Crash damage to fuselage segments with underfloor structure of cargo compartments.

The peak acceleration at the passenger position is still used as an indicator for crashworthiness analysis. The acceleration versus time curves of the four fuselage section models without stanchions, with four stanchions, with four stanchions plus a thin-walled tube, and with a thin-walled tube filled with foam aluminum were selected for comparison, as shown in Fig. 6.

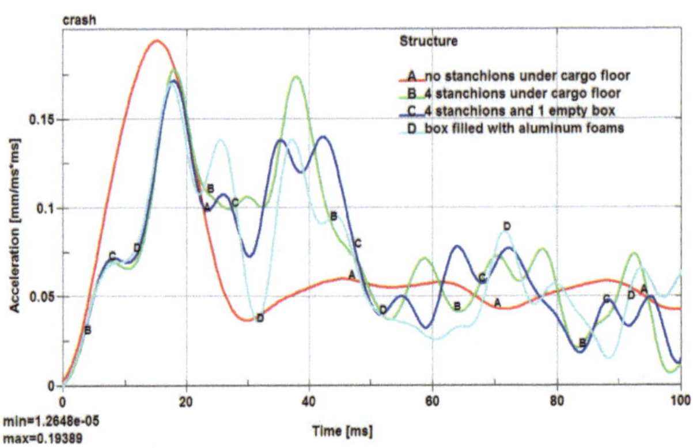

Fig. 6. Acceleration vs. time curves at passenger position for four fuselage segments

As can be seen in Fig. 6, the addition of the cargo floor stanchions increases the second acceleration peak, but there is a significant reduction in the acceleration peak. The addition of the energy absorbing structure also reduces the peak acceleration while increasing another peak acceleration, but there is a significant reduction in the second and third peak acceleration velocities. After filling the thin-walled tube with aluminum foam, the peak acceleration is further reduced. In conclusion, the energy-absorbing structure achieves the design purpose and can improve the crashworthiness of the fuselage section.

4 Conclusion

According to the structural design of civil aircraft fuselage, this paper designs and analyzes the crashworthiness of different forms of fuselage segments, and obtains the following conclusions.

(1) From the requirement of crashworthiness, a smaller angle can be considered when designing the passenger cabin floor strut.
(2) The cargo compartment floor stanchions has a significant effect on the crashworthiness of the fuselage section.
(3) The energy-absorbing structure designed in this paper improves the crashworthiness of the fuselage section.

References

1. Clifford, W.: Method of improving the crashworthiness of an aircraft. U.S. Patent Application No. 09/769,334
2. Mou, H., Xie, J., Feng, Z.: Research status and future development of crashworthiness of civil aircraft fuselage structures: an overview. Prog. Aerosp. Sci. **119**, 100644 (2020)
3. Pengfei, Z.H.A.N.G., Haibo, J.I.N.: Stacking sequence design of composite thin-walled structure based on low energy impact damage resistance. Acta Materiae Compositae Sinica. **31**(1), 18–25 (2014)
4. Fasanella, E.L., et al.: Crash simulation of a Boeing 737 fuselage section vertical drop test (2004)
5. Guida, M., et al.: Review on the design of an aircraft crashworthy passenger seat. Prog. Aerosp. Sci. **129**, 100785 (2022)
6. Yiru, R., et al.: Effect of cabin-floor oblique strut on crashworthiness of typical civil aircraft fuselage section. Acta Aeronautica Et Astronautica Sinica. **2**, 012 (2010)
7. Ren, Y., et al.: Crashworthiness analysis of aircraft fuselage with sine-wave beam structure. Chin. J. Aeronaut. **29**(2), 403–410 (2016)
8. Banhart, J., Seeliger, H.-W.: Aluminium foam sandwich panels: manufacture, metallurgy and applications. Adv. Eng. Mater. **10**(9), 793–802 (2008)

Research on Hypervelocity Impact on Honeycomb Sandwich Panel with Multiple Insulation Layers

Songlin Pang[1], Jinsheng Xu[1(✉)], Xiong Chen[1], Ivan Tkachenko[2], Danhe Chen[1], and Ivanushkin[2]

[1] School of Mechanical Engineering, Key Laboratory of Special Engine Technology, Ministry of Education, Nanjing University of Science and Technology, Nanjing, China
xujinsheng@njust.edu.cn

[2] Institute of Aviation and Rocket and Space Technology, Samara National Research University named after academician S.P. Korolev (Samara University), Samara, Russia

Abstract. In order to study the impact of small space debris on the sandwich honeycomb structure panel of micro/Nano satellite with multi-layer insulation, taking the AIST satellite shell of Samara University as a reference, the deceleration effect of the sandwich honeycomb panel on the fragments when the debris hits the honeycomb core at different positions was studied by numerical simulation method, and the comparison with previous experiments was made. The simulation results showed that, compared to hitting the edge of the honeycomb or hitting the intersection of honeycomb core, when a 4 km/s projectile with a diameter of 2 mm hit the hollow of the honeycomb, the damage area to the honeycomb panel was much smaller than that of the former. The absorbed kinetic energy of the honeycomb panel was less, and the residual kinetic energy of the debris cloud was higher, leading a greater debris damage to the sensors inside the satellite. At the same time, the thickness and side length of the honeycomb core were adjusted while the areal density and the thickness of the panel partition remain unchanged, and the debris kinetic energy and the average debris velocity after deceleration were obtained by numerical simulation. Combining the probability of debris hitting each position of the honeycomb core and the corresponding residual kinetic energy, through mathematical expectation, the original design scheme is a relatively high-quality design scheme under the condition that the areal density and the thickness of each layer are unchanged.

Keywords: micro/Nano satellite · sandwich panel · impact behavior

1 Introduction

As the number of spacecraft in Earth orbit increases, the amount of space debris generated due to aging and scrapping of spacecraft is also increasing day by day, becoming one of the risk factors for the normal operation of spacecraft in orbit [1, 2]. In order to avoid huge losses caused by impact, various spacecraft adopt some special structures for

Y. Zhao (Ed.): ICAMM 2025, SPM 96, pp. 192–202, 2025.
https://doi.org/10.1007/978-3-032-09862-7_19

protection, such as Whipple structures [3]. Micro/Nano satellite is one of the light and convenient type among all types of satellites. They have a high degree of integration and have broad development prospects. Due to weight restrictions, no additional protective structure is generally designed and only the shell is used for protection. The shell mostly adopts honeycomb sandwich structure with multiple insulation layers. The honeycomb sandwich structure, one of advanced Whipple structures [4, 5], has a high strength-weight ratio [6]. Usually the honeycomb sandwich structure is composed of front and rear panels and a honeycomb core. The panel materials can be aluminum alloy and composite materials; the honeycomb can be aluminum alloy, titanium alloy or Nomex. The multi-layer insulation (MLI), which functions as satellite temperature control, is typically consisting of multiple layers of metallized polymer film separated by polyimide, fiberglass or polyester fabrics mesh.

Taylor et al. [7, 8] carried out impact tests of different velocities, angles and sandwich structures for honeycomb sandwich structures, improved honeycomb damage equations, and verified them by numerical simulation using AUTODYN-2D and 3D dynamics software, and compared them with typical Whipple structures. Huang et al. [9] carried out hypervelocity speed impact test research and numerical simulation for honeycomb sandwich structure with MIL. The ballistic limit equation of honeycomb sandwich structure with MIL was obtained by experiments, and the equivalent thickness of the front plate with MIL was calculated by using the equation. Chaturvedi et al. [10] established a three-dimensional finite element model of the metal honeycomb sandwich structure and projectile high-speed impact. In the explicit calculation, the material nonlinearity and geometric nonlinearity of the metal honeycomb were taken into account, and the simulation accuracy of the numerical model was verified by comparison with the test. Turan et al. [11] examined the effects of various hole profiles with equal areas on dynamic mechanical performances of Al 6063-T6 rectangular crash boxes. Mao et al. [12] proposes a soft/hard dual-material honeycomb structure. Hai et al. [13] proposes a modified double re-entrant honeycomb (MDRH) structure and explores its Young's modulus and Poisson's ratio through theoretical derivation and finite element analysis.

In this paper, the AIST series micro/Nano satellite shell designed by Samara National University is selected as the simulation model. AUTODYN-3D software [14] is used to simulate the damage and residual velocity of the projectile when it impacts the honeycomb core at different positions. And according to calculating the probability and mathematical expectation of the hit different position, under the premise of constant surface density and thickness of each layer, various working conditions were simulated and compared.

2 Numerical Simulation

2.1 Numerical Simulation Model

According to the paper [15, 16] and the relevant data obtained from the laboratory of Samara University, the panels of the AIST series satellite adopts a honeycomb sandwich structure. Honeycomb core, front and rear panels are made of aluminium alloy. The thickness of the front panel (t_f) and the thickness of rear panels (t_r) are 0.4 mm. The side length (l) of the honeycomb core is 2.75 mm, with 17 mm of the core height (H), and the thickness of the honeycomb core wall(t_h) is 0.04 mm. The equation of areal density of honeycomb core (ρh) is

$$\rho_h = \frac{M_1 + M_f + M_r}{S_1} = \frac{6\rho_{Al}Hl\frac{t_h}{2} + \rho_{Al}S_1t_f + \rho_{Al}S_1t_r}{3\sqrt{3}l^2/2} = 2\rho_{Al}\left(\frac{\sqrt{3}Ht_h}{3l} + t_f + t_r\right) \quad (1)$$

where M_1 is the mass of single honeycomb core with half thickness, g; M_f is the mass of front panel covering the single honeycomb core, g; M_r is the mass of rear panel covering the single honeycomb core, g; S_1 is the area of single honeycomb, cm^2; ρ_{Al} ρ_{Al} is the density of aluminium, 2.70 g/cm^2. The areal density of honeycomb core ρ_h is 0.5090 g/cm^2.

According to the paper [9], the equivalent front plate thickness of MIL is about 0.15 mm. The overall thickness of the front panel is 0.55 mm. The diameter of projectile is 2 mm, and the initial velocity of the projectile is 4 km/s.

The numerical simulation model was established by AUTODYN-3D software is shown in Fig. 1. SPH units were used for the projectile with front and rear panels, which can simulate the impacting process and the damage of debris cloud to the structure more accurately. Honeycomb core was modeled by shell units. The material of projectile is Al 7039. The material of panels is Al 2024 T351, and the material of honeycomb core is Al5083H116. Johnson Cook mechanical model was used for all three materials, Shock equation of state was used for Al 7039 and Al2024T351, and Linear equation of state was used for Al5083H116. The material parameters are from the AUTODYN material library. In the three simulation conditions, the projectile impacted the hollow of the honeycomb core (HH), the single edge of honeycomb core (EH) and the intersection of the honeycomb core (IH), respectively, as shown in Fig. 2.

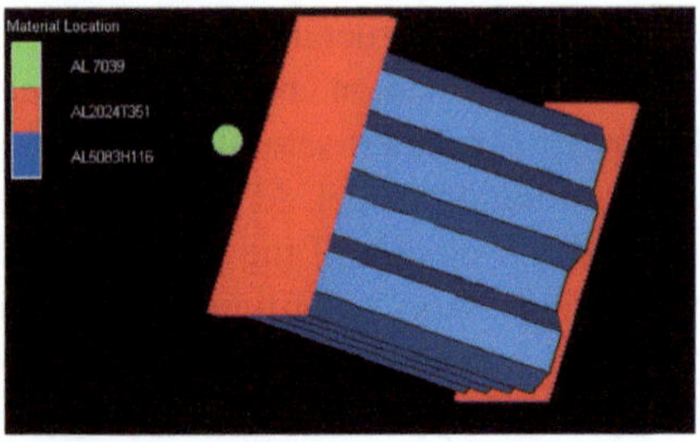

Fig. 1. Numerical simulation model

The mesh density of SPH particles and Shell elements must be controlled reasonably during simulation modeling. In general, it is suggested that at least 10 layers of SPH particles should be divided to ensure accuracy in hypervelocity collision simulation of single-layer plates. If this principle is also adopted in this paper to divide the grid density for the front and rear panels, the number of SPH particles in the whole model will be too large, resulting in a large scale of the calculation model, which exceeds the computing power of the computer. Therefore, the diameter of each SPH particle in this paper was divided into 0.1 mm [17].

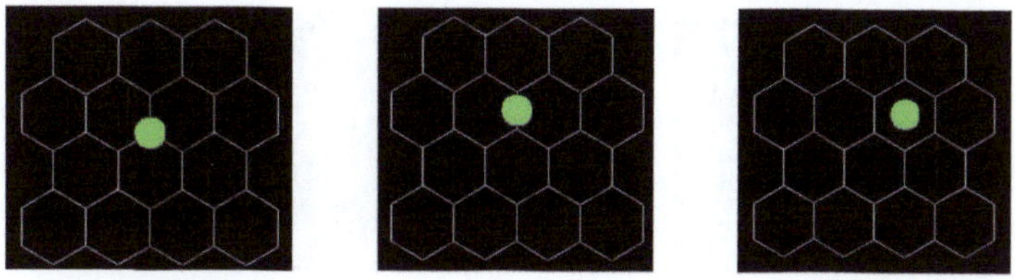

(a) IH model (b) EH model (c) HH model

Fig. 2. The positions between the projectile and the honeycomb core

2.2 Numerical Simulation Result and Analysis

The three simulation results are shown in Fig. 3. It can be obtained from the simulation results that the 4 km/s projectile penetrated the sandwich structure panel in three cases, and a hole with a diameter of about 4.5 mm was formed on the front plate, as shown in Fig. 4.

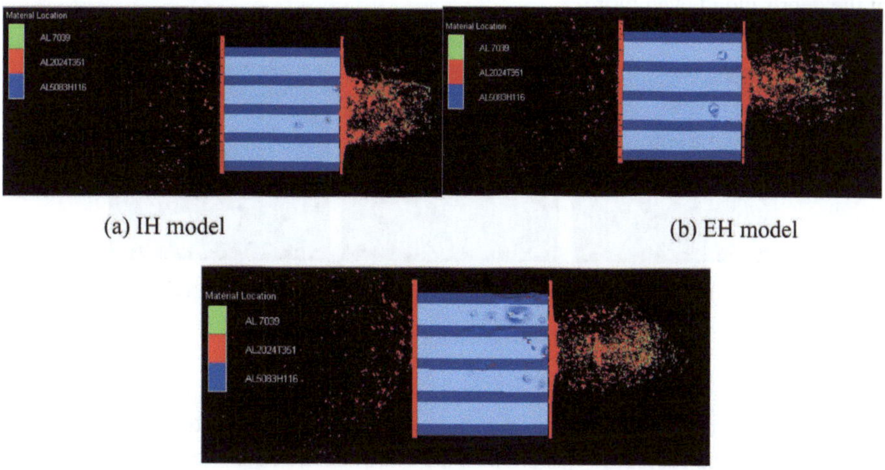

(a) IH model (b) EH model

(c) HH model

Fig. 3. Simulation of perforating impacts.

The debris cloud generated by the fragmentation of the projectile and the front plate will cause great damage to the honeycomb core, as shown in Fig. 5. According to the simulation results of different working conditions, in the IH model, because the projectile directly hit the intersection of the three edges of the honeycomb core, the debris cloud entered the three nearby honeycomb holes, and mainly caused great damage to the walls of the three honeycomb holes, as shown in Fig. 5(a). In the EH mode, the damaged honeycomb walls were more concentrated on the walls of the two honeycomb holes located on both sides of the edge, as shown in Fig. 5(b). In HH mode, because there was no edge division, more fragments swarmed into a honeycomb hole and completely destroyed 6 walls of the honeycomb hole, as shown in Fig. 5(c).

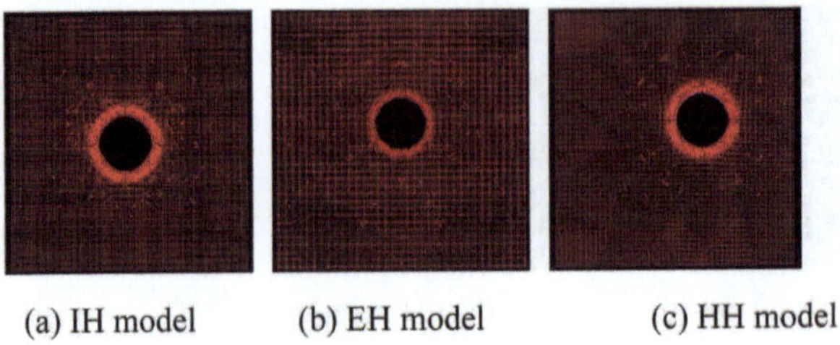

Fig. 4. Damage on the front panels.

It can be seen that no matter where the projectile hits, there were a large lateral velocity of debris, through the multilayer wall. The honeycomb wall absorbed the energy of the debris cloud and reduced its kinetic energy by restraining the lateral diffusion of the debris cloud. At the same time, due to the existence of the edge, the wall directly collides with the debris cloud and splits the debris cloud into several honeycomb holes. Edges can increase the number of walls colliding with them, reducing the kinetic energy of the debris cloud even more.

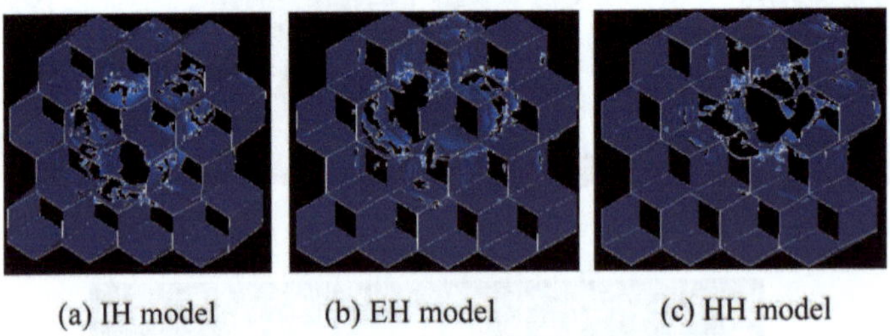

Fig. 5. Damage on the honeycomb cores

The damage of the projectile to the rear plate is shown in Fig. 6. In the IH and EH model, because the projectile was divided, the mass and diameter were reduced, causing the kinetic energy reduced. But the damage area on the rear panel was increased, and the perforation area was three and twice times that of the HH model, respectively. The diameter of the hole on the rear panels are 8.93 mm, 7.03 mm and 5.87 mm.

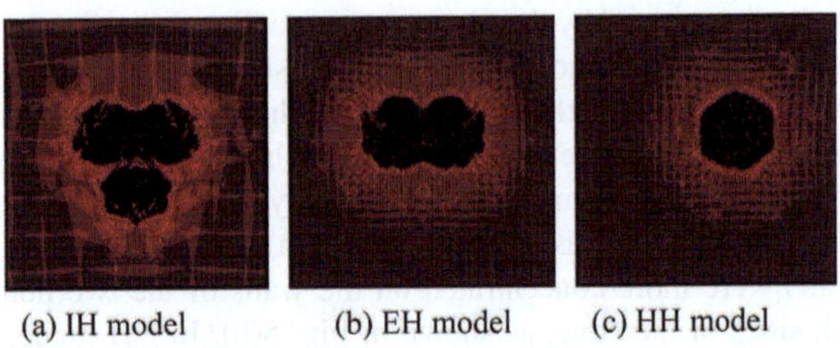

Fig. 6. Damage on the rear panels.

Collision of high-speed particles with a single dense panel of sufficient thickness can lead to the appearance of a crater on the surface of the obstacle, and at a higher speed and with a thinner thickness of the obstacle, to a breakdown of the panel. When a thin barrier is broken, the particle is destroyed and not only fragments of the particle fly behind the panel (the largest fragment is called the "leading element"), but also fragments of the panel [18]. The debris cloud formed after the projectile penetrated the front plate at 5 μs is shown in Fig. 7.

In the IH model, the debris cloud was more dispersed, evenly among the three honeycomb holes, as shown in Fig. 7(a). In EH mode, the debris cloud diffused unhindered in the two honeycomb holes on both sides of the edge, but in the vertical direction of the distribution of the two honeycomb holes, the debris cloud was constrained by the honeycomb walls, and the spherical distribution became a rectangular distribution, as shown in Fig. 7(b). In HH mode, the debris cloud was constrained in all directions, and the debris cloud that diverges around had collided with the honeycomb wall, destroying each other or changing its direction, and moving more intensively in the direction of the backward plate.

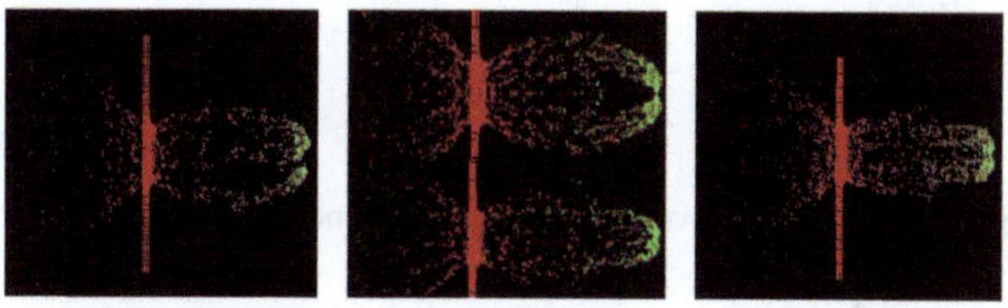

(a) IH model (b) EH model (c) HH model

Fig. 7. Distribution of debris cloud behind front panel

The distribution of position and velocity of the debris cloud at 14 μs is shown in Fig. 8, and the velocity range marked in the Fig. 8 is 0–2000 m/s. As can be seen in Fig. 8, the debris with the highest velocity was at the head of the debris cloud among the debris that has penetrated the panel. In the IH model, there were almost no debris with velocities greater than 2000 m/s, with an average velocity of about 900 m/s. In the EH model, there were still fewer debris with velocities greater than 2000 m/s, with an average velocity of about 1100 m/s. In the HH mode, the debris clouds with velocities greater than 2000 m/s were the most numerous and densely distributed. High-velocity fragments were not only distributed in the head of the debris cloud, but also sporadically in the middle of the debris cloud. The average velocity of the fragments was about 1350 m/s.

The change of velocity and kinetic energy of the projectile is shown in Fig. 9. Figure 9(a) shows that the average velocity of the projectile in the IH model can be reduced to about 750 m/s, which is slightly lower than the average velocity of 880 m/s in the EH model and much lower than the average velocity of 1015 m/s in the HH model. It can be concluded from Fig. 9(b) that, in the IH model, the final residual kinetic energy of the projectile dropped to 4.955 J, which was about 5.171 J lower than that in the HH model and was about 1.862 J lower than that in the EH model.

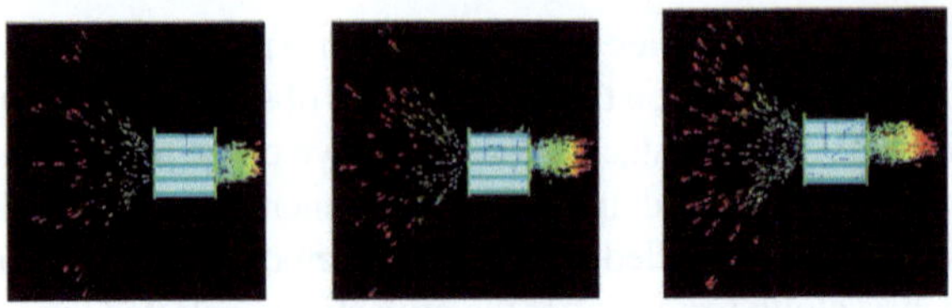

(a) IH model (b) EH model (c) HH model

Fig. 8. Distribution of particle velocities

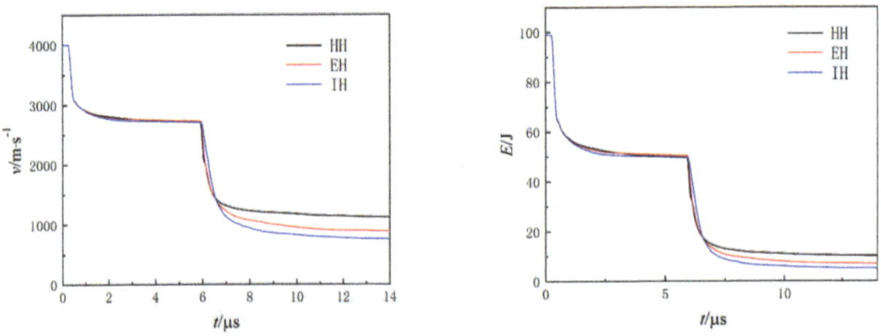

(a) The velocity (b) The kinetic energy

Fig. 9. Curves of the velocity and kinetic energy of the projectiles

Through the above analysis, the conclusion can be drawn that compared with the EH and IH model, when projectiles impact the honeycomb hollow, the damage area of the honeycomb sandwich panel was one-third or half of the former. However, the honeycomb panel absorbed less kinetic energy, resulting in the debris cloud residual kinetic energy is much higher than the former, and the debris damage to the sensor inside the satellite would be more serious. Therefore, while ensuring lightweight, the side length of the honeycomb should be increased as much as possible to increase, so that the probability of space debris hitting the honeycomb is increased.

3 Optimization and Comparison

3.1 Optimization Schemes Design

In the process of design and optimization, while improving the probability of hitting the honeycomb edge, the light weight of the panel should be maintained, that is, the surface density of the panel should not be increased. At the same time, because the volume of the satellite and the position of the sensor and other devices are relatively fixed, the thickness of the panel should not be changed. Therefore, in this paper, the thickness of each layer of the panel was not changed while the areal density was kept unchanged. According to Eq. (2), in the optimization process, the ratio of side length to honeycomb wall thickness remains unchanged, that is

$$\frac{t_h}{1} = \text{constant} \tag{2}$$

Therefore, the optimization was to multiply the length of the side and the honeycomb wall by the same factor to keep the ratio constant. Six optimization schemes were designed and numerically simulated. The side length and honeycomb wall thickness of the optimization scheme are shown in Table 1.

Table 1. The optimization test schemes

Number	l/mm	t_h/mm
1	5.5	0.08
2	4.125	0.06
3	3.4375	0.05
4	2.2	0.032
5	1.833	0.0267
6	1.375	0.02

In the IH model that hits the intersection of honeycomb cores, the panel showed the best deceleration effect and the projectile reduced to the least residual kinetic energy. Therefore, IH model was used for numerical simulation in all optimized working conditions, and the material and parameter settings of numerical simulation were consistent with those mentioned above.

3.2 Simulation Results and Comparison

The final velocity and residual kinetic energy of the projectile obtained under various working conditions are shown in Table 2.

Table 2. The final velocity and residual kinetic energy

Number	l/mm	t_h/mm	v/m•s^{-1}	E_r/J
1	5.5	0.08	766.984	5.667
2	4.125	0.06	735.792	5.329
3	3.4375	0.05	729.423	4.876
4	2.75	0.04	751.630	4.955
5	2.2	0.032	822.847	5.700
6	1.833	0.0267	956.224	7.732
7	1.375	0.02	1124.601	10.730

Combined with Table 2, it can be concluded that when the honeycomb wall thickness was greater than 0.032 mm, with the increase of side length and honeycomb wall thickness, the kinetic energy consumption of the projectile tend to be stable, and the remaining kinetic energy of the projectile remained from 4.8 J to 5.7 J. However, when the wall thickness of the honeycomb was too thin, even if the side length was shortened and the debris cloud penetrated more walls, it cannot consume more kinetic energy, and the remaining kinetic energy rose greatly. In addition, in cases 1–5, although the remaining kinetic energy was almost same, the longer the side length, the lower the probability of the projectile directly hitting the edge. From the previous section, it can be seen that the residual kinetic energy of the EH model was about 1.376 times that of the IH model. The residual kinetic energy of the HH model was about 2.044 times that of the IH model. Therefore, we should directly increase the probability of hitting the honeycomb ridge while ensuring that the residual kinetic energy under the same model is low.

The probability of a projectile hitting the intersection of the honeycomb cores, an edge, and the hollow are as follows:

$$P_1 = \frac{S_i}{S_1} = \begin{cases} \frac{4\sqrt{3}\pi r^2}{9l^2}, l \geq 2r \\ \frac{2\pi r^2 - 3\left[r^2 \cos^{-1}(l/r) - \left(r^2 - l/2\sqrt{4r^2 - l^2}\right)\right]}{3\sqrt{3}l^2/2}, l < 2r \end{cases} \tag{3}$$

$$P_2 = \frac{S_e}{S_1} = \frac{3\sqrt{3}l^2/2 - 3\sqrt{3}(l-r)^2/2 - S_i}{3\sqrt{3}l^2/2} \tag{4}$$

$$P_3 = \frac{S_h}{S_1} = \frac{3\sqrt{3}(l-r)^2}{3\sqrt{3}l^2} \tag{5}$$

where P_1 is the probability of the IH model; P_2 is the probability of the EH model; and P_3 is the probability of the HH model; r is the radius of the projectile, cm; S_i is the set of projected area at the center of the sphere under the IH model, cm^2; S_e is the set of projected area at the center of the sphere under the EH model, cm^2; S_h is the set of projected areas at the center of the sphere under the HH model.

Assuming that under the IH model, the residual kinetic energy is E_r, then under the EH model, the residual kinetic energy is 1.376 E_r, and under the HH model, the residual kinetic energy is 2.044 E_r. The radius of the projectile was set as 0.1 cm. Table 3 lists the probabilities of the various models and calculates the mathematical expectations:

$$E(x) = P_1 E_r + 1.376 P_2 E_r + 2.44 P_3 E_r \tag{6}$$

As can be seen from Table 3, the original design scheme has the lowest mathematical expectation, and is a relatively high-quality design scheme under the condition that the surface density and thickness of each layer are unchanged.

Table 3. Probabilities of the various working condition and the mathematical expectations

Number	l/mm	t_h/mm	P_1	P_2	P_3	E_r/J	E(x)/J
1	5.5	0.08	0.080	0.251	0.669	5.667	9.556
2	4.125	0.06	0.142	0.284	0.574	5.329	9.667
3	3.4375	0.05	0.205	0.293	0.503	4.876	7.971
4	2.75	0.04	0.320	0.275	0.405	4.955	7.563
5	2.2	0.032	0.500	0.203	0.298	5.700	8.042
6	1.833	0.0267	0.689	0.105	0.207	7.732	9.704
7	1.375	0.02	0.896	0.029	0.074	10.730	11.682

4 Conclusion

In this paper, the process of different positions of the honeycomb sandwich structure with MIL of the projectile impact was simulated, and the probability and mathematical expectation of the hit different position were calculated. Under the premise of constant surface density and thickness of each layer, various working conditions were simulated and compared, and the following conclusions are obtained:

1. The impact effects of the projectile on the front plates were almost the same, and the projectile with a diameter of 2 mm and 4 km/s formed a hole with a diameter of about 4.5 mm on the front plate.
2. When the projectile hit the edge of the honeycomb core, it was split into multiple broken projectiles, resulting in a larger area of damage and perforation on the rear panel. But the projectile lost even more kinetic energy.
3. On the premise that the surface density and thickness of each layer remain unchanged, the honeycomb side length and honeycomb wall thickness were changed in equal proportion. When the honeycomb wall thickness was thicker than 0.032 mm, the kinetic energy consumption of the projectile tended to be stable with the increase of the side length and honeycomb wall thickness, and the residual kinetic energy of the projectile remained from 4.8 J to 5.7 J, and the increase of the side length reduced the probability of hitting the honeycomb edge. However, when the wall thickness of the honeycomb was too thin, even if the side length was shortened, with the debris cloud penetrating more walls, it cannot consume more kinetic energy, and the remaining kinetic energy increased greatly.
4. By calculating the mathematical expectation method, the original design scheme is a relatively high-quality design scheme under the condition that the areal density and the thickness of each layer are unchanged.

Acknowledgment. The research was supported by the State Scholarship Fund from China Scholarship Council.

References

1. Katz, V.D., Snyder, D.: Mechanism for spacecraft charging initiated destruction of solar arrays in GEO. In: Presented at the AIAA Paper 98-1002, 36th Aerospace Science Meeting, Reno, USA (1998)
2. Liu, S., Li, Y.: The damage and countermeasures of M/OD on space station. Manned Spaceflight. **3**, 15–19 (2003)
3. Whipple, F.L.: Meteorites and space travel. Astron. J. **52**, 131 (1947)
4. Taylor, E.A.: Computational study of hypervelocity impact onto Whipple bumpers and sandwich plates with honeycomb core. Eur. Space Agency/Eur. Space Res. Technol. Centre. (1999)
5. Ryan, S., Schäfer, F., Destefanis, R.: A ballistic limit equation for hypervelocity impacts on composite honeycomb sandwich panel satellite structures. Adv. Space Res. **41**(7), 1152–1166 (2008)
6. Yuan, J.: Satellite Structure Analysis and Design. Aerosp. Press, Beijing (2004)
7. Taylor, E.A., Glanville, J.P., Clegg, R.A., Turner, R.G.: Hypervelocity impact on spacecraft honeycomb: hydrocode simulation and damage Laws. Int. J. Impact Eng. **29**, 691–702 (2003)
8. Taylor, E.A., Herbert, M., Baughan, V., Melonnell, J.: Hypervelocity impact on carbon fibre reinforced plastic/aluminium honeycomb: comparison with Whipple bumper shields. Int. J. Impact Eng. **23**, 883–894 (1999)
9. Huang, J., Ma, Z., Lan, S., Li, Y., Liu, S.: Study on hypervelocity impact characteristics for honeycomb sandwich with multi-layer insulation. J. Astronaut. **31**(8), 2043–2049 (2010)
10. Chaturvedi, R., Trikha, M., Simha, K.: Impact penetration through spacecraft honeycomb panels analytical, numerical and experimental study. Mater Today Proc. **21**, 1050–1058 (2020)
11. Turan, M.K., Ensarioglu, C., Bakirci, A., Karpat, F.: Impact performance of unconventional trigger holes. Mater. Test. **66**(3), 389–396 (2024)
12. Mao, G., Xu, W., Wang, C., Liu, B.: Experimental research on crush response and impact energy absorption of dual-material biomimetic honeycomb structures. Mech. Adv. Mater. Struct., 1–19 (2024)
13. Hai, H., Chen, C., Wang, W., Xu, W.: Impact resistance of a double re-entrant negative poisson's ratio honeycomb structure. Phys Scr. **99**(2), 025919 (2024)
14. Century Dynamics: AUTODYN User Manual (2002)
15. Kaurov, V., Tkachenko, I.S., Salmin, V.V.: Methodology of designing a system for ensuring the thermal regime of small spacecraft and verification of mathematical models based on telemetry data. Des. Constr. Manuf. Flying Veh. **28**(3), 113–129 (2021)
16. Nosatenko, P.Y., Bobrov, A.V., Baranov, M.L., Shlyapnikov, A.N.: Experimental determination of acoustic loads during "strela" launches and computational determination of experimental testing modes of launched spacecraft. Bull. Samara State Aerosp. Univ. **2**, 112–123 (2010)
17. Yuan, X., Ma, T.: Research on structural optimization design of hypervelocity Impact on honeycomb sandwich panel. J. Ordnance Equip. Eng. **42**(01), 179–184 (2021)
18. Kurenkov, V.I., Lukashev, L.G., Yumashev, L.P.: Assessment of the Meteor and Technogenic Danger of Spacecraft Flight. Samara State Aerosp. Univ., Samara (2004)

In-Service Behavior and Failure Analysis of Engineering Materials and Structures

Evaluation of Corrosion Resistance of an Underwater Robot Joint through Accelerated Testing in Simulated North Sea Marine Conditions

Ying Qu[1(✉)], Michael Blom Hermansen[2], and Thomas Ebel[1]

[1] University of Southern Denmark, Odense, Denmark
yingqu@sdu.dk
[2] SubBlue Robotics ApS, Odense, Denmark

Abstract. This study investigates the corrosion resistance of a hard anodized aluminum joint enclosure used in an underwater robotic arm. An accelerated corrosion test, based on Arrhenius equation, is designed to simulate 10,000 h of operational exposure in North Sea conditions (salinity 38 ppt, seawater typical temperature 15 °C) within 1,043 h by increasing simulated seawater temperature (50 °C). A sacrificial magnesium anode is applied to provide cathodic protection against galvanic corrosion to critical areas where are exposed directly to the seawater. Throughout the test, comprehensive insights, including visual inspections, weight measurements, and vacuum pressure evaluations, are conducted to monitor corrosion effects on the joint enclosure surface and ensure sealing performance. This vacuum pressure test methodology provides a non-destructive, quantitative assessment of the integrity of the joint enclosure before and after corrosion testing, which offers insight into the corrosion resistance and functional reliability of the joint. The results confirm that the joint enclosure performs high corrosion resistance and reliable sealing under operational marine environment, validating the effectiveness of accelerated test for corrosion evaluation. Additionally, practical recommendations are proposed to enhance the corrosion resistance and sealing performance of the joint enclosure.

Keywords: Corrosion Resistance · Underwater Robot · Accelerated Reliability Test · Galvanic Corrosion · Anodized Aluminum

1 Introduction

Corrosion resistance is crucial for ensuring the long-term reliability and durability of underwater robotic components, particularly joints that serve structural and protective roles. Aluminum alloys, especially 6082-T6 with anodized coatings, are frequently utilized in underwater robotics due to their favorable strength-to-weight ratio and inherent corrosion resistance [1]. However, despite anodization, corrosion vulnerabilities persist at geometrically complex regions, interfaces of dissimilar metals, and locations damaged during operation [2, 3].

Y. Zhao (Ed.): ICAMM 2025, SPM 96, pp. 205–218, 2025.
https://doi.org/10.1007/978-3-032-09862-7_20

Accelerated corrosion testing methodologies involving elevated temperatures have emerged as an effective approach for rapidly simulating long-term marine exposure conditions, allowing reliable prediction of material behavior within shorter test durations. Additionally, effective cathodic protection methods are necessary for galvanic corrosion risks arising from interactions between dissimilar materials, such as aluminum alloys and stainless-steel fasteners [4]. For example, [5] shows the effectiveness of magnesium sacrificial anodes in reducing corrosion on aluminum robotic components by preferentially corroding to protect vulnerable areas.

This paper presents an accelerated corrosion test designed to evaluate the corrosion resistance and sealing integrity of a newly developed joint enclosure. The methodology combines elevated temperature testing, sacrificial magnesium anodes, and periodic evaluation, including visual inspections, weight measurements, and vacuum pressure tests, to provide comprehensive insights into the performance and reliability of an underwater robot arm joint in a simulated typical operational marine environment.

1.1 Joint Dimension and Material Specification

The joint under investigation is a critical mechanical component in the underwater robotic arm system, serves both as a structural link in the robotic arm and as a protective enclosure for key electronic and mechanical components. As illustrated in Fig. 1, this joint specifically designed to house and protect the motor positioning controller and motor assembly.

The aluminum enclosure housing forms the primary structural body of the joint, with overall external dimensions of 443 mm in length and 243 mm in height. The cylindrical portion of the enclosure, which accommodates the motor positioning controller and motor assembly, has an outer diameter of 168 mm. This design ensures both a compact form factor and adequate internal space for component integration.

The enclosure is fabricated from 6082 Aluminum Alloy, which is widely adopted for underwater robots for its mechanical properties, such as corrosion resistance, strength-to-weight ratio, etc. To enhance its performance in the harsh marine environment, the external enclosure surface has a hard anodizing treatment, which provides a thick, uniform oxide layer to improve typically resistance against pitting and crevice corrosion in high-salinity underwater conditions.

Inside the enclosure, the motor positioning controller PCB is mounted securely to provide precise control over the motor's function, enabling accurate positioning of the robotic arm segments. The motor itself is installed within the cylindrical enclosure and directly interfaces with the robotic arm's drive mechanisms.

Several components within the joint assembly utilize different materials from the primary enclosure:

- **Plugs:** allow trapped pressure to escape from an enclosure after it is closed and serve as pressure test ports for sealing property check of individual cellars of the joint enclosure. Manufactured from an alternative aluminum alloy (7075-T6 Anodized Aluminum).
- **Fasteners:** All screws on the joint are made from stainless-steel, providing high mechanical strength and corrosion resistance.

Additionally, despite the application of a hard anodizing treatment to enhance the corrosion resistance of the enclosure, there are certain geometrically complex areas, such as screw holes, threaded interfaces, recessed contact surfaces, where anodizing treatment is less effective or entirely inaccessible. Besides, scratches may occur on the joint surface during the robot working, which can cause the anodizing layer falls off and make the Aluminum Alloy directly expose to seawater.

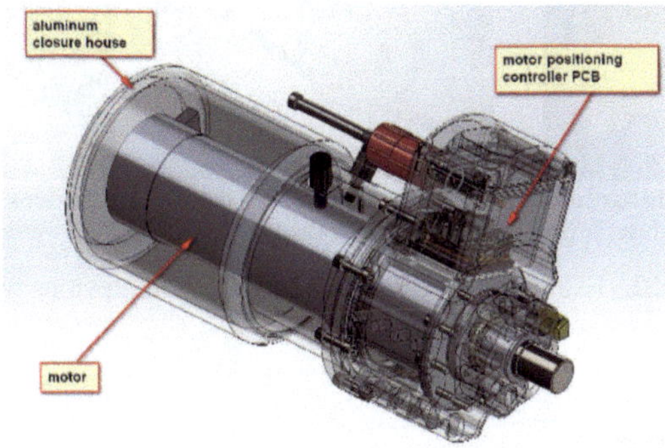

Fig. 1. Mechanical assembly model with joint outline and dimensions.

1.2 Corrosion Description

As shown in Fig. 2, by the experience of the previous joints, which are used in multiple seawater operation, the corrosion on the joint mainly include:

- pitting
- surface discoloration
- material erosion
- rust formation
- scratches

As the joint for the corrosion test is newly produced, there is no scratches on the surface. Therefore, the corrosion that is caused by scratches is not expected or considered in this test, and the related picture is only for reference.

1.3 Study Emphasis

The dissimilar materials of the plugs and the fasteners on the joint satisfy the balance between structural requirements, corrosion resistance, and manufacturability. However, the presence of dissimilar metals introduces the risk of galvanic corrosion where seawater acts as an electrolyte. Additionally, the uncoated regions, where is with no treatment, are inherently more susceptible to localized corrosion when exposed to seawater.

This study concerns these risks and particularly monitored the potential phenomenon in the corrosion test, in order to analyze the corrosion resistance properties of the mechanical design of the proposed joint in a typical seawater working environment, and provide the recommendations for improving the mechanical design.

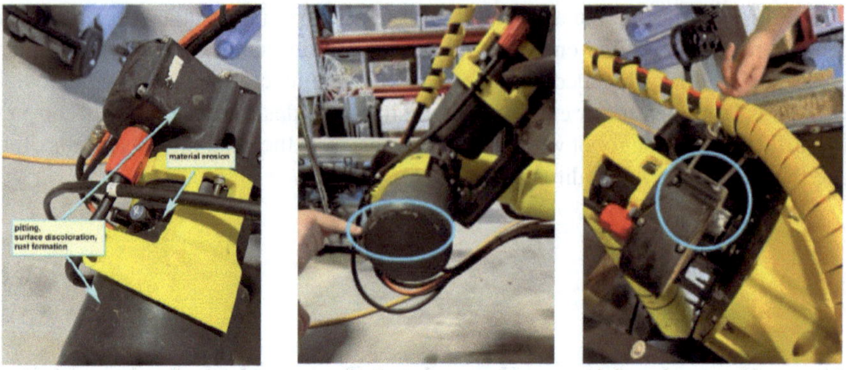

(a) pitting, surface discoloration, material erosion, and rust formation

(b) Corrosion caused by scratches

(c) Corrosion caused by material erosion and pitting

Fig. 2. Examples of corrosion phenomena in previous work

2 Methods

2.1 Accelerated Corrosion

In order to shorten the testing duration while maintaining the reliability and relevance of the corrosion performance evaluation, this study employed an accelerated corrosion testing methodology to simulate the long-term effects of the corrosive environment.

The relation between temperature and corrosion rate is generally exponential, where an increase in temperature leads to a significant acceleration in the corrosion rate. This is primarily due to the increased kinetic energy available for chemical and electrochemical reactions.

Corrosion rate equation (Arrhenius equation), which describes how the rate of a chemical reaction increases with temperature. In physical chemistry, the Arrhenius equation is a formula for the temperature dependence of reaction rates [6, 7]:

$$k = A \cdot \exp\left(-\frac{E_a}{R \cdot T}\right) \tag{1}$$

k: corrosion rate.

A: pre-exponential factor, which is specific to the material and environment.

E_a = 50.000 (J / mol): activation energy for the corrosion process.

$R = 8.314$ $(J / mol \bullet K)$: universal gas constant.

T: temperature (K).

2.2 Magnesium Sacrificial Anode Application

Galvanic corrosion occurs when two dissimilar metals come into electrical contact in the presence of an electrolyte solution, such as seawater. The metal with a lower electrode potential, anodic metal, corrodes faster, while the metal with a higher electrode potential, cathodic metal is protected. This process is driven by the electrochemical potential difference between the two metals.

In this study, the primary structural material of the joint assembly is Aluminum Alloy 6082, while the plugs are made from Aluminum Alloy 7075-T6. While both alloys are aluminum-based, 7075-T6 has a significantly different electrochemical potential compared to 6082, due to its higher zinc content. This difference in composition, coupled with their direct contact within the joint assembly, introduces an increased risk of galvanic corrosion on the seam of the two parts.

Additionally, the interface regions between the stainless-steel fasteners and the screw holes of the underwater robotic arm joint are with no protective hard anodized coating, where the aluminum alloy becomes the anode and increases localized corrosion risk.

To mitigate the risk of galvanic corrosion in the joint assembly, a magnesium sacrificial anode is applied to the surface of the aluminum enclosure. Magnesium, with a much lower electrode potential than aluminum, acts as a sacrificial anode in this study. Magnesium anodes preferentially corrode and thereby cathodically protect the aluminum components from corrosion in an electrolyte. This galvanic protection system extends the lifetime of the joint in seawater [8, 9].

2.3 Leakage Inspection by Vacuum Test

A vacuum pressure test is conducted to establish the baseline pressure property of the joint enclosure prior to corrosion testing, in order to verify the sealing integrity and ensure no leakage occurs on the joint enclosure, which is critical for underwater applications.

As shown in Fig. 3, the sealing integrity of the joint enclosure is tested through a vacuum pressure by a hand operated vacuum pump. To check the sealing integrity, connecting the vacuum pump to one of the plugs, and pump the joint inner pressure to approximately -15 inHg, and maintain for 15 min. The evaluation criteria for the sealing integrity are as follows:

- **PASS:** If the gauge shows that the pressure does not have an obvious decline, or less than 0.5 inHg, that means no leakage happens and evaluated PASS. The test can continue.
- **FAIL:** If the pressure declines by more than 0.5 inHg in 15 min (faster than the baseline), repeat the vacuum operation for another two times to ensure the decline. If the pressure results of the 3 times are similar, with a consistent pressure decline, leakage is confirmed, and the test is terminated.

The baseline vacuum test is shown in Fig. 4. The plot illustrates the pressure decline tendency within the sealed joint enclosure over time. Under normal leak-free conditions, the decrease rate of pressure remains gradual and consistent.

This baseline pressure trend serves as a reference for subsequent pressure tests conducted after the accelerated corrosion testing phase. the pressure retention performance is expected to be consistent with the established baseline if the joint enclosure maintains structural integrity and no leakage occurs. Conversely, if corrosion damage leads to a degradation of the joint enclosure and leakage happens, the rate of pressure decrease will exhibit a steeper decline.

This vacuum pressure test methodology provides a non-destructive, quantitative assessment of the joint's sealing performance during the corrosion test, which offers an insight into the corrosion resistance of the joint enclosure under seawater conditions.

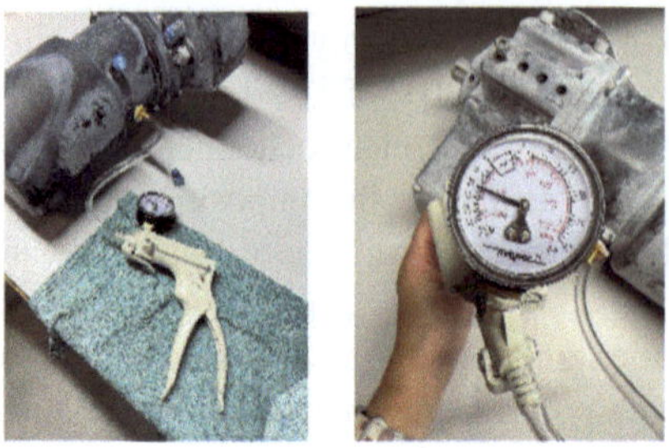

Fig. 3. Illustration of vacuum pressure test

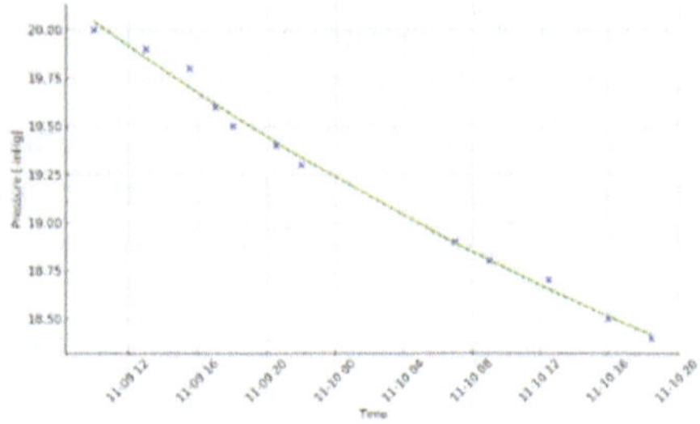

Fig. 4. Baseline of pressure feature of joint by vacuum test

3 Results

3.1 Test Configuration

Environmental Configuration. In this study, accelerated corrosion testing was conducted by increasing the temperature to expedite the corrosion process. The typical temperature T_{typ} of North Sea, where the robot works normally, is 15 °C, and to 288.15 K. The acceleration test temperature T_{acc} is elevated to 50 °C:

$$T_{typ} = 15°\mathrm{C} = 288.15K$$
$$T_{acc} = 50°\mathrm{C} = 323.15K$$

The relationship between the corrosion rates at the typical and accelerated temperatures is determined by Arrhenius equation, as shown in Eq. (1):

$$\frac{k_2}{k_1} = \frac{A \cdot \exp\left(-\frac{E_a}{R \cdot T_2}\right)}{A \cdot \exp\left(-\frac{E_a}{R \cdot T_1}\right)} = \exp\left(\frac{E_a}{R}\left(\frac{1}{T_1} - \frac{1}{T_2}\right)\right)$$
$$= \exp\left(\frac{50{,}000}{8.314}\left(\frac{1}{288.15} - \frac{1}{323.15}\right)\right) \approx 9.58$$

In conclusion, the accelerated test time t_{acc} corresponding to an equivalent exposure time t_{exp} under 15 °C typical condition (Table 1):

$$t_{acc} = \frac{t_{exp}}{f_{acc}} = 1,043\ hours$$

Environmental Setup. A custom test was designed to simulate operational marine conditions which provides a physical environment. The test setup is shown in Fig. 5.

Test Apparatus. The test system consists of two concentric containers:

Inner Box: is filled with simulated seawater to replicate marine conditions and the joint enclosure is fully submerged in the seawater.

Outer Box: is filled with normal water and serves as thermal insulation and additional containment. It ensures temperature stability during the test process and avoids any corrosion damage to the heater if directly heating the simulated seawater.

Table 1. Environmental configuration.

Parameter	Value
Simulated seawater salinity	38 *ppt*
Normal water and simulated seawater temperature	50 °C
expected lifetime	10,000 h under 15 °C
Test time (wall clock)	1,043 h

Temperature Control and Monitoring. Temperature plays a critical role in accelerating the corrosion process. Therefore, precise temperature control and monitoring is significant:

Heater: installed in the outer box, setups and maintains the normal water temperature at 50 °C, which indirectly elevates the temperature of the simulated seawater in the inner box.

Thermometer: Positioned in the inner box to monitor the temperature of the simulated seawater to ensure it remains consistent throughout the experiment.

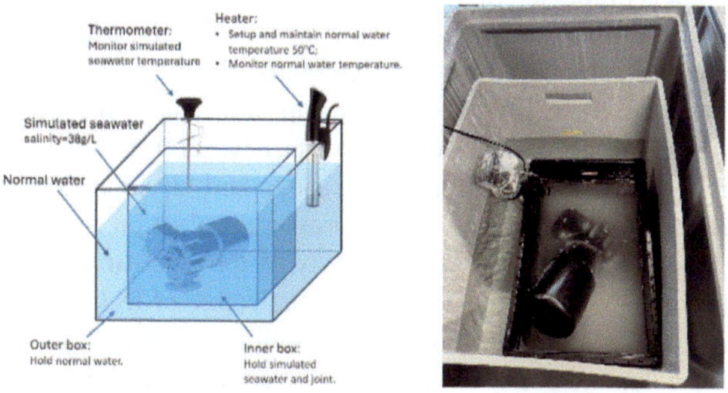

Fig. 5. Illustration of corrosion test on SBR joint

Magnesium Sacrificial Anode. To provide electrochemical potential protection to the exposed surface, a sacrificial magnesium anode was installed by bolting into one available screw hole on the joint enclosure, as illustrated in Fig. 6.

3.2 Test Procedure

The corrosion testing procedure for the underwater robotic arm joint is illustrated in the flowchart in Fig. 7. This process is designed to evaluate the corrosion resistance and sealing integrity of the joint under an accelerated simulating typical seawater environment.

First of all, the test needs to be ready by setting up the hardware and ensuring the joint is correctly assembled with all plugs securely fastened and magnesium sacrificial

Fig. 6. Magnesium sacrificial anode assembly

anodes properly positioned. Simulated seawater with 38 *ppt* salinity should be prepared in the immersion tank.

Once the setup is complete, the initial condition of the joint is documented by recording pictures, measuring weight, and conducting a vacuum pressure test to establish a baseline for sealing integrity. The simulated seawater is heated to 55 °C. A 5 °C higher temperature is configured to compensate for the temperature drop upon immersion the joint.

Following this, a decision point is reached where the test either proceeds or concludes, depending on whether the predetermined test duration has elapsed. If the test continues, the joint is immersed in the heated seawater, and the temperature is stabilized and maintained at 50 °C.

The joint remains submerged for the accelerated duration t_{acc}. During the test, the joint is inspected regularly by lifting out, drying up, and visually inspecting for any signs of corrosion, such as pitting, discoloration, erosion, or rust formation, etc. If no corrosion is observed, the regular inspection undergoes a standard visual evaluation. However, if corrosion is detected, an additional vacuum pressure evaluation is needed to assess the joint's sealing integrity. If the pressure property of the joint fulfils the PASS criteria, the test continues. If abnormal pressure trends indicative of leakage are observed, the test is terminated. In such cases, final data—including visual records, weight changes, vacuum pressure data, and corrosion details—are collected and analyzed to assess the joint's corrosion resistance performance and identify potential failure modes.

3.3 Test Results

The results of the corrosion test on the joint will be elaborated from three perspectives, including visual, weight and vacuum pressure evaluation.

Visual Evaluation Result. The visual inspection of critical areas of the joint during the corrosion is summarized in Table 2.

Joint Surface Condition.

(a) Throughout the entire test duration, no significant damage was observed on the joint surface, indicating that the cathodic protection layer effectively prevented corrosion.
(b) The joint maintained its structural integrity with no visible degradation.

Corrosion Conditions.

(a) representative exposed areas, showed no significant rust formation. This is attributed to the effective protection from the magnesium anode, which was directly assembled to contact with the housing.
(b) Minimal rust spots were observed in some specific areas, including the motor shaft keyseat, and the seam of the motor driver cellar cover.
(c) These rust spots remained localized and did not expand, suggesting that the overall corrosion resistance of the joint was maintained.

Magnesium Anode Performance.

(a) The magnesium anode actively corroded throughout the test due to sacrificial reactions with the seawater, ensuring that the joint's exposed surfaces remained protected.
(b) As a result of these reactions, deposits of $Mg(OH)_2$ and $MgCl_2$ accumulated on the joint surface, leading to a whitish discoloration.

Magnesium Anode Degradation.

(a) The first magnesium anode was severely damaged after approximately 2,682 accelerated hours of testing.
(b) The second magnesium anode, which was installed to replace the first, remained in good condition at the end of the test, lasting 7,328 accelerated hours without significant wear.

Weight Evaluation Results. The weight comparison of the joint with the second anode is shown in Table 3. The weight of the joint set after the test is 0.2 kg lighter than the original situation.

Pressure Evaluation Results. Vacuum pressure tests were conducted regularly to monitor the sealing integrity of the joint enclosure. Throughout the test period, the vacuum pressure of the joint enclosure can consistently maintain -15 inHg after 15 min without obvious drop in the regular inspections, which indicates the joint had no leakage occurs and maintained its sealing performance under accelerated corrosion conditions during the entire test.

4 Analysis

This section elaborates the analysis of the test results, consistent with visual, weight and pressure evaluations.

4.1 Visual Evaluation

Corrosion Resistance.

(a) The robot joint is coated with a cathodic protection layer, which effectively prevents significant corrosion damage throughout the test.

Table 2. Summary of visual evaluation during corrosion test.

Real time [hours]	**baseline**	**185.65**	**280**	**471.25**	**638.3**	**876**	**1045.6**
Accelerated time [hours]		**1,778.5**	**2,682**	**4,514.1**	**6,114.9**	**8,373**	**10,010**
Side view (left)							
Side view (right)							
Plug area	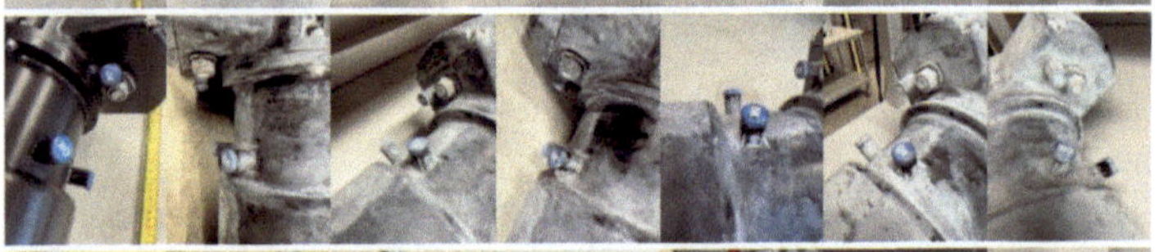						
Exposed surface (screw holes)							
Motor shaft							
Seam of motor driver cellar seam	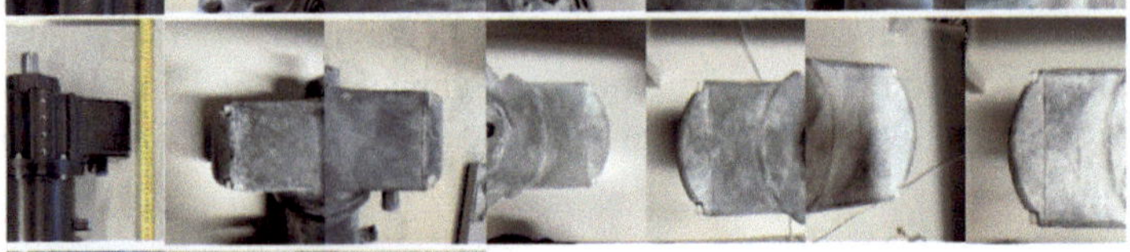						
Magnesium anode	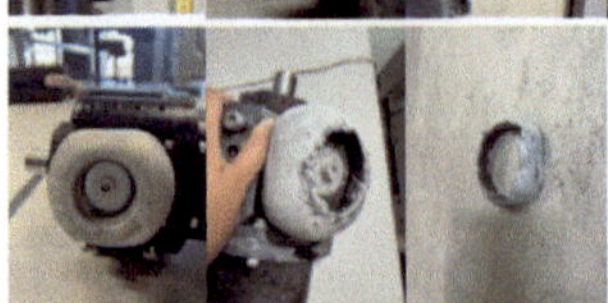						

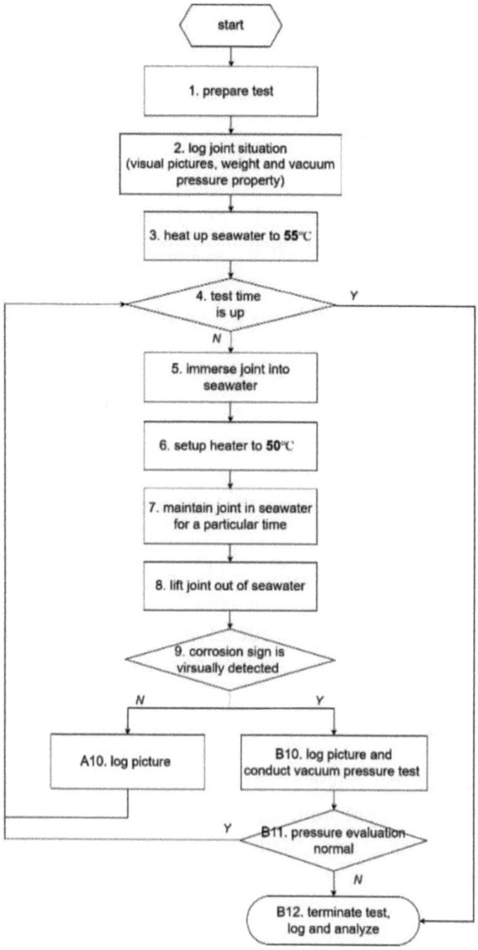

Fig. 7. Flow chart of corrosion test on the joint enclosure

Table 3. Weight comparison

Weight Measurement	Total Weight (joint + second anode) [kg]
Before test	9.87
After test	9.67
Weight loss from corrosion test	0.2

(b) Throughout the test, the joint surface remains intact with no visible structural degradation.
(c) The reason of the corrosion occurs on the motor shaft and the seam of the motor driver cellar cover is due to the contact between the main housing and the shaft or

the cellar cover is not electrically continuous, i.e., poor galvanic coupling to the magnesium anode.

Plug Seam Sealing Performance.

(a) no signs of crack at the plug seam, and remains well-sealed, which indicates that the sealing withstands the prolonged saltwater exposure, preventing water ingress into the internal structure.
(b) Deformation caused by the accumulated deposits is visible near the plug, but it does not affect its sealing functionality.

Magnesium Deposition.

(a) Over time, deposits are observed on the surface of the joint, particularly in areas near the magnesium anode.
(b) These deposits are composed of $MgCl_2$ and $Mg(OH)_2$, which are sacrificial reactions products of the magnesium anode in seawater.
(c) The presence of these deposits suggests that the magnesium anode is actively functioning to protect the exposed surface of the joint.

4.2 Weight Evaluation

During the corrosion test, no obvious reduce occurs on the weight of the joint set, including the joint and the magnesium anode.

The minimal weight loss can be attributed primarily to the material degradation of the magnesium anode, which is caused by sacrificial reactions to protect the joint from corrosion.

This result indicates that no significant loss of the joint material from the simulated seawater corrosion.

4.3 Pressure Evaluation

The pressure evaluation confirms that the joint housing can sustaining 10,000 working hours in a simulated North Sea environment. The test conditions are designed to replicate the North Sea environment.

5 Conclusion

The test results confirm that the robot joint performs high corrosion resistance and sealing performance under simulated North Sea environment (38 *ppt* salinity, 15 °C temperature) without failure. The combination of cathodic protection, effective sealing, and magnesium anode protection ensures long-term reliability (10,000 accelerated hours) in the simulated marine conditions.

- The cathodic protection coating on the robot joint is effective, as there is no observable damage to the surface.
- The plug sealing remains fully intact, preventing leakage during the entire test period.
- The deposits on the surface, primarily composed of $MgCl_2$ and $Mg(OH)_2$, are mainly corrosion by-products from the magnesium anode, rather than damage to the joint itself.
- The test confirms that the current protective system successfully prevents direct corrosion of the joint, ensuring long-term durability in a marine environment.

5.1 Application Recommendations

Regular Monitor Joint Surface, Critical Sealing and Leakage Situation.
Visual inspect is a critical and fast way to ensure the long-term durability and reliability of the joint in a seawater environment. Visual inspection can detect early signs of corrosion, material degradation, or deposit accumulation; ensure the cathodic protection coating remains intact, and identify any changes in surface texture due to saltwater exposure. If any damage on the joint surface is inspected, there are some preventive actions can be taken to avoid further degradation:

(a) apply a localized corrosion inhibitor, for example, sealants or adhesives, if coating wear is detected, such as pitting, scratches, etc. Otherwise, reapply cathodic coatings on the joint surface if the exposure area significantly affects the joint strength.
(b) clean with fresh water rinsing or a mild cleaning solution If excessive deposits form.
(c) apply additional marine-grade sealants on the plug sealing if any degradation, cracks, or deformation signs occurs.

Leakage situation of the joint chamber can be evaluated through monitoring the pressure loss in a vacuum test after seawater working if obvious visual damage can be inspected on the joint surface, especially the critical sealing area.

If pressure loss is observed, apply sealants or adhesives on the plug sealing first, and test the chamber pressure once again. The leakage crack can be confirmed to occur at the sealing if no pressure loss this time. Otherwise, find the leakage crack using visual inspection, coloring or ultrasound methods.

Adopt Magnesium Anode.
This corrosion study confirms the efficacy of magnesium anodes in protecting aluminum components from saltwater-induced degradation. No significant rust formation on critical areas, particularly the joint head and screw holes, indicates that sacrificial anode technology can be an effective and practical method for extending the lifespan of the exposed surface.

The mechanical design of the joint should consider the assembly of more magnesium anodes, which can contact independent critical housings, to ensure good galvanic coupling between the exposed surface and the magnesium anode.

Monitor Magnesium Anode Situation and Exchange Promptly.
Regularly monitor the situation and the **rate of** Magnesium **anode consumption,** and

replaced periodically or when it is significantly depleted, to avoid severely corroded anodes cannot provide protection to the exposed surface.

Highly Recommend Deep Cleaning Promptly After Working in Seawater Environment.
Rinse the joint with fresh water after prolonged exposure to seawater, and the joint enclosure life tends to reduce if no deep clean after seawater working, due to the NaCl attached to the joint surface can continue the corrosion on the cathodic coating, pitting and scratching exposure surface and magnesium anodes.

The accumulated $MgCl_2$ and $Mg(OH)_2$ deposits should also be periodically removed to prevent excessive buildup, which could affect performance.

Acknowledgment. The authors would like to thank Kurt Godiksen for his valuable assistance in setting up the test equipment to ensure the successful execution of the test procedure.

This work is supported by the project "Fuel savings and emission reductions in the shipping industry through innovative propeller polishing robot technology", funded by EUDP Project no. 64022-1009.

References

1. Davis, J.R.: Aluminum and aluminum alloys. In: Alloying: Understanding the Basics. ASM International (2001)
2. Szklarska-Smialowska, Z.: Pitting corrosion of aluminum. Corros. Sci., Elsevier, 1743–1767 (1999)
3. Vargel, C.: Corrosion of Aluminium. Elsevier (2004)
4. Andrei, M., Andrei, M., di Gabriele, F., Bonora, P.L., Scantlebury, D.: Corrosion behaviour of magnesium sacrificial anodes in tap water. Mater. Corros., 5–11 (2003)
5. Knudsen, O.Ø., Vada, E., Krieger, W., Bertram, J., Jevremovic, I., Wilson, H.: Cathodic protection of aluminium in seawater. Mater. Corros. (2024)
6. Bullough, R.P., Edwards, W.H., Harrison, A.I.: Stationary valve regulated sealed lead-acid batteries can the Arrhenius equation be used to predict product life. In: INTELEC '87 - The Ninth International Telecommunications Energy Conference (1987)
7. Yang, S., Xiang, D., Bryant, A., Mawby, P., Ran, L., Tavner, P.: Condition monitoring for device reliability in power electronic converters: a review. IEEE Trans. Power Electron., 2734–2752 (2010)
8. Ahmad, A., Yahya, Z., Daud, N.A.Q.M., Daud, M.: Electrochemical behaviour of aluminum alloys in natural seawater. In: 2009 3rd International Conference on Energy and Environment (ICEE), pp. 110–116 (2009)
9. Desmouliere, L.: Cable trays offshore: a choice of materials. In: 2009 Conference Record PCIC Europe, pp. 28–34 (2009)

Mechanical Performance of Knee Implant Components: A Study of Flexion and Rotation Angle Variations

Deepak Kumar(✉) and Rina Maiti

Human Engineering Research Laboratory, Department of Design and Manufacturing, Indian Institute of Science, Bangalore, Bengaluru 560012, India
deepakkumar4@iisc.ac.in

Abstract. This study investigates the biomechanical behaviour of Ti-6Al-4V, Co-Cr-Mo alloy, and Stainless Steel in knee implants using Finite Element Method (FEM) analysis across various flexion angles (−8°, 5°, 15°, 30°, 60°, 90°, and 120°) and internal/external rotational conditions (5°, 6°, and 7°). Key parameters assessed include Von Mises stress, deformation, contact area, and contact pressure under physiologically relevant loads. Deep flexion angles (90° and 120°) produce peak stress concentrations in the femoral and tibial components, especially at their interfaces, increasing the risk of wear and fatigue. The polyethylene (PE) insert exhibited localized strain peaks, highlighting its role in load transfer. Maximum deformation occurred at 120°, raising concerns about implant stability. In contrast, moderate flexion (−5° to 30°) showed uniform stress distributions. Even minor internal/external rotations significantly altered stress profiles, stressing the need for precise rotational alignment in implant design. These findings advocate for material-specific optimization, especially for PE inserts, and highlight flexion-related loading as a critical factor in implant fatigue. Future work will include experimental validation and long-term fatigue studies across varied patient anatomies and activity levels.

Keywords: Finite element analysis · knee joint implant · Flexion and Extension angles · stress distribution · contact pressure · joint kinematics

1 Introduction

The knee joint is one of the most functionally significant and biomechanically complex joints in the human body. It facilitates a combination of flexion-extension and rotational movements essential for locomotion, weight-bearing, and dynamic stability. Flexion and rotational movements occur through intricate interactions between bones, ligaments, and cartilage, influencing joint loading, stability, and overall kinematics [1]. Accurately assessing these movements is crucial in orthopaedic research, prosthetic development, sports biomechanics, and injury prevention strategies.

Orthopaedic implants, particularly in joint replacements, endure complex biomechanical loads varying with flexion, extension, and rotation. Material selection critically

Y. Zhao (Ed.): ICAMM 2025, SPM 96, pp. 219–228, 2025.
https://doi.org/10.1007/978-3-032-09862-7_21

influences implant performance, affecting stress distribution, fatigue resistance, and wear [2]. Commonly used materials, Ti-6Al-4V, Co-Cr-Mo, and Stainless Steel, are valued for their mechanical strength and biocompatibility. Ti-6Al-4V offers a high strength-to-weight ratio and corrosion resistance, Co-Cr-Mo provides superior wear resistance, while Stainless Steel is cost-effective but may lack long-term fatigue durability. Finite Element Method (FEM) analysis is a key computational tool to simulate real-life mechanical loads on implants, offering detailed insights into stress, strain, deformation, and load transfer [3]. This study evaluates how these materials respond under flexion angles ranging from −8° (hyperextension) to 120° (deep flexion) and internal/external rotations of 5°–7°, which significantly affect stress patterns and joint stability. These movements influence ligamentous function, particularly the anterior cruciate and collateral ligaments [4].

Previous studies show that flexion angles impact anterior-posterior stress distribution, and internal/external rotation governs stability and wear [5]. FEM helps assess ligament strain and prosthetic design under dynamic conditions. This comparative FEM study aims to optimize implant materials for improved joint stability and longevity, enhancing clinical outcomes and rehabilitation strategies in orthopaedic care.

1.1 Knee Joint Contact

Under loading, the knee's articular surfaces show increased contact area, providing insights into joint stability and load-bearing capacity. Typically, the patella-femur contact area is smaller than that of the tibia-femur. During knee flexion, the tibiofemoral contact center shifts posteriorly; for example, it moves from 25 mm to 38.5 mm from the front joint line as the knee bends [6]. Figure 1(a) and (b) illustrate this contact shift based on flexion angles. Additionally, a contact strip on the patella travels upward, reaching its top at 90° flexion. Researchers have extensively studied pressure distribution across knee regions in various flexion positions to assess biomechanical behaviour.

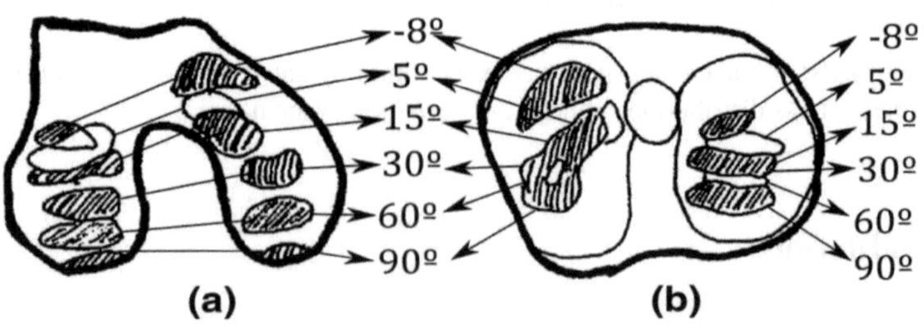

Fig. 1. Tibio-femoral contact area as a function of knee flexion angle: (a) for the femur and (b) for the tibia [7].

This study investigates how bony deformities, and rotational movements affect knee joint contact behaviour under compression, using advanced FEM analysis. By evaluating anterior-posterior displacements and internal-external rotations at flexion angles (−8° to 120°) and rotational angles (5°–7°), the research aims to enhance understanding of knee mechanics. The model uses realistic material properties and boundary conditions. Findings will support improved prosthetic knee designs, rehabilitation strategies, and Sports injury prevention, offering valuable insights for clinicians, engineers, and researchers in orthopaedic biomechanics [8].

2 Materials and Methods

2.1 Material Properties of Implant Materials

Each component is modelled using its respective material properties, including elastic modulus E, Poisson's ratio v, and density ρ (Table 1) [10].

Table 1. Material properties of Knee Implant

Material	Elastic Modulus E(GPa)	Poisson's Ratio v	Density ρ (kg/m^3)
Ti-6Al-4V	110	0.34	4430
Co-Cr-Mo Alloy	210	0.30	8500
Stainless Steel	200	0.29	7850
UHMWPE	0.8–0.15	0.42	950

For isotropic materials, the linear elastic stress-strain relationship is given by Eq. 1

$$\sigma_{ij} = C_{ijkl}\varepsilon_{kl} \tag{1}$$

Where σ_{ij} is the stress tensor, C_{ijkl} is the elastic stiffness tensor, and ε_{kl} is the strain tensor.

Equation 2 defines metallic components (Ti-Alloy, Co-Cr-Mo, Stainless Steel) as isotropic linear elastic. UHMWPE is modelled hyperelastically using strain energy functions: Mooney-Rivlin (Eq. 3), Neo-Hookean (Eq. 4), and Ogden (Eq. 5).

$$\sigma = \frac{\partial W}{\partial \varepsilon} \tag{2}$$

Neo-Hookean Model

$$W = \frac{C_1}{2}(I_1 - 3) \tag{3}$$

Mooney -Rivlin Model

$$W = C_1(I_1 - 3) + C_2(I_2 - 3) \tag{4}$$

Where C_1 and C_2 is the material constant, and I_1 and I_2 are the first invariants of the deformation tensor

Ogden Model

$$W = \sum_{i=1}^{N} \frac{\mu_i}{\alpha_i}\left(\lambda_1^{\alpha_i} + \lambda_2^{\alpha_i} + \lambda_3^{\alpha_i} - 3\right) \tag{5}$$

Here, $\lambda_1, \lambda_1, \lambda_1$ are the principal stretch ratios and μ_i, α_i are material constants.

2.2 Model of Knee Implant

Comprehending how people move and the forces at work on the body are critical when doing tiresome jobs [9]. In training for sporting events and daily life routines, knee rotation angles change from *−8°* to *120°* in anterior and posterior, and *5°* to 8° in internal and external angles. Table 2 Lists the approximate dimensions of the knee implant model.

Table 2. Estimated knee implant dimensions

Name of Components	Length X (mm)	Breadth Z (mm)	Height Y (mm)
Femur Tray	60	76	73
PE component	62	36	69
Tibial Tray	62	28	62
Model Assembly	62	132	73

As shown in Fig. 2 An artificial knee joint comprises three primary components: (a) the femoral component, (b) the tibial segment, and (c) the cushion pad (plastic spacer). A compressive load $F_y = 700$ N was applied to the femoral tray with boundary conditions $U_1 = 0$ and $U_3 = 0$.

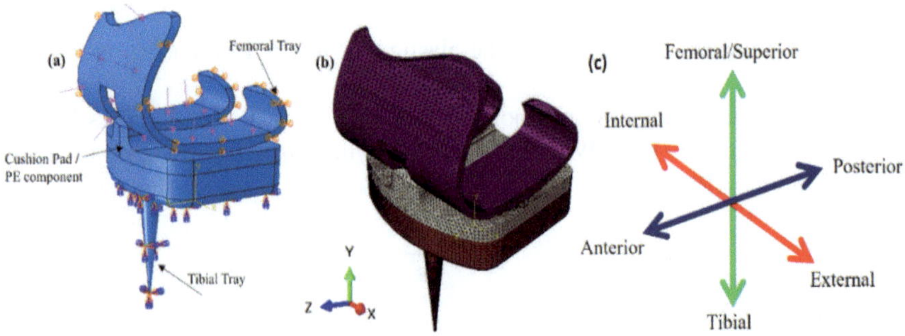

Fig. 2. (a) The artificial knee joint consists of femoral, tibial, and cushion components. A compressive load (700 N) with fixed boundary conditions is applied and analyzed via the FE meshed model (b), FE Meshed model of knee implant (c), and loading force directions.

Surface-to-surface contact with friction was modelled between the femoral tray and the plastic pad, while the tibial component was constrained using the CSTER boundary condition. ($U_1 = U_2 = U_3 = 0$, $U_{R1} = U_{R2} = U_{R3} = 0$), Effectively fixing it in place. The tibial component acts as the fixed reference point in the force analysis of the knee prosthesis, with the resultant contact forces being critical to implant performance.

Figure 3 illustrates Finite Element Analysis (FEA) of the knee joint at flexion angles from −8° (hyperextension) to 120° (deep flexion), highlighting stress distribution, contact pressure, and contact area. At −8°, minimal contact occurs with ligaments resisting hyperextension. Flexion from 5° to 30° shows stable contact and rising ligament tension. At 60° and 90°, contact pressure increases, especially during sitting or squatting. At 120°, extreme flexion leads to significant deformation and compression. This analysis enhances understanding of knee biomechanics, aiding prosthetic design and injury prevention strategies.

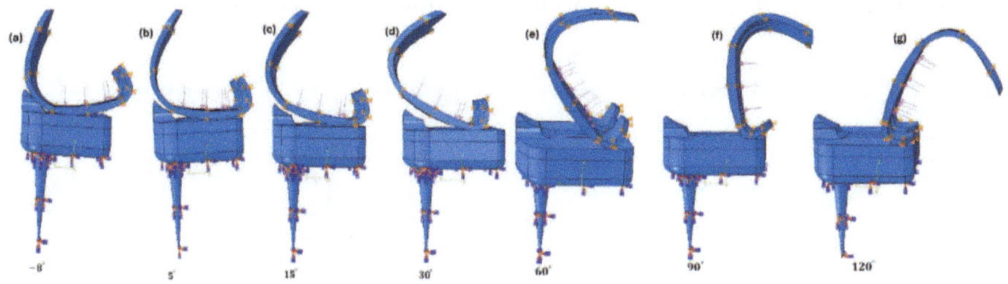

Fig. 3. Finite Element Model of Knee Joint under Various Flexion Angles (−8°, 5°, 15°, 30°, 60°, 90°, and 120°) at the Anterior and Posterior planes.

Figure 4 The knee joint implant is analysed under rotational movements at 5°, 6°, 7°, and 8°, simulating internal and external rotations. These angles assess implant stability, wear, and alignment. At 5°–6°, strain remains within tolerable limits, while 7°–8° introduces higher stress, risking ligament strain and instability.

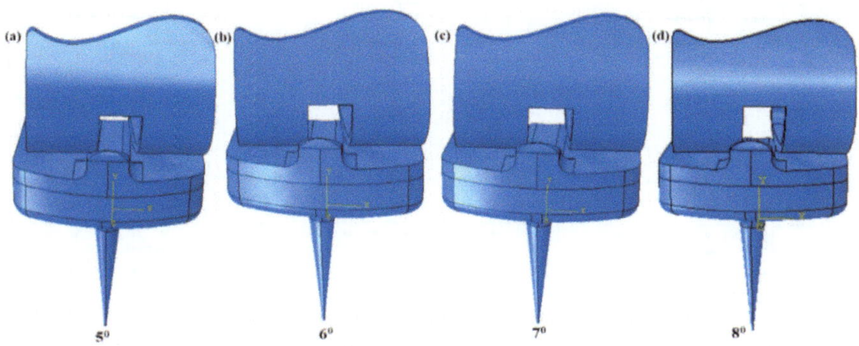

Fig. 4. Model of the Knee Implant under rotational angles 5°, 6°, 7°, and 8° in Internal and external planes.

Such analysis helps evaluate implant performance during dynamic movements, especially in sports. Tested across Ti-alloy, Co-Cr-Mo alloy, and Stainless Steel, the findings guide optimized implant design for durability, reduced wear, and improved joint function.

3 Results and Discussion

Stress variation across femur, polyethylene (PE), and tibial components was analysed at different flexion angles. In Fig. 5(a–c), Femoral stress is minimal at $-8°$ and $5°$, peaking near $60°$, with Stainless Steel (SS) exhibiting the highest stress, followed by Co-Cr-Mo and Ti-Alloy, the latter showing the most favourable distribution. In Fig. 5(d–f), The PE component peaks at $30°$, again with SS inducing the highest stress.

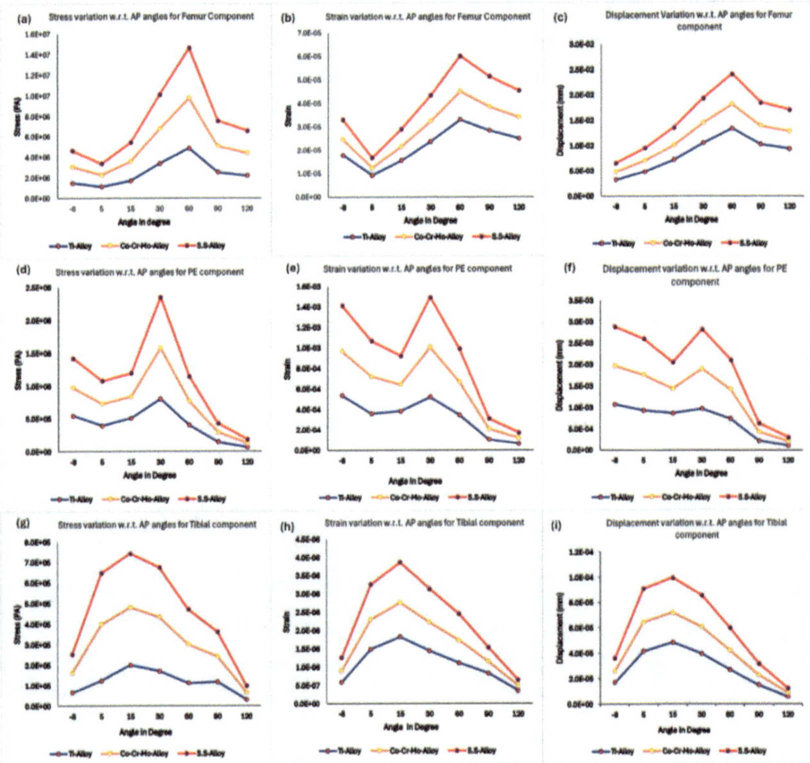

Fig. 5. Stress, strain, and displacement variations with respect to anterior-posterior (AP) flexion angles for the femur, polyethylene (PE), and tibial components using Ti-Alloy, Co-Cr-Mo Alloy, and Stainless-Steel materials.

Figure 5(g–h) shows tibial stress increasing to $60°$, then slightly decreasing. Ti-Alloy consistently performs best. Figure 6 reveals that peak stress, strain, and displacement occur at $6°$ rotation. Co-Cr-Mo shows maximum stress, while Ti-6Al-4V offers lower stress but higher displacement, raising concerns about potential deformation. The polyethylene (PE) inserts, essential for shock absorption, exhibited maximum strain when paired with stiffer alloys like stainless steel, indicating a higher risk of fatigue under rotational misalignment. Figure 7 shows contact area variations between femur-PE and tibia-PE interfaces across flexion angles ($-8°$ to $120°$). Femur-PE contact increased with flexion, peaking at ~25 mm^2 at $60°$, while tibia-PE contact remained stable (~5 mm^2), suggesting that femur-PE bears most of the load. Figure 8 illustrates contact area changes under internal-external (IE) rotation (5°–8°), showing a rise in femur-PE contact from ~35 mm^2 to ~40 mm^2, indicating increased articulation and compressive stress. Tibia-PE contact remained steady (~25 mm^2), highlighting stable load transfer. Figure 9 presents contact pressure (CPRESS) over time across flexion angles, with the highest pressures observed in the PE component, especially at 120° (~7.0×10^6 Pa). These findings emphasize the need for optimized implant designs to minimize stress concentrations and reduce wear under high flexion and rotational loads.

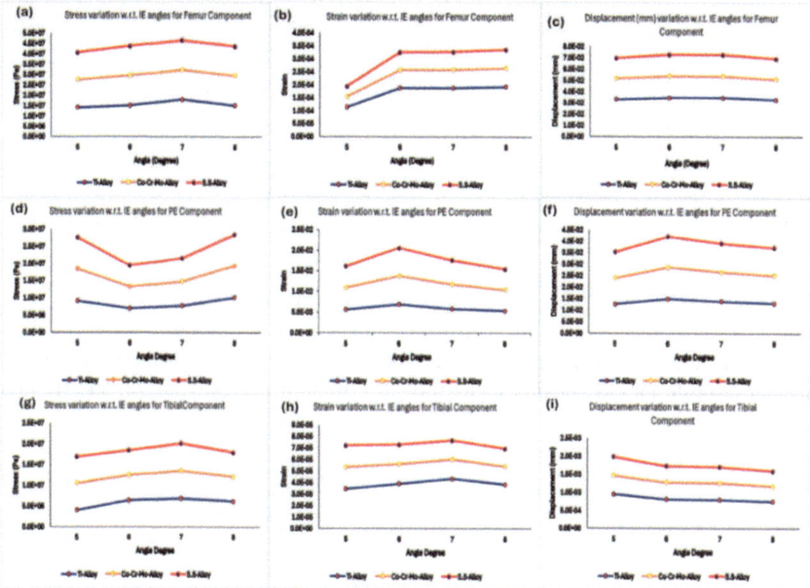

Fig. 6. Stress, strain, and displacement variations with respect to internal-external (IE) rotation angles for the femur, polyethylene (PE), and tibial components using Ti-Alloy, Co-Cr-Mo Alloy, and Stainless-Steel materials.

Figure 10 illustrates rising contact pressure (CPRESS) with increasing internal-external rotation (5°–8°), peaking at ~7.0×10^6 Pa on the PE component and elevated femoral pressures. Tibial pressure remains lower but steady. These results stress the need for optimized implant designs to reduce wear and ensure joint stability.

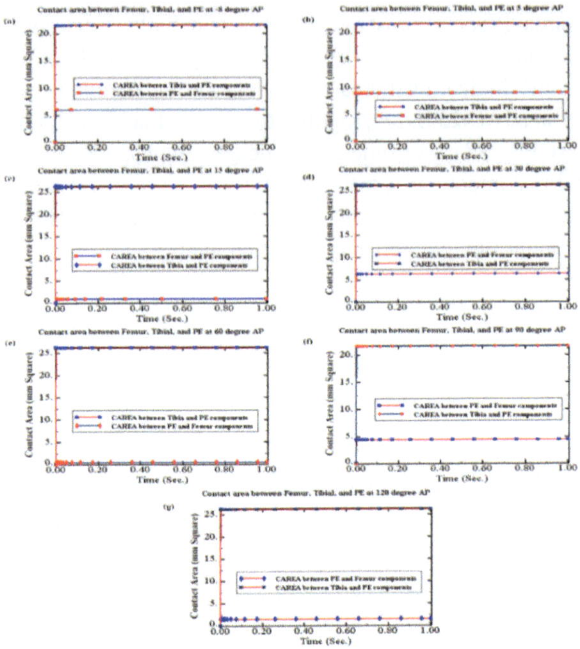

Fig. 7. Contact area variation between femur, polyethylene (PE), and tibial components at different anterior-posterior (AP) flexion angles: (a) −8°, (b) 5°, (c) 15°, (d) 30°, (e) 60°, (f) 90°, and (g) 120°.

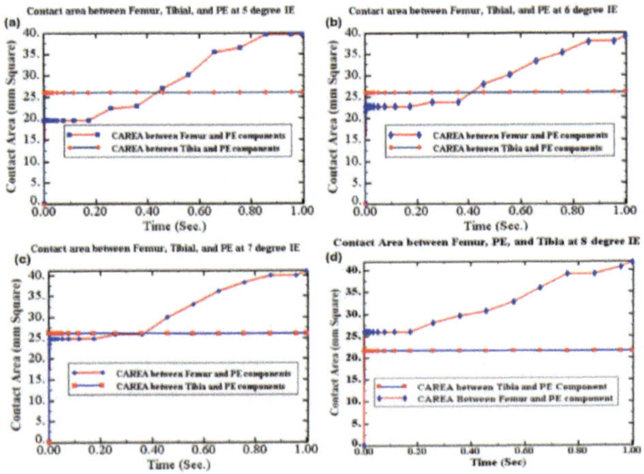

Fig. 8. Contact area variation between the femur, polyethylene (PE), and tibial components under internal-external (IE) rotation angles: (a) 5°, (b) 6°, (c) 7°, and (d) 8°.

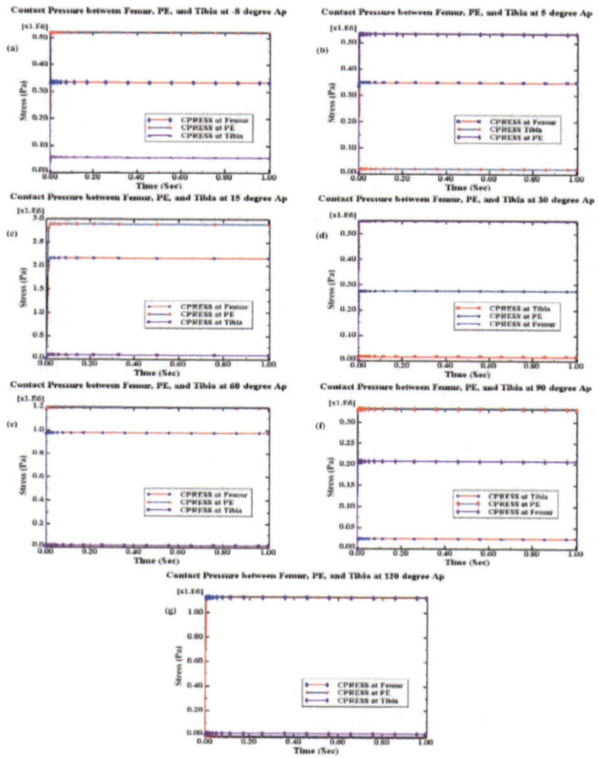

Fig. 9. Contact pressure (CPRESS) variation between the femur, polyethylene (PE), and tibial components at various anterior-posterior (AP) flexion angles: (a) −8°, (b) 5°, (c) 15°, (d) 30°, (e) 60°, (f) 90°, and (g) 120°.

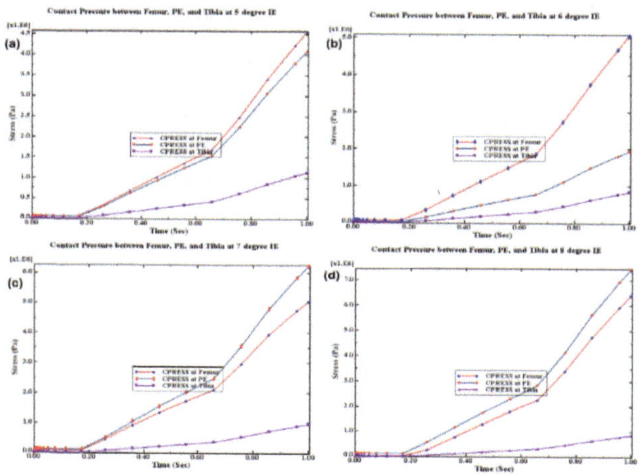

Fig. 10. Contact pressure (CPRESS) variation between the femur, polyethylene (PE), and tibial components under internal-external (IE) rotation angles: (a) 5°, (b) 6°, (c) 7°, and (d) 8°.

4 Conclusion

This study highlights that high flexion (60°–120°) and rotation (7°–8°) angles significantly raise contact pressure, especially on the PE insert, increasing wear risk. Ti-6Al-4V reduces stress concentrations, while Co-Cr-Mo and stainless steel elevate localized stress. Optimized material combinations improve implant longevity. Future research should target adaptive prosthetics and long-term performance in high-impact activities.

References

1. Piazza, S.J., Cavanagh, P.R.: Measurement of the screw-home motion of the knee is sensitive to errors in axis alignment. J. Biomech. **33** (2000). https://doi.org/10.1016/S0021-9290(00)00056-7
2. Taylor, M., Prendergast, P.J.: Four decades of finite element analysis of orthopaedic devices: where are we now and what are the opportunities? J. Biomech. **48**(5), 767–778 (2015). https://doi.org/10.1016/j.jbiomech.2014.12.019
3. Baldwin, M.A., Laz, P.J., Stowe, J.Q., Rullkoetter, P.J.: Efficient probabilistic representation of tibiofemoral soft tissue constraint. Comput. Methods Biomech. Biomed. Engin. **12**(6), 651–659 (2009). https://doi.org/10.1080/10255840902822550
4. Hallab, N.J., Jacobs, J.J.: Orthopedic Applications. https://doi.org/10.1016/B978-0-12-816137-1.00070-2
5. Li, G., DeFrate, L.E., Rubash, H.E., Gill, T.J.: In vivo kinematics of the ACL during weight-bearing knee flexion. J. Orthop. Res. **23**(2), 340–344 (2005). https://doi.org/10.1016/j.orthres.2004.08.006
6. P., T., Scott, S.A.T., Galante Jorge, O.: Knee biomechanics and total knee replacement. J. Arthroplast. **1**(3), 211–219 (1986). https://doi.org/10.1016/s0883-5403(86)80033-x
7. Iseki Taisuke, F.T.: The biomechanics of the knee joint with special reference to the contact area. Keio J. Med. **25**(1), 37–44 (1976). https://doi.org/10.2302/kjm.25.37
8. Pandy, M.G., Shelburne, K.B.: Dependence of cruciate-ligament loading on muscle forces and external load (1997). https://doi.org/10.1016/S0021-9290(97)00070-5
9. Meenakshi, C.V.R., Shivendra Kumar, K., Ramana, S.V.: Tribological aspects on human knee joint – a review. Mater Today Proc. **22**, 3100–3105 (2020). https://doi.org/10.1016/j.matpr.2020.03.446
10. Carr B.C., Goswami, T.: Knee implants - review of models and biomechanics. Mater. Des. **30**(2), 398–413 (2009).

Study on Aging and Life Prediction of One Type Passenger Oxygen Mask

Huo Liqin[1(✉)] and Dong Zhongbao[2]

[1] Shanghai Aircraft Design and Research Institute, Shanghai 201210, China
huoliqin@comac.cc

[2] Handan Purification Equipment Research Institute, Handan 056027, China

Abstract. One type passenger oxygen mask used in civil aircraft is made of polymer materials such as rubber and plastic. This paper has an analysis of these materials aging. These materials will age and then fail influenced by external factors such as heat, light, mechanical stress and ozone. After the analysis of the mask installation environment, it is determined that the thermal environment is the main factor that causes mask age accelerating. Accelerated thermal aging and the Arrhenius Equation is a common mean to study the life of rubber and plastics. This paper brings up an efficient method to predict mask life. It is to use accelerated heat aging test to predict life and evaluate the degree of deterioration of mask function. The results showed that at normal storage and use conditions (21 °C), the life of the mask is 8.38 years, which is similar to the congeneric products.

Keywords: passenger oxygen mask · polymer materials · accelerated thermal aging · life prediction

1 Introduction

Passenger oxygen masks used in civil aircraft are oronasal type masks which are mainly made of polymer materials such as rubber and plastic. Passenger oxygen masks are generally used as oxygen distribution devices for passenger oxygen supply units and portable oxygen cylinder assemblies which are installed in the passenger cabin. As one kind of onboard equipment, passenger oxygen masks need to withstand harsh natural and mechanical environment to ensure that their performance will not degrade [1]. Rubber and plastic will be influenced by heat, ozone, ultraviolet and other environmental factors and gradually aging. This will result in material performance decline. As the material aging degree aggravating, it will eventually make the equipment failure. One type of passenger oxygen mask used in civil aircraft uses silicone rubber as mouth cup and breathing valves material, polycarbonate as cup panel, plunger and other supporting parts material, polyvinyl chloride as oxygen hose and air storage bag material, rubber mixed polyester as headband material. The typical structure of the mask is shown in Fig. 1. Due to the lack of service experience in civil aircraft of these materials, whether the life of these materials can meet the technical requirements of the product can't be verified by analysis.

Y. Zhao (Ed.): ICAMM 2025, SPM 96, pp. 229–236, 2025.
https://doi.org/10.1007/978-3-032-09862-7_22

With the development of industrial technology, polymer materials are more wildly used in the field of civil aircraft. Civil aviation examination is a very strict process, and only the materials that have passed airworthiness verification can be used in civil aircraft. The passenger oxygen mask in this paper uses a variety of polymer materials, and the verification of each type material will undoubtedly consume a lot of manpower and resources. This paper brings up a fast and efficient life evaluation method to show airworthiness conformity. This paper has an analysis of the aging reasons of the mask materials and has a consideration of actual environmental conditions of the mask, then selects thermo-oxygen aging factor as the key factor which mainly cause the aging and performance decline of the mask. Accelerated thermal aging test is carried out and the life of the mask is predicted.

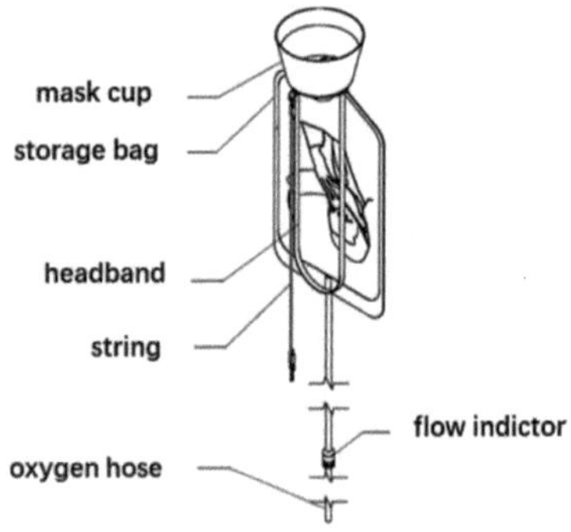

Fig. 1. Passenger Oxygen Mask

2 Analysis of Material Aging Reasons

2.1 Silicone Rubber

Silicone rubber has the characteristics of good resilience, high plasticity, high and low temperature resistance, and the storage life is more than 20 years at room temperature. The mask uses silicone rubber as the material of the mouth cup and the breathing valves. The cup serves to seal the user's nose, mouth and maintain the shape of the mask. Suction and exhalation valves act as intake and exhaust air. As the main material of the mask, silicone rubber undertakes the important functions of the mask, like inhalation, exhaust and be fixed on the user's face. The aging performance of silicone rubber should be regard as the most important indicator for predicting the life of the mask.

Silicone rubber aging will be affected by heat, light, mechanical stress, ozone, contact media and other external factors [2], and it will show irreversible physical or chemical changes like softening, hardening or cracking. Its performance will deteriorate until lose its use value. The passenger oxygen mask used in civil aircraft is stored in the mask box or mask package, there is no light, mechanical stress either pollution medium conditions, but will experience high and low temperature alternating thermal environment, as well as the impact of ozone. The thermal environment will accelerate the aging of silicone rubber. Due to the extremely low concentration of ozone in the atmosphere, ozone in the cabin will not cause additional aging effects of the silicone components of the mask. Therefore, the aging of silicone rubber parts of the mask is mainly caused by thermal heating, thermal aging is also the most important factor [3, 4] causing the aging of silicone rubber.

2.2 Polycarbonate

Polycarbonate has the characteristics of impact resistance, flame retardant and good mechanical properties, and its storage life is more than 10 years [5] under normal temperature conditions. The structural parts of the mask use polycarbonate, which mainly plays a supporting role. The aging of polycarbonate mainly shows color discoloration and brittle cracking. Like the mouth cup of the mask, considering the actual environment where masks are stored, thermal aging is also the main reason for the aging of the structural parts of the mask.

2.3 PVC

PVC has the advantages of adjustable hardness, easy coloring and good flame retardancy, and is usually used as a material for hard pipes and films. The storage life of PVC is more than 10 years [6] under normal temperature conditions. PVC is used as storage bag and oxygen hose material, the main role is to store oxygen from the oxygen source and transport oxygen to the mask cup. The aging of PVC mainly shows color discoloration and brittle cracking [7]. Like the mask cup, thermal aging is also the main factor of aging of the air storage bag and oxygen hose.

2.4 Rubber and Polyester

Rubber is woven with polyester silk to make a headband which is used to fix the mask to the user's face. Under normal temperature conditions, the storage life of ordinary rubber is 3–5 years, and the storage life of polyester braided materials is more than 10 years. Polyester is one of the most durable materials in textile fibers. Rubber aging behavior is similar to silicone rubber and will show softening, hardening or cracking phenomena until loses its use value. Polyester will fade and harden as it ages. Like the mask cup, thermal aging is the major factor of headband aging.

2.5 Brief Summary

Passenger oxygen mask is stored in a mask box or package and will not be affected by light, mechanical stress, either pollution medium, but will experience harsh thermal environments and ozone in the atmosphere. Due to the extremely low concentration of ozone in the atmosphere, ozone in the cabin does not cause additional aging effects on the polymer material components of the mask. When the aircraft is passing airport station or parking at airport station, the cabin environment is not controlled, and the cabin temperature is affected by environment. The cabin temperature may range from −55 °C to 85 °C affected by extreme natural environment [1]. When the aircraft is in operation, the cabin environment is controlled, and the temperature is kept within the human comfort range of 17 °C ~ 27 °C. Based on the above scenarios, it can be concluded that the mask is mainly subjected to high and low temperature changes when it is stored in the cabin. Therefore, the alternating thermal environment mainly causes the aging of the materials of each part of the mask.

Accelerated thermal aging and Arrhenius equation are a common means [8–10] to study the life of rubber and plastic. In this paper, the life of the type passenger oxygen mask is predicted by thermal aging test with the assessment of the degree of functional performance decline or deterioration of the mask, and the mask life is preliminarily obtained.

3 Accelerated Thermal Aging Test

3.1 Part to Be Tested

20 passenger oxygen masks. The masks have passed the product acceptance test and the performance meets the technical requirements.

3.2 Test Equipment

The main test equipment is 3 ovens, mask performance test bench and digital display tension meter.

3.3 Accelerated Thermal Aging Test

Oven accelerated thermal aging is usually considered to be the closest research method to the actual natural aging results. According to GB/T 20028-2005, the service life of the mask can be calculated by applying Arrhenius equation [8]. The aging of the mask is accelerated by exposing the mask to high temperature. Considering that the maximum storage temperature of the mask is 85 °C, the three test temperatures are determined according to the temperature adjustment downward, which are 65 °C, 75 °C and 85 °C respectively. The appearance, strength, respiratory resistance and oxygen supply performance of the mask are tested after temperature exposure. The test steps are as follows:

1. Set the temperature of 3 ovens to the specified temperature;
2. Put 10 masks in the 65 °C oven, 6 masks in the 75 °C oven and 4 masks in the 85 °C oven;
3. Take out a mask from the 85 °C drying oven every 48 h and place it at room temperature for 2 h, and carry out the following checks until non-conformance occurs. After test, the test mask which passes test will be put back to the oven.
   (a) Visual inspection: the mask components should not be damaged or degraded, the elastic parts have no obvious elastic changes, the oxygen hose and the oxygen storage bag can be freely extended, no cracking, no cracks, no adhesive connection, and the mask can be fixed and maintained on user's face.

(b) Strength test: apply the specified tension force to the mask along the oxygen hose axis and maintain the specified time, and then visually check the parts and interfaces, the mask should be without any damage. The tension force is defined 89 N and the implementation time is defined 3.5 s [11];
(c) Respiratory resistance: Use the respiratory resistance test bench to test the mask respiratory resistance should not exceed the specified value when providing the specified respiratory flow rate. When the respiratory flow rate is 30 L/min, the respiratory resistance should not exceed 0.2 kPa; when the respiratory flow rate is 70 L/min, the respiratory resistance should not exceed 0.25 kPa.
(d) Oxygen supply performance test: the respiratory resistance of the test mask meets the requirements of the product, and the oxygen concentration obtained under the specified oxygen supply flow meets the requirements of the product.

4. Take out a mask from the 75 °C drying box every 96 h and place it at room temperature for 2 h, and repeat the test steps in step c;
5. Take out a mask from the 65 °C drying oven every 168 h and place it at room temperature for 2 h, repeat the detection steps in step c;
6. Repeat steps 3) to 5) until the test is stopped in all oven chambers.

The mask in each oven is taken according to sequence of serial number from small to large. After the completion of one cycle testing, mask will be taken according to the same sequence. The limit exposure time of the mask under the aging conditions of 65 °C, 75 °C, 85 °C is the accumulated exposure time of the last test before the test which the unqualified items appear.

3.4 Test Results

All the masks completed the aging test. Test result is shown below:

1. 85 °C thermal accelerating test

After masks exposure at 85 °C for 288 h, the appearance inspection, strength, respiratory resistance and oxygen supply performance of test masks met the requirement of test. After 366 h thermal aging acceleration, the air stowage bag slightly adhere and the appearance inspection didn't meet the requirement, but strength, respiratory resistance and oxygen supply performance still met the requirement, the result of this time is judged to be unqualified. Therefore, it is determined that the limit exposure time of mask under aging condition of 85 °C is 288 h.

2. 75 °C thermal accelerating test

After masks exposure at 75 °C for 672 h, the appearance inspection, strength, respiratory resistance and oxygen supply performance of test masks met the requirement of test. After 768 h thermal aging acceleration, the oxygen hose is slightly yellow and the appearance inspection didn't meet the requirement, but strength, respiratory resistance and oxygen supply performance still met the requirement, the result of this time is judged to be unqualified. Therefore, it is determined that the limit exposure time of mask under aging condition of 75 °C is 672 h.

3. 65 °C thermal accelerating test

After masks exposure at 65 °C for 1344 h, the appearance inspection, strength, respiratory resistance and oxygen supply performance of test masks met the requirement of test. After 1512 h thermal aging acceleration, the oxygen hose is slightly yellow and the appearance inspection didn't meet the requirement, but strength, respiratory resistance and oxygen supply performance still met the requirement, the result of this time is judged to be unqualified. Therefore, it is determined that the limit exposure time of mask under aging condition of 65 °C is 1344 h.

In conclusion, the mask limit exposure time at 65 °C, 75 °C, 85 °C respectively is 1344 h, 672 h and 288 h.

4 Results and Discussion

4.1 Rationale Analysis

The Arrhenius Equation is used to expresses the temperature and speed of chemical reaction. When the temperature rises, the rate of chemical reaction generally increases. For some organic chemical reactions, increasing the temperature by 10 °C, the rate of reaction will increase two or three times. The relationship between temperature and chemical reactions can be expressed using Arrhenius Equation shown in Eq. (1). When temperature is the main environment factor, the Arrhenius Equation can be used to predict the failure performance of the product.

$$K(T) = A \cdot e^{-E/RT} \tag{1}$$

K(T)—reaction rate constant, /h.
T—temperature, K.
A—exponential factor, /h.
E—activation energy, J/mol.
R—molar gas constant, 8.314 J/mol/K.

The chemical reaction relation function is expressed by Eq. (2). In the mask aging test, *F(t)* could be defined as the performance of mask appearance, strength, respiratory resistance and oxygen supply performance. If one item is unqualified, it means that the mask performance reaches its critical condition, and the exposure time is the limit exposure time.

$$F(t) = K(T) \cdot t \tag{2}$$

F(t) — reaction relation function.
t — aging time, h.

Life prediction method: To detect the performance of the mask after aging and it is considered that the performance of the mask fails when the exposure time reaches the limit time *ti* at temperature *Ti*. Equation (1) is substituted into Eq. (2) and expressed by chemical reaction relation function as follows:

$$F(ti) = A \cdot e - E/RT \cdot ti \tag{3}$$

Shift the phase, take the logarithm of both sides, merge the constant terms to become B, and get Eq. (4):

$$\log(ti) = E/RTi + B = A/Ti + B \quad (4)$$

Equation (4) shows that the logarithm of aging time log(ti) is linearly related to the reciprocal of temperature T. The slope A is a constant determined by the activation energy E. The storage life corresponding to the desired temperature is obtained by extrapolating from the Eq. (4) which can also be called the aging life equation.

4.2 Mask Life Prediction

Under the aging conditions of 65 °C, 75 °C and 85 °C, the limit exposure time of the mask is shown in Table 1.

Table 1. Limit Exposure Time of Mask

Temperature (°C)	65	75	85
Limit exposure time (h)	1344	672	288

Draw the Arrhenius diagram according to the test data recorded in Table 1. Plot and linearly fit the *logt* of every limit exposure time with reciprocal of temperature 1/T of the corresponding test to obtain the fitting liner equation (Fig. 2):

$$logt = 3920/T - 8.467 \quad (5)$$

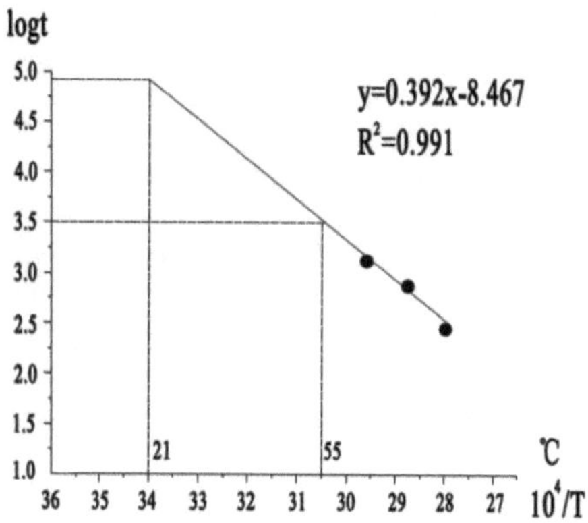

Fig. 2. Mask Life versus Temperature

According to Eq. (5), the life of the oxygen mask is 8.38 years at normal storage and use conditions (21 °C).

5 Conclusion

This paper has an analysis of the mask installation environment and materials aging reasons, then determines the thermal environment is the main environment factor which causes mask age accelerating. Based on the Arrhenius Equation, the prediction equation of mask life between temperature and exposure time is established. Using the equation, it is calculated that at normal storage and use conditions (21 °C), the life of oxygen mask is 8.38 years, which is similar to the congeneric products. The mask installation environment is actually a temperature change environment, although most of the time is normal temperature condition, but there also will be a short time to experience extreme environment temperature. For this complex working condition, it is necessary to evaluate the mask life iteratively by collecting the feedback data of the route. The mask prediction method proposed in this paper improved the efficiency of product life verification and passed airworthiness certification. It is also a reference for polymer materials to be verified and applied in the field of civil aircraft.

References

1. RTCA: Environmental Conditions and Test Procedures for Airborne Equipment: RTCA DO-160G. U.S. (2010)
2. Sijing, W., Jinping, X., Zuo, Y.: Progress of aging mechanism and research methods of rubber. Aging Appl. Synth. Mater. **38**(2), 23–33 (2009)
3. Yan, X., Wei Borong, D., Maoping.: Accelerated aging test and storage life estimation method of rubber. Aging Appli. Synth. Mater. **36**(1), 40–43 (2007)
4. Xuan, X.: Prediction of storage life of silicone rubber sealing materials. Aerospace Mater. Technol. **2**, 83–87 (1996)
5. Xueliang, W., Faming, Z., Wei, L., et al.: Study on service life prediction of polycarbonate based on thermogravimetric analysis. Smart Rail Transit. **60**(3), 55–58 (2023)
6. Qiongying, C., Rongfu, J., Ming, P.: Study on thermal degradation of polyvinyl chloride. Polyvinyl Chloride. **6**, 12–14 (1988)
7. Wenjun, P., Jun, G., Sixin, L., et al.: Study on thermal aging properties of rigid polyvinyl chloride. J. Southwest Univ. Naty. Nat. Sci. Ed. **35**(3), 559–561 (2009)
8. Standardization Administration of China. Estimation of life and maximum service temperature of vulcanized rubber or thermoplastic rubber using Arrhenius Diagram: GB/T 20028–2005, China (2005)
9. Huaqing, G., Xiuyin, Q., Fuwei, J., et al.: Effect of accelerated aging at high temperature and high humidity on structure and properties of polycarbonate. Plastics Ind. **45**(1), 22–31 (2017)
10. Weibin, G., Libo, M., Rixin, L.: Prediction of storage life of polycarbonate by extrapolation of Arrhenius equation. Plastics. **39**(1), 120–121 (2010)
11. SAE International: Passenger Oxygen Mask: AS8025-1999[S]. SAE International, US (1999)

A Systematic Review and Meta-analysis of Vibratory Stress Relief (VSR) Effectiveness of Steel Components

L. N. Frandsen(✉) and J. H. Høyer

Department of Business Development and Technology, Aarhus University, Birk Centerpark 15, 7400 Herning, Denmark
201906642@post.au.dk

Abstract. Vibratory stress relief (VSR) is a long-established method for reducing residual stress (RS) in metal components, yct its industrial adoption remains limited due to a lack of comprehensive process understanding. This study presents the first meta-analysis on VSR effectiveness in steel specimens, systematically reviewing 25 peer-reviewed studies and analyzing data from 30 specimens. The results indicate an average RS reduction of 42%, though with significant variability among samples. A correlation analysis of eight parameters: frequency, volume, stress source, carbon content, the sum of alloying elements, RS before VSR, RS after VSR, and yield strength revealed that only stress source exhibited a statistically significant correlation with RS reduction. The findings highlight the potential of VSR as a viable stress-relief technique while emphasizing the need for standardized procedures to optimize its effectiveness. This study provides valuable insights for both researchers and industry practitioners, offering a data-driven perspective on VSR's expected performance and influential parameters.

Keywords: Vibratory Stress relief (VSR) · Steel specimens · Systematic literature review · Meta-analysis

1 Introduction

Vibratory stress relief (VSR) is a process in which components are vibrated to relieve the residual stresses (RS) introduced during manufacturing. RS can be induced in numerous ways, such as plastic deformation from machining, cutting, and grinding, phase transformation leading to differences in microstructural properties, or steep temperature gradients seen in welding or other heat-inducing operations [1, 2]. It is important to relieve the RS as these can lead to dimensional changes, cracking, or reduced fatigue and strength properties [3, 4].

The effects of vibrations leading to stress relief have been known for almost a century but were first described in the literature as late as 1943 by McGoldrick and Saunders [5] when the phenomenon was discovered in shipbuilding. It was later noted that the development of the technology has been almost stagnant since 1980 [6]. Despite being an old technology, the underlying mechanisms of VSR are still not fully understood.

Y. Zhao (Ed.): ICAMM 2025, SPM 96, pp. 237–251, 2025.
https://doi.org/10.1007/978-3-032-09862-7_23

Despite the lack of knowledge about VSR, there is a growing consensus that its effects occur at multiple structural scales. During VSR, alternating stresses are superimposed onto existing residual welding stresses, leading to localized microplastic and, in some cases, macroplastic deformations. These deformations facilitate the redistribution and reduction of residual stresses within the material structure [7]. Early foundational work by Dawson and Moffat [8] demonstrated that cold-worked specimens could only be stress-relieved through plastic deformation, observed as a permanent strain change at the surface. Additionally, they identified a critical vibrational threshold, equivalent to the material's yield strength, that must be exceeded for VSR to have any effect. This supports the dominant theory that the applied vibratory stress combines with residual stresses to locally exceed the yield point, thereby activating dislocation motion and enabling strain redistribution [9].

The industrial application of the technology also appears to be limited, which may seem odd as VSR should be advantageous in certain scenarios compared to annealing, which is a more common and widely studied thermal stress relief method [10, 11]. Annealing is usually performed in large ovens, requiring a lot of time and energy and restricting the maximum size of structures that can be treated. On the other hand, VSR equipment is cheaper and more portable [11], which further suggests increased usage.

1.1 Aim

Currently, isolated experimental VSR studies exist, but no collection has been made to shed light upon the effectiveness of a VSR treatment. This study is, to the authors' knowledge, the first meta-analysis made within the field of VSR. The aim of this study is two-fold:

1. To systematically review VSR literature to shed light on the expected effectiveness of VSR treatment in steel specimens.
2. To identify any influential factors on the effectiveness of selected VSR and material parameters.

The aim will be explored through a systematic literature review and meta-analysis. This work will have relevance not only in the field of study but also in an industrial setting, as it should also be viewed as a practitioner's guide to VSR, acting as an indicator of the expected effectiveness of a VSR treatment for a given steel type and as a clarification of influential VSR parameters.

1.2 Scope

The study is limited to VSR treatment in steel and cast-iron components. The study will focus on the metric "reduction of RS" as an indicator of an effective treatment. Other VSR techniques exist, such as torsional VSR, where components are stress-relieved by cyclic torsional excitation, and ultrasonic VSR treatments where frequencies are in the range of 20 kHz. This study will solely focus on "standard" VSR, which is done in the frequency range of 0–200 Hz [11].

2 Methodology

The systematic review consisted of literature from three databases: Scopus, Web of Science, and ProQuest. The resulting papers were handled using EndNote 21. The search was limited to papers published from 1943 to 2023, as this is the first time VSR is described in the literature by McGoldrick and Saunders [5]. The papers were gathered on the 15th of December, 2023.

The search phrases were identified from keywords and papers in existing literature and can be seen in Table 1.

Table 1. Search phrase

VSR **OR** "Vibrat* stress reli*" **OR** "post-weld* vibrat*" **OR** "post-heat* vibrat*" **OR** Vibrat*	**AND**	Steel **OR** Iron	**AND**	"Residual stres*"

The search phrase was used to identify relevant papers through title, abstract, and keyword searches, with a primary focus on vibratory stress relief (VSR). VSR, also referred to as post-weld or post-heat vibration, was explored in relation to iron and steel-based materials, and residual stress (RS). Wildcards (e.g., *vibrat**) were used to capture the many phrasings of key terms, such as vibration, vibratory, and vibrational.

After retrieving the papers, references were managed using EndNote. Duplicates were removed automatically, followed by manual screening. One author reviewed all records by title, abstract, and full text using predefined inclusion criteria: peer-reviewed sources (journal articles or conference papers), experimental studies with primary data (excluding simulations and reviews), and quantitative results on stress reduction. Only studies using post-production VSR (excluding torsional or ultrasonic variants) on clearly identified steel or iron specimens were included. Papers also had to be in English.

Full texts were accessed via institutional databases, Google Scholar, ResearchGate, or purchased directly. Included studies were analysed and summarised in an evidence table.

2.1 Data Extraction

For the meta-analysis, data was extracted from the included papers. It was preferred to have the data in either in-text or tabular form. When this was not available, data was extracted by reading the graphs. To increase precision, the image analysis software ImageJ was used. The software allowed tables to be inserted. A scale was set based on the graph axis, and a measurement was taken.

2.2 Reporting Measures

RS reduction was selected as the measure of VSR effectiveness. When investigating the influential factors for RS reduction, two sets of variables were categorised, namely, the dependent and independent variables. The effect of eight independent variables was tested on the dependent variables. The two variable types can be seen in Table 2.

Table 2. Variables used for statistical analysis

Parameters	Description	Unit	Data type
Dependent variable RS reduction	The highest measurement of RS must be the most critical in terms of impact on the specimen. RS reduction was calculated: RS reduction $= \sigma_{before}^{peak} - \sigma_{before}^{after}$. The peak RS before minus peak RS after VSR was reported as RS reduction	MPa	Continuous
Independent variable Stress source	The stress sources had 2 levels: 1 = *Heat induced,* i.e. Welding, quenching, or casting 2 = *Strain induced,* i.e. Hot-rolled, cold-rolled, or cold-drawn	–	Categorical
Frequency	The frequency had 2 levels: 1 = *Resonant* 2 = N*on-resonant*	–	Categorical
Peak RS before	The maximum RS value recorded before VSR	MPa	Continuous
Peak RS after	The maximum RS value recorded after VSR	MPa	Continuous
Volume	The volume of the specimen	m^3	Continuous
Yield strength	The yield strength of the specimen material	MPa	Continuous
C content	The carbon content in the specimen	wt%	Continuous
Sum of AE	The sum of all alloying elements (AE) in the specimen	wt%	Continuous

In addition to these variables, other descriptive measures were also reported and presented in the evidence table but excluded from the statistical analysis. The descriptive measures are seen in Table 3.

Table 3. Additional descriptive measures in the evidence table

Parameters	Description
Specimen #	The specimen number is a unique identifier
Relative RS reduction	The relative RS reduction is calculated as: Relative RS reduction = $\frac{\sigma_{before}-\sigma_{after}}{\sigma_{before}} \cdot 100$

2.3 Statistics

The statistical analysis for the meta-analysis was conducted with JMP Pro 15 statistical software. A simple pairwise correlation analysis was done using the Pearson correlation coefficient. All categorical variables were converted into coded variables. The "multivariate" JMP function was used, where all parameters were inserted. From this, the Pearson correlation coefficients and the respective p-values were calculated. All correlations were tested at a significance level of 5%. The parameters included in the correlation analysis are seen in Table 2. As only the magnitude of the reduction in RS was important and not the direction, the absolute value of RS reduction was used in the correlation analysis.

3 Results

The systematic review resulted in 926 papers from the three databases. The duplicates were removed, which left 656 papers for a title and abstract screening. This was followed by a full-text screening, where a total of 104 were removed for not meeting the inclusion criteria. The remaining 25 papers were found to be eligible for inclusion. The systematic literature approach is visualised in Fig. 1.

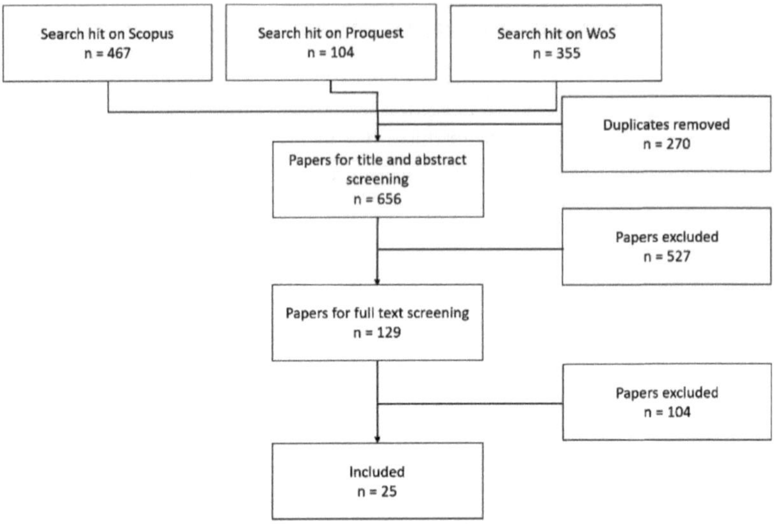

Fig. 1. Results from systematic review

3.1 Review Results

The result of the systematic review is presented in the evidence table sorted by specimen type, see Table 4.

Table 4. Evidence table

#	RS before [MPa]	RS after [MPa]	RS Red. [MPa]	Rel. RS Red. [%]	Freq. [−]	Spec. type [−]	YS [MPa]	Vol. [m^3]	SS [−]	C [wt%]	AE [wt%]
1 [12]	−197	−99	−98	49.7	R	BS 060A35	325	7E-2	SI	0.36	2.1
2 [13]	250	124	126	50.4	R	09G2S	440*	7E-3	HI	0.12	3.3*
3 [13]	248	126	122	49.2	R	20 GML	235*	7E-3	HI	–	–
4 [14]	216	58	158	73.1	R	A36	296	1E-1	HI	0.15	0.87
5 [15]	295	98	197	66.8	–	AISI 304 L	320	4E-5	HI	0.02	29
6 [16]	316	164	152	48.1	R	AISI 304 L	350	–	HI	0.03	31*

(*continued*)

Table 4. (*continued*)

#	RS before [MPa]	RS after [MPa]	RS Red. [MPa]	Rel. RS Red. [%]	Freq. [–]	Spec. type [–]	YS [MPa]	Vol. [m^3]	SS [–]	C [wt%]	AE [wt%]
7 [17]	526	475	51.3	9.7	NR	AISI 316	205*	2E-5	HI	0.46	37
8 [18]	409	392	17	4.1	R	ASTM A131 EH36	383	5E-3	HI	0.18	3.1*
9 [19]	585	247	338	57.8	R	BS 970 080A15	607	1E-4	HI	0.18	1.3
10 [20]	370	251	119	32.2	–	BS 970 080A15	607	1E-4	HI	0.16	1.3*
11 [21]	932	890	42	4.5	R	Cr-Ni-Mo steel	400*	1E-4	SI	0.25	6.3
11 [22]	99	74	25	25.1	R	D406A	825*	9E-4	HI	0.32	4.7
13 [22]	−99	−49	−50	50.3	R	D6AC	1720*	9E-4	HI	0.47	4.3
14 [23]	416	250	166	39.9	R	DH36	355*	6E-3	HI	0.12	2.0
15 [24]	356	221	135	37.9	R	DH36	355	0.9	HI	0.12	2.1
16 [25]	293	228	65	22.2	R	DH36	355*	2E-3	HI	0.18	2.1
17 [26]	−266	−187	−79	29.7	R	DOMEX 700 MC	700	2E-3	HI	0.079	2.6
18 [27]	287	237	50	17.4	NR	DP600	390	3E-5	HI	0.049	2.4
19 [28]	108	51	57	52.8	R	E26–4	420*	1.3	HI	0.18	2.3
20 [4]	453	−124	577	127.4	R	EN3b	250	–	SI	0.25	1.7
21 [29]	395	307	88	22.2	NR	IS2062 E250BR	290	5E-4	HI	0.16	1.3
22 [8]	265	16	249	94.0	R	Mild steel	607	–	SI	0.11	1.1
23 [8]	116	22	94	81.0	R	BS 4360 Grade 43A	290	–	SI	0.25	2.5*

(*continued*)

Table 4. (*continued*)

#	RS before [MPa]	RS after [MPa]	RS Red. [MPa]	Rel. RS Red. [%]	Freq. [–]	Spec. type [–]	YS [MPa]	Vol. [m^3]	SS [–]	C [wt%]	AE [wt%]
24 [30]	264	209	55	20.7	R	Q235	235	3E-3	HI	0.2	2.0
25 [31]	−189	−161	−28	14.8	R	Q235B	235*	–	HI	0.2	2.0*
26 [30]	260	270	−10	−3.8	R	Q345	325	3E-3	HI	0.18	2.1
27 [16]	311	168	143	46.0	R	Q345	325	1.6	HI	0.23	2.1
28 [32]	−197	−110	−87	44.2	R	QT400–15	400*	–	HI	–	–
29 [33]	250	117	133	53.2	R	S355J2	461	5E-4	HI	0.16	2.2
30 [34]	343	210	133	38.8	NR	SS41	235*	2E-3	HI	–	–

All alloy compositions and yield strength values noted with * were not identified through the papers but through other sources like datasheets from steel manufacturers. The abbreviations used in the table are: Red. = reduction, Freq. = frequency, R = resonant, NR = Non-resonant, Spec. = Specimen, YS = Yield strength, Vol. = volume, SS = Stress source, SI = Strain induced, HI = Heat induced, C = Carbon content, AE = Sum of all alloying elements contents.

Descriptive Statistics

Descriptive statistics of the data are presented in Table 5.

Table 5. Descriptive statistics

	Count	Mean	Max	Min	Std.
RS reduction	30	112 MPa	338 MPa	−10 MPa	83 MPa
Relative RS reduction	30	42%	127%	−4%	28%

In total, 30 specimens were identified from the included papers. The mean reduction was found to be 112 MPa or 42%. The maximum reduction exceeds 100% because the VSR operations turned positive RS to negative RS. The minimum reduction is negative, meaning that VSR induced a higher level of RS compared to before VSR. The standard deviation of 83 MPa or 28% indicates that the RS reduction of different specimens varies greatly.

3.2 Correlation Analysis

The results of the correlation analysis can be seen in Table 6. When examining the significance of the correlations, only the correlation between RS reduction and stress source is significant, while the remaining correlations are not.

Table 6. Result from the correlation analysis

Response	Predictor	Correlation coefficient	P-value
RS reduction	Stress source	0.37	0.045
RS reduction	RS before	0.23	0.21
RS reduction	RS after	−0.23	0.22
RS reduction	C content	−0.15	0.45
RS reduction	Frequency	−0.13	0.49
RS reduction	Yield strength	−0.11	0.57
RS reduction	Volume	0.049	0.82
RS reduction	Sum of AE	−0.031	0.88

The significant relationship between RS reduction and stress source is visualized as seen in Fig. 2 where the plot is overlayed by a line fit.

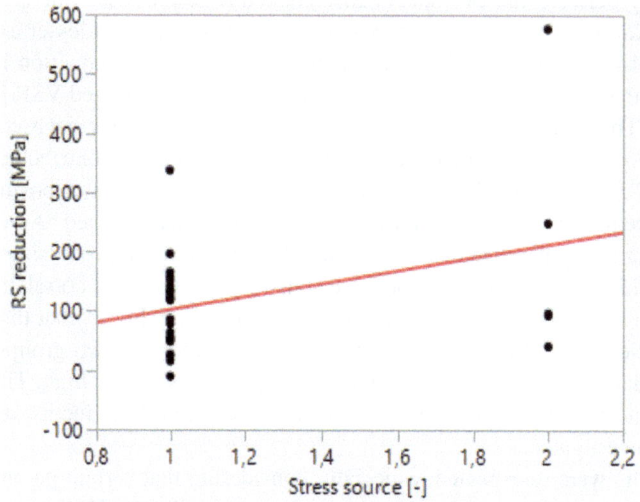

Fig. 2. Plot of RS reduction by stress source 1: Heat, 2: Strain

From the line fit, the positive correlation is shown. The distribution of samples with RS induced by either heat or strain is uneven. In total, 25 samples were subjected to heat stress, while only five samples were subjected to strain stress. This is also reflected in the standard deviation, which demonstrates that strain-stressed specimens exhibit a standard deviation approximately three times that of heat-stressed specimens. The mean RS reduction is higher for strain-induced specimens compared to heat-induced ones. The results are summed up in Table 7.

Table 7. Sample size and mean RS reduction by stress source category

	Heat induced	Strain induced
Count [−]	25	5
Mean RS reduction [MPa]	103	212
Standard deviation [MPa]	73	218

4 Discussion

From the 30 specimens analysed, the mean stress relief achieved from VSR was 42% across a wide selection of steel types and qualities. All but one VSR operation was successful, namely specimen #26. The standard deviation for the data is large at 28%, which suggests that a satisfactory effect is not guaranteed as the results vary. The large standard deviation could indicate that an appropriate procedure is needed to obtain good results from VSR. This tendency is seen in the RS reduction of specimens #5 and #6, both done in similar highly alloyed stainless steels, AISI 304 L. Here, the achieved reduction varies greatly between the two specimens. The same result is seen in specimens #24 and #25 made in Q235, as well as #26 and #27 in Q345, and to a lesser extent, DH36 specimens #14, #15, and #16. This result suggests that the RS reduction is not solely attributed to the specimen type. However, establishing a standardized VSR procedure is challenging. The data presented in Table 4 were derived from studies that used varied, and in some cases poorly described, processing methods, which likely contributed to the wide variation in RS reduction. Key experimental details such as vibrator position, applied force, and frequency were often missing or inconsistently reported. As a result, the findings cannot be generalized based on unified or standardized processing conditions.

Despite this, the study found a positive statistically significant correlation between RS reduction and stress source. It is evident from the graph, Fig. 2, that the result is of limited practical use. The sample size distribution between the two groups is uneven, and the standard deviation of the strain-stressed specimens is very high. Therefore, the study found no meaningful significant correlations between RS reduction and the eight parameters investigated.

The results were unexpected, especially considering that certain parameters, such as frequency, are well-documented in the literature to influence VSR outcomes. Yang [35] found that resonance frequency vibration most effectively relieved RS, likely due to amplified vibration amplitudes. Similar findings were reported by Ebrahimi et al. [19], Wang [36], and Chen et al. [37], who noted the need for higher amplitudes when using non-resonant frequencies. by Ingram et al. [38] suggested that specimen geometry may also affect RS reduction, and Kwofie [9] indicated through simulations that materials with high yield strength respond poorly to VSR.

The selection of parameters in this study was partly informed by Walker's theory [6], which proposed that the interaction between geometry, stress source, and alloy composition governs VSR effectiveness. However, the lack of clear correlations in the data does not support this hypothesis. Minding that geometry was measured pragmatically using volume, which may have limited the ability to detect more nuanced effects.

Other potentially relevant parameters were excluded from the study. For instance, stress amplitude has been shown to influence VSR outcomes. Manh et al. [14] observed increased RS reduction with rising stress amplitude up to a threshold, beyond which further increase led to deformation and higher residual stress. Walker et al. [4] and Munsi et al. [20] found that stress amplitude must reach the material's yield strength to be effective.

The effect of vibration time has also been studied. Dawson and Moffat [8] reported that increasing vibration cycles improved RS reduction only when above the stress amplitude threshold. Beyond a certain number of cycles, the effect diminished asymptotically. Similar observations were made by Walker et al. [4], Hassan [3], Munsi et al. [20] and Yang [35]. Despite their relevance, these parameters were often unreported in the reviewed literature and thus excluded from this study.

As previously noted, this study found the mean RS reduction to be 42% across all included papers. While the reduction is positive, it can be discussed if this reduction is sufficient in a practical setting. Whether the 42% RS reduction is sufficient must be dependent on the use case if VSR is used as either an intermediate or final RS relief treatment. With VSR used as an intermediate treatment between production steps, a 42% RS reduction might not be sufficient. In this case, RS reduction is used to decrease warpage in subsequent machining operations, where high RS reduction leads to less warpage. In other use cases, VSR could be used as a post-production treatment to increase ductility, thereby reducing the risk of premature failure or cracking. In this case, a reduction of 42% might be sufficient. VSR has been used on a larger scale as a post-production treatment for maglev rails in China. Here, the RS was reduced by approximately 30%, which was sufficient for that particular use case [39].

As VSR has a proven positive effect on RS reduction, the question of why VSR is not more extensively used in the industry is interesting. Here, the answer might be three-fold:

1. The expected effectiveness is hard to estimate, unlike in annealing, where equations like the Zener-Wert-Avrami function have been used to model RS relaxation [40]. Additionally, VSR seems inferior when compared to annealing. Here, other researchers commonly find close to 100% stress relief [41–46] from annealing treatments.
2. Many researchers have investigated the underlying mechanism of VSR without coming to a definitive conclusion. Therefore, the lack of knowledge about the physical changes in the materials may discourage some potential practitioners. For example, some authors have speculated on the effect of VSR on fatigue life, with conflicting results. It is known that RS has a negative effect on fatigue life [47]. Therefore, some authors find a good effect on fatigue life by reducing RS with VSR [48, 49], while others have found a vibrational limit where fatigue life is negatively affected [50].

As no clear conclusion is drawn, there might be some degrees of confusion regarding the effects of VSR on material properties.
3. As presented in this study, a successful treatment is by no means guaranteed and requires know-how, which is uncommon both in academia and the industry and acts as a self-reinforcing effect when the usage of VSR is scarce.

The results underline the fact that VSR is still not well understood, characterized as a highly complex topic with many studies reaching non-definitive conclusions. The results indicate that more focused studies are needed to understand the theory behind VSR and to accurately predict the influential variables and the effects of these.

4.1 Limitations

This study has several limitations. Not all included papers aimed to optimise VSR for maximum residual stress reduction, meaning some specimens may not have reached their full potential. Consequently, the actual average RS reduction could be higher than reported. Additionally, not all studies fully stressed the specimens, meaning that higher reductions could have been achieved.

The choice of the RS reduction reporting measure only identified the maximum stress of a single point in the specimens before and after VSR treatment. While informative, this measure does not reflect the overall stress distribution. However, given the self-equilibrating nature of RS [51], some redistribution over time or with additional vibration cycles can be expected.

There was also an uneven distribution of studies across parameter types, many used heat-induced specimens and resonant frequency vibrations. This imbalance introduces statistical uncertainty in the results. Moreover, only linear and pairwise correlations were analysed, limiting the ability to detect non-linear effects or parameter interactions.

Lastly, it is worth noting that a meta-analysis is highly reliant on the quality of the studies included. Most studies reported positive effects of VSR, with only one reporting no effect, suggesting possible publication bias [52]. Additionally, methodological differences between studies such as specimen geometry, vibrator or accelerometer placement, and other unreported variables make direct comparisons difficult. As such, the findings should be interpreted as indicative rather than definitive.

5 Conclusion

It can be concluded that a relative mean RS reduction, calculated from 30 steel specimens, is 42%. Furthermore, the results indicate a significant correlation between RS reduction and stress source. This correlation is likely of no practical use due to imbalanced sample sizes. The correlation between RS reduction and the remaining seven parameters was not statistically significant. Lastly, a great variation in achieved stress reduction was observed in the studies, which could indicate that other parameters could affect the reduction of residual stresses.

6 Future Work

For future work, different material parameters, such as stress amplitude or time, could be tested against RS reduction, possibly yielding other significant correlations. Furthermore, standardised methods of testing stress-induced steel specimens should be developed. This will increase similarities between VSR studies, thereby enhancing the validity of the comparisons.

References

1. Withers, P.J., Bhadeshia, H.K.D.H.: Residual stress. Part 1 – measurement techniques. Mater. Sci. Technol. **17**(4), 355–365 (2001)
2. Callister, W.D., Rethwisch, D.G.: Materials Science and Engineering, vol. 9, SI version edn p. 905. Wiley (in eng), Hoboken (2015)
3. Hassan, A.H.: Fundamentals of vibratory stress relief. Asian J. Appl. Sci. **7**(5), 317–324 (2014)
4. Walker, C.A., et al.: Vibratory stress relief—an investigation of the underlying processes. Proc. Inst. Mech. Eng., Part E: J. Mater. Des. Appl. **209**(1), 51–58 (1995)
5. McGoldrick, R.T., Saunders, H.E.: Experiments in stress relieving castings and welded structures by vibration. J. Am. Soc. Nav. Eng. **25**(4), 589s–609s (1943)
6. Walker, C.: A theoretical review of the operation of vibratory stress relief with particular reference to the stabilization of large-scale fabrications. Proc. Inst. Mech. Eng., Part L: J. Mater. Des. Appl. **225**(L3), 195–204 (2011)
7. Marushchak, P., et al.: Effect of shock and vibration preloading on the deformation and fracture behavior of 17G1S-U steel. Strength Mater. **51**, 418–426 (2019)
8. Dawson, R., Moffat, D.G.: Vibratory stress relief: a fundamental study of its effectiveness. J. Eng. Mater. Technol. Trans. ASME. **102**(2), 169–176 (1980)
9. Kwofie, S.: Plasticity model for simulation, description and evaluation of vibratory stress relief. Mater. Sci. Eng. **516**(1–2), 154–161 (2009)
10. Lv, T., Zhang, Y.: A combined method of thermal and vibratory stress relief. J. Vibroengineering. **17**(6), 2837–2845 (2015)
11. Jurčius, A., et al.: Vibratory stress relieving – it's advantages as an alternative to thermal treatment. J. Vibroengineering. **10**(1), 123–127 (2008)
12. Sun, M.C., et al.: The vibratory stress relief of a marine shafting of 35# bar steel. Mater. Lett. **58**(3), 299–303 (2004)
13. Gevlich, S.O., Knyazev, V.N.: Examination of the relaxation of residual stresses in welded joints between 09G2S and 20GML steels after vibration treatment. Chem. Petroleum Eng. **22**(8), 369–371 (1986)
14. Manh, C.B., et al.: Effects of the vibration amplitude in vibratory stress relief on the fatigue life of structures. Adv. Technol. Innov. **7**(1), 56–65 (2022)
15. Rao, D., et al.: The effectiveness evaluation of 314L stainless steel vibratory stress relief by dynamic stress. Int. J. Fatigue. **29**(1), 192–196 (2007)
16. Delin, R., et al.: Reduce the residual stress of welded structures by post-weld vibration. Mater. Sci. Forum. **490-491**, 102–106 (2005)
17. Mohanty, S., et al.: The residual stress distribution of CO2 laser beam welded AISI 316 austenitic stainless steel and the effect of vibratory stress relief. Mater. Sci. Eng. A. **703**, 227–235 (2017)
18. Serrati, D.S.M., et al.: Effects of heat and vibration treatments for welding on residual stresses and mechanical properties. Int. J. Adv. Manuf. Technol. **128**(3), 1473–1481 (2023)

19. Ebrahimi, S.M., et al.: The influences of the cyclic force magnitude and frequency on the effectiveness of the vibratory stress relief process on a butt welded connection. Int. J. Adv. Manuf. Technol. **102**(5–8), 2147–2158 (2019)
20. Munsi, A., et al.: The effect of vibratory stress on the welding microstructure and residual stress distribution. Proc. Inst. Mech. Eng., Part L: J. Mater.: Des. Appl. **215**(L2), 99–111 (2001)
21. Xu, W., et al.: Residual stress relaxation and microstructure evolution in cold radial forged high-strength steel tube using thermal-vibratory stress relief technique. J. Mater. Eng. Perform. **33**, 2064–2076 (2023)
22. Sun, M.C., et al.: Vibratory stress relieving of welded sheet steels of low alloy high strength steel. Mater. Lett. **58**(7–8), 1396–1399 (2004)
23. Li, S., Fang, H.: Vibration stress relief of DH 36 rectangle welded plates. In: IOP Conference Series: Materials Science and Engineering, vol. 322, p. 042002. IOP Publishing (2018)
24. Li, S., et al.: A combined computational and experimental study on vibration stress relief for large welded DH36 steel tube. J. Vibroengineering. **18**(3), 1486–1496 (2016)
25. Li, S.Q., et al.: Influence of vibration stress relief (VSR) on the residual stress and microstructure in welded plate of DH36 low alloy high strength steel. Adv. Mater. Res. **887–888**, 924–928 (2014)
26. De Moraes, A.G., et al.: Quantification of CTOD fracture toughness in welded joints to evaluate the efficacy of vibration stress relief compared to thermal stress relief. Mater. Res. **23**(4), Art no. e20200157 (2020)
27. Chuvas, T.C., et al.: Residual stress relief of welded joints by mechanical vibrations. J. Braz. Soc. Mech. Sci. Eng. **38**(8), 2449–2457 (2016)
28. Bouhelier, C., et al.: Vibratory stress relief of welded parts. Mech. Relax. Residual Stresses. **993**, 58–70 (1988)
29. Pradhan, R., et al.: An investigation on microstructure-texture of E250BR steel plate subjected to post-weld vibration. Proc. Inst. Mech. Eng. Part B: J. Eng. Manuf. **238**, 514–523 (2023)
30. Li, S.Q., et al.: Relationship between the microstructure of the welded steel plates and the efficiency of vibration stress relief. Adv. Mater. Res. **941**, 2062–2065 (2014)
31. Liang, X.D., et al.: Welding residual stress reduction assessment in steel rings via the spectral harmonic vibration aging test. In: IOP Conference Series: Materials Science and Engineering, vol. 770, 1st edn. IOP Publishing (2020)
32. Dong, X., et al.: Effect of residual stress in ductile iron component on its natural frequency. Adv. Mater. Res. **308**, 967–972 (2011)
33. Jurčius, A., et al.: Effects of vibration energy input on stress concentration in weld and heat-affected zone of S355J2 steel. Solid State Phenom. **165**, 73–78 (2010)
34. Wang, J.S., et al.: The effects of waveform in residual stress relief by vibration technique. In: ASM Proceedings of the International Conference: Trends in Welding Research, p. 427 (2013)
35. Yang, Y.P.: Understanding of vibration stress relief with computation modeling (in en). J. Mater. Eng. Perform. **18**(7), 856–862 (2009)
36. Wang, J.S.: The effects of waveform in residual stress relief by vibration technique. In: Trends in Welding Research 2012, pp. 427–431. ASM International, Chicago (2012)
37. Chen, Y., et al.: Study of vibratory stress relief effect of quartz flexible accelerometer with FEA method. J. Vibroengineering. **15**(2), 784–793 (2013)
38. Ingram, E., et al.: The effect of localized vibration during welding on the microstructure and mechanical behavior of steel welds. Materials. **12**(16), 2553 (2019)
39. Rao, D., et al.: Vibratory stress relief in manufacturing the rails of a maglev system. J. Manuf. Sci. Eng. **126**(2), 388–391 (2004)
40. Lin, Y.-C., et al.: Residual-stress relaxation mechanism and model description of 5052H32 Al alloy spun ellipsoidal heads during annealing treatment. Adv. Manuf. **10**, 1–14 (2022)

41. Epp, J., et al.: Residual stress relaxation during heating of bearing rings produced in two different manufacturing chains. J. Mater. Process. Technol. **211**(4), 637–643 (2011)
42. Feng, Q., et al.: Influence of annealing on the shot-peened surface of duplex stainless steel at elevated temperatures. Nucl. Eng. Des. **255**, 146–152 (2013)
43. Jang, J.-i., et al.: Assessing welding residual stress in A335 P12 steel welds before and after stress-relaxation annealing through instrumented indentation technique. Scr. Mater. **48**(6), 743–748 (2003)
44. Lu, B., Lu, X.: Evolution of residual stress and distortion of cold-rolled bearing ring from annealing to quenched-tempered heat treatment. J. Mater. Eng. Perform. **27**(2), 368–378 (2018)
45. Wawszczak, R., et al.: Residual stress field in steel samples during plastic deformation and recovery processes. Philos. Mag. **91**(17), 2263–2290 (2011)
46. Xu, W., et al.: Evolution of residual stress and microstructure during annealing of 30SiMn2MoVA high-strength alloy steel tube processed by cold radial forging. J. Mater. Eng. Perform. **30**(8), 5889–5897 (2021)
47. Harati, E., et al.: The relative effects of residual stresses and weld toe geometry on fatigue life of weldments. Int. J. Fatigue. **77**, 160–165 (2015)
48. Munsi, A., et al.: The influence of vibratory treatment on the fatigue life of welds: a comparison with thermal stress relief. Strain. **37**(4), 141–149 (2001)
49. Kaçar, S., Yılmaz, O.: Effect of vibratory stress relief on fatigue life of S355J2 steel welded joints. Makina Tasarım ve İmalat Dergisi. **17**(1), 1–8 (2019)
50. Lai, H.-H., et al.: Evolution of internal friction in low-carbon steel during vibratory stress relief. J. Mater. Res. Technol. **9**(3), 5403–5409 (2020)
51. Withers, P.J., Bhadeshia, H.K.D.H.: Residual stress. Part 2 – nature and origins (in en). Mater. Sci. Technol. **17**(4), 366–375 (2001)
52. Felson, D.T.: Bias in meta-analytic research. J. Clin. Epidemiol. **45**(8), 885–892 (1992)

Comparative Analysis of Machine Learning and Deep Learning Fault Detection Across Various Time Segmentation

Nurul Hannah Mohd Yusof[1], Nurul Adilla Mohd Subha[1(✉)], Norikhwan Hamzah[2], Nurulaqilla Khamis[1], Mohamad Amir Shamsudin[1], and Muhamad Fadli Ghani[1]

[1] Faculty of Electrical Engineering, Universiti Teknologi Malaysia, 81310 Johor Bahru, Malaysia
nuruladilla@utm.my

[2] Faculty of Mechanical Engineering, Universiti Teknologi Malaysia, 81310 Johor Bahru, Malaysia

Abstract. The increasing complexity of industrial systems necessitates accurate and timely fault detection mechanisms to maintain reliability and avoid unplanned downtimes. This paper presents a comparative study on fault detection methods applied to a classifier drive system in a cement manufacturing plant. Four models: One-Class Support Vector Machine (OC-SVM), Random Forest (RF), Gated Recurrent Unit (GRU), and Long Short-Term Memory (LSTM) were evaluated across three time-series segmentation strategies: overlapping, non-overlapping, and expert knowledge driven. Current and speed data collected over a one-year period were used and expert-annotated fault dates serving as ground truth. Thirteen statistical and frequency-domain features were extracted from each segmented window and used to train both traditional machine learning (ML) and deep learning (DL) models. Results show that RF achieved the highest accuracy of 98.90%, successfully detecting 11 faults with a relatively low average computational time, ranging from 0.33 to 0.94 s. In terms of computational time, ML (OC-SVM and RF) models demonstrated far lower processing times compared to DL models, but expert knowledge-driven segmentation has significantly enhanced the DL (GRU and LSTM) models' ability to align predictions with actual fault occurrences by achieving an average temporal error of just 0.2 days which is crucial in industrial settings for rapid and accurate fault localization. This study highlights the importance of selecting appropriate segmentation strategies and tuning model parameters for achieving accurate, interpretable, and efficient fault detection in complex industrial environments.

Keywords: Expert Knowledge · Fault Identification · Time Segmentation

1 Introduction

Vast development in manufacturing has led to increasingly complex systems characterized by sophisticated automation and extensive sets of variables [1]. This complexity brings challenges in maintaining smooth operations and preventing unforeseen faults.

Y. Zhao (Ed.): ICAMM 2025, SPM 96, pp. 252–261, 2025.
https://doi.org/10.1007/978-3-032-09862-7_24

Fault detection based on data-driven techniques plays a paramount role in the manufacturing industry to quickly detect and diagnose anomalies while improving the reliability and safety of the processes [1]. The fault detection delay is defined as the amount of time passes between the actual occurrence of a fault and its detection which should be minimized as much as possible [2]. This measure reflects the responsiveness and effectiveness of a system in detecting issues early. Therefore, rapid and precise fault detection tasks have become vital in industrial environments [3].

Time is a crucial metric whenever rapid response is required. The segmentation of time is essential for data mining applied in fault and anomaly detection [4]. Time-series data are being analyzed using window slicing techniques in most data-driven fault detection systems. There are two common types of strategies that can be used: non-overlapping [5] and overlapping (sliding) [6] windows. The signals of time-series data are divided into independent and fix sized segmentation in non-overlapping windows. This approach is computationally efficient but has the potential to miss the short-term anomalies that occur within the segment boundary [7]. In contrast, overlapping can overcome the loss of information problem by effectively capturing subtle variations through shared data points across segmentations. However, the drawbacks may include higher memory consumption, computational complexity, and data redundancy [7].

Considering these limitations, expert-driven segmentation has been developed as a complementary method. Time segmentation windows can be precisely aligned with anticipated fault events by incorporating expert knowledge which can help to enhance detection accuracy and temporal precision such that can be found in fault detection and diagnosis for a cement rotary kiln [1]. This approach provides a more accurate detection without upholding excessive computational overhead.

Despite the time segment considerations, premature detection of subtle fault has always been a challenge since such anomalies gradually developed within vast amount of data that is continuously produced by control systems and sensors [8]. The primary difficulty arises not solely from processing large volumes of data, but also from discovering significant features that distinguish normal from anomalous behaviour. The fault detection techniques in identifying anomalous span from traditional approaches such as rule-based and statistical approach to modern data-driven methods like machine learning (ML) and deep learning (DL).

Compared to traditional manual data analysis methods, ML offers possibility to estimate and foresee malfunctions. Particularly, ML algorithms excel in identifying anomalies, capturing data patterns, and optimizing maintenance schedules [9]. Support Vector Machines (SVM) is a type of classic machine learning technique that has demonstrated high generalization capabilities in fault detection and one popular example of SVM is One-Class SVM (OC-SVM) [10] that has been investigated for welding fault detection. It is compatible for unsupervised or semi-supervised contexts. However, OC-SVM may not perform well in dynamic situations with changing fault patterns and necessitate careful kernel parameter tuning [11]. Random Forests (RF) is another well-known ensemble ML method which is appreciated for its robustness and accuracy in managing outliers and noise [12]. This method pools weak decision trees to form a strong classifier that is effective while having huge dependencies on discriminative power of the extracted features just like in [12, 13].

To further investigate distinct performance of the fault detection, deep learning-based approaches can be considered. It has grown in popularity for the diagnosis of motor faults due to its capacity to automatically extract relevant features from raw data and capture intricate non-linear correlations between inputs and outputs [14]. Notable Recurrent Neural Networks (RNN) methods such as Long Short-Term Memory (LSTM) and Gated Recurrent Units (GRU) have great potential in capturing temporal dependencies in time-series data. GRU is a simplified variant of LSTM that offers comparable performance with reduced complexity in term of gated and parameters usage. Both methods have successfully detected anomalies in diverse industrial applications including electrolytic copper production [15] and coal mining hydraulic system [6].

This study focuses on fault detection in cement manufacturing which discovered lack of expert knowledge integration in the time segmentation process, whereas previous research has concentrated on traditional time windowing. To address these gaps, this study assesses OC-SVM, RF, GRU, and LSTM models on real-world sensor data from a classifier drive system in cement manufacturing, using three segmentation strategies: non-overlapping, overlapping, and expert-driven knowledge as can depicted in Fig. 1. The models are evaluated based on accuracy, fault detection count, fault detection gap with expert real fault occurrence annotations, and computational efficiency to determine the most effective approach for early and reliable fault detection. The key contributions of this study include:

- Employment of feature extraction and usage of ensemble techniques for feature classification of multivariate data of the classifier motor.
- Development of a unified framework for evaluating ML and DL algorithms in fault detection
- Comprehensive comparison of different ML and DL algorithms' performance with a special emphasis on fault detection delays in performance evaluation.

2 Methodology

The fault detection experiments were carried out on a local computing system equipped with an Acer Intel Core i5-7200U processor, 20 GB of RAM, and a 512 GB SSD, operating on Windows 10. All implementations were developed in MATLAB as shown in Fig. 2 which integrates the built-in functions and custom scripts to perform data preprocessing, segmentation, feature extraction, model training, fault detection, and evaluation.

2.1 Data Collection and Preprocessing

The dataset used in this study was collected from a classifier drive system within a cement manufacturing facility over a year period which spans from 1 August 2023 to 1 August 2024. Two parameters were collected for a classifier motor: current and speed. Both parameters are sampled for every two hours with timestamped entries. The timestamp synchronization was performed using a ±60-second tolerance to align both signals. Once aligned, a Savitzky-Golay filter with a frame length of 21 and a polynomial order of 3 is being implemented to denoise the data and at the same time preserve essential signal features. After synchronization and filtering, 4000 samples were successfully retained for further analysis. A set of expert-annotated fault dates was also provided and used as ground truth for evaluation as shown in Fig. 3.

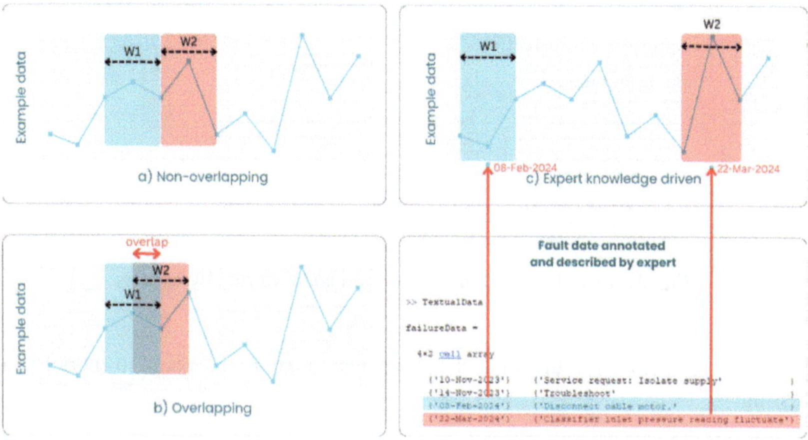

Fig. 1. Time segmentation strategy for a) Overlapping b) Non-overlapping c) Expert knowledge-driven

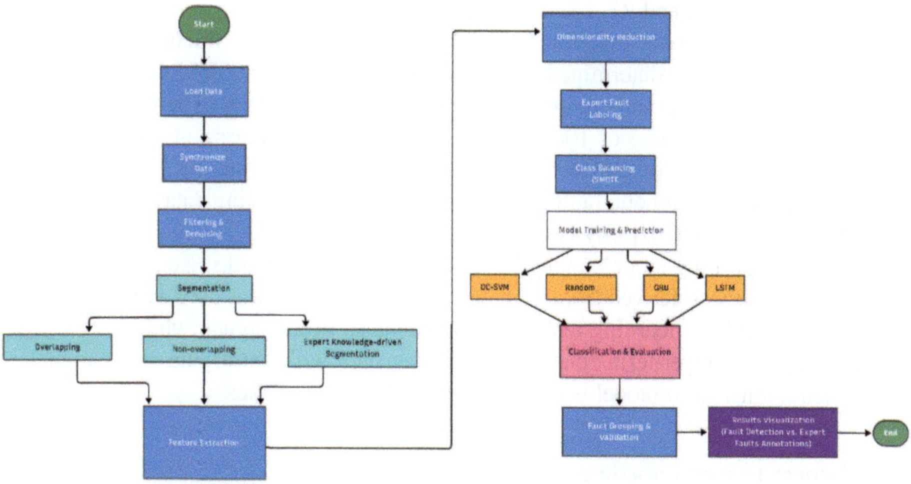

Fig. 2. Flowchart of fault detection methodology integrating multiple segmentation strategies and machine learning models

2.2 Feature Extraction

From each segmented window, a total of 13 features were extracted into a structured dataset to capture the statistical and frequency-domain characteristics of the current and speed signals. Statistical features included the mean, standard deviation, variance, skewness, kurtosis, root mean square (RMS), median, and range. Signal shape descriptors such as crest factor, impulse factor, and shape factor were also calculated. In the frequency domain, the dominant frequency and maximum magnitude were extracted from the one-sided Fast Fourier Transform (FFT) of the signal. These features from both signals were

No	Fault Date	Description
1	{'10-Nov-2023'}	{'Service request: Isolate supply' }
2	{'12-Nov-2023'}	{'Service request: Isolate supply' }
3	{'14-Nov-2023'}	{'Troubleshoot' }
4	{'14-Nov-2023'}	{'Service request: Isolate supply' }
5	{'08-Feb-2024'}	{'Isolate supply' }
6	{'08-Feb-2024'}	{'Disconnect cable motor.' }
7	{'10-Feb-2024'}	{'Service request: Isolate supply' }
8	{'13-Feb-2024'}	{'Service request: Isolate supply' }
9	{'13-Feb-2024'}	{'Service request: Isolate supply' }
10	{'15-Feb-2024'}	{'Maintenance routine' }
11	{'22-Mar-2024'}	{'Classifier inlet pressure reading fluctuate'}

Fig. 3. List of actual faults described and reported by experts

concatenated to form a feature matrix, which was normalized prior to model training and testing.

2.3 Fault Detection Models

Four fault detection models were employed: two for ML and two for DL. The ML models are OC-SVM and RF. OC-SVM was trained exclusively on normal (non-faulty) data using a radial basis function (RBF) kernel and an outlier fraction of 0.05. An adaptive threshold was determined from the 5th percentile of the SVM decision scores. Windows scoring below this threshold were labeled as anomalies. The RF model was trained using an ensemble of 100 decision trees. Due to the class imbalance between normal and fault samples, SMOTE (Synthetic Minority Oversampling Technique) was applied to generate synthetic fault samples and balance the training dataset. Predictions were then made using majority voting.

The deep learning models included GRU and LSTM networks. Both models were trained on sequences derived from segmented windows. GRU was used as a lightweight RNN that capable of learning temporal dependencies in the signal, while LSTM was used to capture longer-term relationships in the data with extra gate and parameter implementations. Each model was configured to detect anomalies.

2.4 Expert-Driven Labelling and Evaluation

To evaluate model performance, each segmented window was labeled based on whether its midpoint fell within a ±5-day range of any expert-annotated fault date. Windows that met this criterion were labeled as faults. To prevent redundant detections of the same fault, windows flagged as faults were grouped using a 5-minute threshold to ensure that only distinct fault events were counted.

Model performance was evaluated using four practical metrics. First, accuracy was computed as the percentage of correctly classified windows (both fault and non-fault). Second, total faults detected measured the number of distinct fault events identified by the model. Third, total time difference quantified the cumulative deviation (in days) between the detected fault times and their nearest expert annotations, offering a measure of temporal precision. Lastly, computational time captured the model's runtime in seconds, including training and inference, providing insight into its practicality for real-time applications.

2.5 Iterative Parameter Tuning

Each model in this study underwent iterative tuning to enhance its fault detection performance and temporal alignment with expert-annotated events. At the segmentation stage, window size was fixed at 30 samples and overlap percentage was tuned. For overlapping segmentation, 75% overlap was selected as suggested in [16] which enables the model to capture subtle, short-lived transitional behaviors. On the other hand, the non-overlapping configuration was tested as a baseline. For expert-driven segmentation, the window boundaries were aligned based on ±5-day tolerance around the fault dates.

For the OC-SVM model, an adaptive threshold was dynamically set at the 5th percentile of the decision scores from normal-labeled samples. This quantile-based strategy was iteratively tested to balance sensitivity and reduce false positives.

In the RF model, the number of decision trees was set to 100, and its performance was tuned by varying the SMOTE parameters to balance the skewed distribution of fault and non-fault classes. The number of synthetic fault samples was determined based on the size of the majority class, and the neighborhood parameter ($k = 5$) in SMOTE was retained as a default for diversity. Predictions were made using majority voting, and model behavior was analyzed through confusion matrices and fault match comparisons.

For the DL models (GRU and LSTM), hyperparameters were tuned through repeated experiments with different random seeds. Both models were trained using a sequence input of size $[2 \times 30]$ (2 channels for current and speed, and 30 timesteps). The network architecture consisted of a single recurrent layer (either GRU or LSTM) with 50 hidden units, followed by a dropout layer (dropout rate = 0.2), a fully connected layer, and a softmax output. The models were trained using the Adam optimizer with an initial learning rate of 0.001 for 50 epochs. 80:20 split was applied for training and validation, and model tuning was guided by training progress plots and validation accuracy metrics. Confusion matrices and fault timestamp comparisons were used to evaluate not only classification accuracy but also how closely the detected faults aligned with expert fault dates.

Final tuning cycle included the grouping of predicted fault windows using a 5-minute threshold and comparison against expert faults with a ±7-day matching window. This approach allowed all models to be evaluated not just on standard metrics like accuracy, but also on more application-relevant metrics such as the number of faults detected and the cumulative time difference from expert diagnoses.

2.6 Model Evaluation Strategy

The evaluation strategy emphasized fault detection relevance over traditional classification metrics. The performance was evaluated using the total number of detected faults, the temporal difference from expert annotations, computational time, and the overall classification accuracy. These metrics were chosen to reflect the practical requirements of industrial fault detection systems, where timely and reliable detection is often more important than achieving perfect classification balance. The final comparison helped assess the trade-offs between detection resolution, computational load, and alignment with expert knowledge across the four models and three segmentation strategies.

3 Result and Discussion

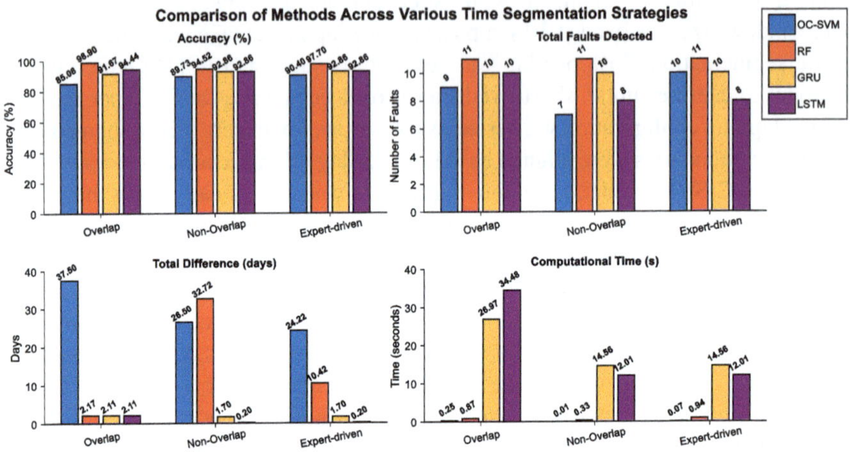

Fig. 4. Comparative method of ML and DL across various time segmentation strategies

The four fault detection models of OC-SVM, RF, GRU, and LSTM were evaluated with three distinct time segmentation: overlapping, non-overlapping, and expert knowledge-driven. The performance of each model is presented in Fig. 4 based on accuracy, number of faults detected, temporal difference from expert fault dates, and computational time.

In terms of accuracy, RF consistently outperformed all other models across entirely segmentation strategies. Under the overlapping strategy, RF achieved the highest accuracy at 98.90%, while OC-SVM recorded the lowest at 85.08% due to its sensitivity to the complex decision boundaries introduced by overlapping segments. LSTM and GRU also performed well, achieving accuracies of 94.44% and 91.67%, respectively. With non-overlapping segmentation, all models demonstrated reasonably high accuracy, but RF maintained its lead at 94.52%. In the expert-driven setup, the performance of all models converged, with accuracies ranging from approximately 90.40% to 92.86%. These results indicate that RF is the most robust model in terms of classification accuracy. The model benefits from the ensemble learning which enhanced by the SMOTE oversampling due to its ability to handle high-dimensional feature spaces.

The number of faults detected provides an insight into the sensitivity of each model. RF was the most sensitive with 11 detected faults consistently across all segmentation methods. In contrast, OC-SVM detected fewer faults, especially under non-overlapping segmentation whereby only 7 faults were identified. Both GRU and LSTM showed balanced detection behavior by identifying 8 to 10 faults depending on the segmentation strategy. While consistent fault detection of RF suggests strong coverage, it also raises the potential concern of over-detection, which must be validated against false positives. The lower fault counts for OC-SVM in non-overlapping segments suggest its sensitivity is highly dependent on the segmentation window and distribution of training data.

Fault detection delays is measured by the total difference in days between detected faults and expert-annotated fault dates has revealed interesting patterns. OC-SVM exhibited high total time differences under the overlapping strategy with a deviation of 37.50 days. This situation indicates poor temporal alignment. However, its performance improved significantly under the expert-driven segmentation where the total difference significantly reduced to just 1.70 days. This suggests that OC-SVM performs best when fault boundaries are tightly defined. On the other hand, GRU and LSTM showed very minimal fault detection delays, particularly under expert-driven segmentation, where their total differences were as low as 0.20 and 1.70 days, respectively. RF demonstrated moderate temporal precision across all strategies. These findings highlight the strength of DL models (GRU & LSTM) in capturing fault timing more precisely due to their ability to learn from patterns in sequential time-series data.

From a computational overhead perspective, OC-SVM was the most efficient model, with processing times as low as 0.0095 s for non-overlapping data. RF also performed well by maintaining computational times below one second for all strategies. OC-SVM is faster than RF due to its simpler decision boundary evaluation, whereas RF's ensemble of trees can increase fault detection time. In contrast, GRU and LSTM required significantly more time, ranging from 12 to 34 s depending on the segmentation method. This is expected due to their complex architectures and recurrent computations. While DL models may offer superior temporal alignment, their high computational cost could limit their application in real-time environments.

These results suggest that DL models (GRU and LSTM) demonstrated the lowest deviation in fault detection delay, which is very good for rapid fault detection. However, this advantage comes at the expense of computational efficiency. In contrast, RF offered a more balanced trade-off by achieving high accuracy with significantly lower computational time and fewer fault detection delay which make it more compelling and practical for real-time or resource-limited applications. OC-SVM proved to be lightweight and effective when faults are well-labeled, but less robust under loose or ambiguous segmentation and large data volume.

Expert-driven segmentation enhances model performance by narrowing the learning scope to periods surrounding actual fault events. This approach notably benefits ML models like OC-SVM and RF which improved detection accuracy compared to non-overlapping that is loosely segmented but still does not outperform overlapping RF that has been performed excellently with precise tuning and SMOTE oversampling. Interestingly, although DL models may not detect as many total fault instances as ML models, they consistently produced detection times that were closer to the ground truth.

4 Conclusion and Future Works

Based on the findings, a trade-off must be considered. ML models generally perform better in terms of total fault coverage, while DL models excel in temporal precision, accurately capturing the timing of fault events due to their ability to model sequential patterns. This discrepancy might be due to the inherent differences in learning mechanisms. ML models rely on feature-based generalization, whereas DL models learn deeper temporal dependencies from sequential data. Therefore, the selection of both segmentation strategy and model type should be guided by the specific priorities of the application by balancing detection accuracy, timing precision, and computational efficiency.

As a matter of fact, hybrid and ensemble approaches of ML like combining OC-SVM and RF can enhance classification capability to improve accuracy and robustness. Furthermore, DL models like GRU, LSTM, and autoencoders can also be explored for temporal sequence modeling and compare their performance against traditional classifiers. Finally, dynamic segmentation strategies can be investigated to make it more adaptive by dynamically adjusting window sizes based on signal characteristics rather than fixed segmentation strategies.

Acknowledgment. This work was supported by the grants provided by Universiti Teknologi Malaysia (UTM), under the UTM Fundamental Research (Q.J130000.3823.22H48) and Digital-Infused Research (Q.J130000.5123.00L77). Simultaneously, the authors would also like express their sincere thanks to the Cement Industries of Malaysia Berhad (CIMA) for their technical support.

References

1. Bencheikh, F., Harkat, M.F., Kouadri, A., Bensmail, A.: New reduced kernel PCA for fault detection and diagnosis in cement rotary kiln. Chemom. Intell. Lab. Syst. **204**, 104091 (2020). https://doi.org/10.1016/j.chemolab.2020.104091
2. Stoorvogel, A.A., Niemann, H., Saberi, A.: Delays in fault detection and isolation. Proc. Am. Control Conf. **1**, 459–463 (2001). https://doi.org/10.1109/ACC.2001.945587
3. Šprogar, M., Verber, D.: Accuracy is not enough: optimizing for a fault detection delay. Mathematics. **11**(15), 3369 (2023). https://doi.org/10.3390/math11153369
4. Lu, S., Huang, S.: Segmentation of multivariate industrial time series data based on dynamic latent variable predictability. IEEE Access. **8**, 112092–112103 (2020). https://doi.org/10.1109/ACCESS.2020.3002257
5. He, B., Li, L., Zhang, Q.: PNC-based trend feature extraction method for time series data. Neurocomputing. **618**(December), 2025 (2024). https://doi.org/10.1016/j.neucom.2024.129174
6. Zhan, K., et al.: Seq2Seq-based GRU autoencoder for anomaly detection and failure identification in coal mining hydraulic support systems. Sci. Rep. **15**(1), 1–19 (2025). https://doi.org/10.1038/s41598-024-84130-8
7. Ni, Q., Patterson, T., Cleland, I., Nugent, C.: Dynamic detection of window starting positions and its implementation within an activity recognition framework. J. Biomed. Inform. **62**, 171–180 (2016). https://doi.org/10.1016/j.jbi.2016.07.005
8. Edozie, E., Shuaibu, A.N., Sadiq, B.O., John, U.K.: Artificial intelligence advances in anomaly detection for telecom networks. Artif. Intell. Rev. **58**(4), 100 (2025). https://doi.org/10.1007/s10462-025-11108-x
9. Farooq, U., Ademola, M., Shaalan, A.: Comparative analysis of machine learning models for predictive maintenance of ball bearing systems. Des. Electron. **13**(2), 438 (2024). https://doi.org/10.3390/electronics13020438
10. Melakhsou, A.A., Batton-Hubert, M., Casoetto, N.: Welding fault detection and diagnosis using one-class SVM with distance substitution kernels and random convolutional kernel transform. Int. J. Adv. Manuf. Technol. **128**(1–2), 459–477 (2023). https://doi.org/10.1007/s00170-023-11768-5

11. Goel, A., Srivastava, S.K.: Role of kernel parameters in performance evaluation of SVM. In: Proc. - 2016 Second International Conference on Computational Intelligence & Communication Technology (CICT), pp. 166–169. IEEE (2016). https://doi.org/10.1109/CICT.2016.40
12. Wu, F., Chen, K., Qiu, G., Zhou, W.: Robust open circuit fault diagnosis method for converter using automatic feature extraction and random forests considering nonstationary influence. IEEE Trans. Ind. Electron. **71**(10), 13263–13273 (2024). https://doi.org/10.1109/TIE.2023.3344816
13. Thakur, D., Biswas, S.: An integration of feature extraction and guided regularized random forest feature selection for smartphone based human activity recognition. J. Netw. Comput. Appl. **204**(October 2021), 103417 (2022). https://doi.org/10.1016/j.jnca.2022.103417
14. Sikinyi, L.A., Maina, C.M., Ngoo, L.: Condition monitoring and fault diagnosis of induction motor faults using deep learning: a review. IEEE AFRICON Conf., 1–6 (2023). https://doi.org/10.1109/AFRICON55910.2023.10293578
15. Yoo, J., Song, S., Chang, K., Baek, J.G.: Fault detection and diagnosis using two-stage attention-based variational LSTM in electrolytic copper manufacturing process. Int. J. Adv. Manuf. Technol. **129**(3–4), 1269–1288 (2023). https://doi.org/10.1007/s00170-023-12356-3
16. Kshirasagar, S., Guntoro, A., Mayr, C.: Impact of sliding window variation and neuronal time constants on acoustic anomaly detection using recurrent spiking neural networks in automotive environment. Algorithms. **17**(10), 440 (2024). https://doi.org/10.3390/a17100440

Evaluation of Open Face Helmet Head Foam to Investigate Critical Injury Location

Hammad Shaikh[1], Muddassar Sharif[1], Fayyaz[2], Sikandar Khan[2,3](✉), Numan Khan[4], Ashfaq Khan[5], and Salem Bashmal[2,3]

[1] Department of Mechanical Engineering, Pakistan Institute of Engineering and Applied Science, Islamabad, Pakistan
mudasar@pieas.edu.pk

[2] Department of Mechanical Engineering, King Fahd University of Petroleum & Minerals, Dhahran 31261, Saudi Arabia
{g202202260,sikandarkhan,bashmal}@kfupm.edu.sa

[3] Interdisciplinary Research Center for Intelligent Manufacturing and Robotics, King Fahd University of Petroleum & Minerals, Dhahran 31261, Saudi Arabia

[4] Department of Engineering, University of Campania "L. Vanvitelli", Via Roma 29, 81031 Aversa, Italy
numan.khan@unicampania.it

[5] School of Engineering and Built Environment, Sheffield Hallam University, Sheffield, UK
ashfaq.khan@shu.ac.uk

Abstract. According to medical surveys, bike riders are more at risk of road accidents than other modes of transportation. Among injuries sustained by a bike rider, head injuries are more frequent and are a major cause of fatality. Medical research showed that about 60% of the injuries in an accident are likely to the head. This causes skull fracture or brain damage due to high acceleration upon impact. Also, several research showed that the use of helmets decreases the fatality rate due to head injury by half. This research investigates the impact analysis of realistic open-face helmet head foam (OFHHF) at different locations. The helmet has an expanded polypropylene (EPP) liner and acrylonitrile butadiene styrene (ABS) shell. First, the numerical modeling of a simple helmet head foam with ABS and EPP was modeled in design modeler ANSYS, analyzed in explicit dynamics ANSYS using an autodyn solver, and validated with available published work. Then the analysis of realistic OFHH at a velocity of 7.5 $\frac{m}{s}$ was carried out as per European safety standard ECE 22.05. The OFHHF was analyzed at four different locations, i.e. top, front, rear, and side locations using a flat anvil that mimics a hard surface, such as road or pavement. OFHHF acceleration results at each location were calculated using ANSYS Explicit dynamics (Autodyn Solver) and compared to the 275g limit in ECE 22.05 standard. The impact investigation of OFHHF results showed that there was a higher acceleration at the top and side locations. Thus, the impact of OFHHF on top and side locations is critical to human head injury.

Keywords: Open Face Helmet · Head Foam · Head Injury · Explicit Dynamics ANSYS · SOLIDWORKS · ECE 22.05 Standard

Y. Zhao (Ed.): ICAMM 2025, SPM 96, pp. 262–272, 2025.
https://doi.org/10.1007/978-3-032-09862-7_25

1 Introduction

Approximately 1.35 million people in the world die each year from road-related traffic injuries, with motorcyclists and bicyclists traffic injuries, representing 31% of all traffic-related deaths worldwide and 45% in South-East Asia [1]. Further, head injuries are the main cause of death and disability among users of motorcycles and bicycles in traffic accidents [2]. The annual data of domestic road traffic accidents from 2019 to 2022 released by the National Bureau of Statistics shows that the total number of bicycle and motorcycle traffic accidents are the second highest road accidents after the car traffic accidents [3]. For cyclists, helmets are used for protection against injuries such as traumatic brain injury [4–7]. The use of helmets is an evidence-based, preventive intervention that saves lives [2], continued improvements in helmet assessment and design can create additional opportunities to further reduce injuries and deaths. In a road accident, the motorcyclist is more prone to fatal injuries than in any other mode of transport. A helmet is a common protective element that is worn to protect the head thus protecting the skull (a natural protective layer for the brain) and preventing brain injuries such as skull fracture; hematoma and brain haemorrhage [8–10]. Medical surveys proved that the major cause of death in a bike accident is fatal injuries to the head and neck. The usage of helmets greatly reduces the risk of fatal injuries to the bike rider [11–14]. Wearing a motorcycle helmet correctly can reduce the risk of death by almost 40% and the risk of severe injury by over 70% [15]. Motorbike helmets come in different types, such as full-face, open-face, etc.; however, a typical helmet consists of two basic components; shell and liner [16–18]. The shell is a hard outer layer that protects the head from being penetrated by sharp objects. Its main function is to distribute the impact energy over a large area. The shell is generally made of thermoplastics (ABS, polycarbonate) but composite materials are also used in advanced helmets [19]. The liner is a soft, deformable layer that provides sufficient stopping distance for gradual deceleration in an accident. The liner compresses and absorbs the energy of the impact, preventing damage to the rider's head [20, 21]. Most helmet liners are made of polymers such as EPS [22]. The energy-absorbing ability of the liner is directly proportional to the liner thickness and inversely proportional to the density of the material [23]. Other components of the helmet are cushion pads that are used for fitting and a retention system with a strap that keeps the helmet on the head.

There has been a need to improve the performance of helmets by introducing better material that enhances the energy absorption of helmets. Helmets were initially experimentally tested to study their performance characteristics. This process was somewhat cumbersome, and the destructive analysis made the research expensive. The advancement in computer technology has created ease in the numerical study of helmets and FE models were introduced. Mustafa et al. [26] performed the computational analysis by taking a 3D scan of a helmet followed by pre-processing.

Gilchrist and Mills [27] introduced an accurate one-dimensional computer model for simulating an impact test on a helmet using the experimental results of the shell and liner. Kostoupolos et al. [28] performed a study and examined the performance of composite shells properties and development of delamination and damage on the impact response of a motorcycle helmet using LS-DYNA. Caserta et al. [29] used an experimental approach to present a novel design for a helmet liner with a reinforced aluminum honeycomb.

Coelho et al. [30] presented a new helmet liner configuration composed of EPS and micro-agglomerated cork (MAC) to enhance the performance of the helmet liner after it had undergone its first impact in LS-DYNA. A critical review of the previous works carried out and different aspects regarding analysis was studied by Cernicchi et al. [23]. Blanco et al. [31] introduced a new helmet liner that consists of ABS lamina shaped into deformable cones, which absorb energy by folding and collapsing cones.

Teng et al. [32] studied two oblique impact tests on open-face helmets and measured the linear and rotational acceleration along with the impact force experienced by the head foam. The numerical results were compared with the experimental results, which showed good agreement. Pinnoji et al. [33] investigated the validity of the use of metal foam as a suitable helmet liner. The research comprised a numerical study in which an open-face helmet was analyzed. The result showed that metal foam helmets have less acceleration and low HIC value as compared to ABS shell helmets. Shuaeib et al. [34] concluded that EPP can be a viable substitute for EPS as a helmet liner. The helmet was analyzed on both flat and hemispherical anvil, followed by optimization of EPP with respect to density, the thickness of liner, and the thickness of ABS outer shell in LS-DYNA. A simplified helmet head foam model was used with the impact carried out only at the top of the helmet. Also, no real dimensions of the human head were used.

The current research work is mostly focused on the study of liner material thickness for only simple helmet head foam models. Impacts were studied at only one location, i.e. impact of helmet head foam top to road ground. In this study, a more realistic helmet head foam model was used that mimics the actual human head and helmet combination. Also, the impact of helmet head foam is studied at four different locations to simulate road ground impact scenarios. At first, a simplified helmet head foam was modeled in the ANSYS design modeler. The model was analyzed in the Autodyn and results were validated with the published literature. EPP was used as a helmet liner in this study. Afterward, an actual realistic model was created in SolidWorks and analyzed in ANSYS Explicit dynamics using Autodyn Solver at different impact locations. The results showed that the top and side helmet head foam impact is critical for human head injury as per European standards (ECE 22.05).

2 Materials and Methods

The research is started from the CAD modeling of the available simple helmet head foam model. The design and analysis capabilities of CAD models have been greatly expanded by developments in computing technologies [35–37]. The developed numerical model was later used for realistic OFHHF model analysis. In the next step, the numerical modeling results were validated with the already existing published work. The CAD model of an available simple helmet head foam model was created using ANSYS design modeler. Then ABS and EPP materials were defined in the ANSYS material library and assigned to the liner and shell of the helmet, respectively. The meshing of the developed model was done in ANSYS's built-in mesh tool. Afterward, the simulations were performed in ANSYS Explicit Dynamics using Autodyn solver. The results were then compared with the available published results from the literature and initial model was validated. After validation of a simple helmet head foam model, subsequently, a realistic scale open faced helmet head foam model was developed in SOLIDWORKS software. The developed model was imported into ANSYS software for subsequent impact investigation at different locations. The steps for realistic OFHHF model analysis are like simple helmet head foam, except for its CAD modeling. Finally, the results of validation and impact investigation at different locations of OFHHF are analyzed and discussed.

2.1 Head Injury Criterion (HIC)

HIC is a quantitative value to judge the severity of the accident relating to the deceleration experienced in a crash [29]. The HIC calculates the average acceleration over time. The average value is increased by a power of 2.5 (experimentally deduced for head impact). Though highly used to assess the effectiveness of safety equipment, it is criticized for not considering the probability of injury sustained to the brain without the skull. The HIC also does not account for the effect of rotational acceleration of the head. This is because no proper limit of rotational acceleration is proposed that can quantify the severity of accident. The HIC, however, has been consistently used in standards used for validating the performance of helmets.

2.2 European Safety Standard ECE 22.05

The ECE 22.05 is the European safety standard brought out by the Economic Commission of Europe [30, 38]. The ECE 22.05 states that the helmet, together with the head foam, should be dropped towards the anvil at a speed of 7.5 $\frac{m}{s}$. Two types of anvils, i.e. Flat and Curbstone are used. The helmet must not exceed the maximum acceleration value of 275g and the HIC criterion of 2400.

2.3 Material Model for Helmet Shell and Liner

The ABS material was used for the shell and EPP was used for the liner. Unlike conventional helmets, the comfort padding was excluded from the model as it was used for ergonomics only and has an insignificant role in the safety characteristics. Thus, the focus of this study was on the mechanical properties of the ABS shell and EPP liner. The material properties were studied, and an accurate material model was imported to ANSYS engineering data. The material model was validated by comparing published results from the literature.

3 Numerical Modeling

The numerical modeling of helmet head foam is performed for both the simple helmet head foam and the realistic OFHHF. The geometry of simple helmet head foam is taken from the available published literature.

3.1 Available Simple Helmet Head Foam Modeling

The geometry was modeled in the ANSYS design modeler and consists of hemispherical head foam with an outer diameter of 180 mm and height of 220 mm. The helmet comprised of liner and shell with thicknesses of 25 mm and 5 mm, respectively. A flat anvil of 130 mm diameter was created to produce similarity to being fallen onto a road ground or pavement during an accident. A velocity of 6 $\frac{m}{s}$ was used for simulation. Body sizing was used to generate tetrahedral mesh by the ANSYS mesh tool. The mesh size of head foam, liner, and shell was 10 mm, 10 mm, and 5 mm, respectively. The geometric model and the mesh are shown in Fig. 1

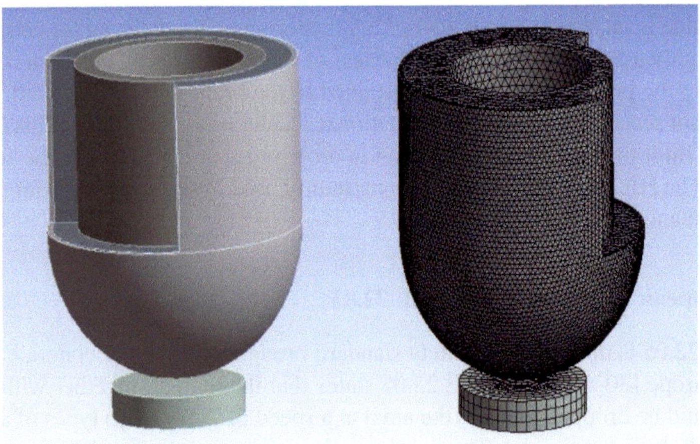

Fig. 1. OFHHF CAD model and mesh for validation

3.2 Realistic OFHHF Model

Dynamic analysis of the realistic OFHHF model was also carried out in the current study. The model was first generated in SOLIDWORKS and was then exported into ANSYS Explicit Dynamics for simulation. The helmet was analysed on possible four different locations and the results were compared with the acceleration threshold specified in the ECE 22.05 safety standard. The human head, also known as head foam, was created using the Cartesian coordinates that were available from the World Wide Web. Additionally, the liner and shell were created with thicknesses of 30 mm and 5 mm, respectively. The material properties for the EPP liner and ABS shell were the same as those used for the simple helmet head foam model. The stiffness behavior of both head foam and anvil was set as rigid so that the deformation in these parts is neglected and to capture the impact performance (acceleration) of the open-faced helmet head foam model. As dynamic analysis was simulated using an explicit method, the mesh generated was tetrahedral due to the arbitrary complex shape of the used head foam assembly. The mesh body sizing was applied to maintain a uniform mesh size for better simulation by eliminating the disadvantage of small-time steps that would increase the simulation time. The mesh size for head foam, liner, and shell was 6 mm. The total number of elements created was 205,578. Figure 2 shows the CAD and meshing of the open-face helmet head foam model.

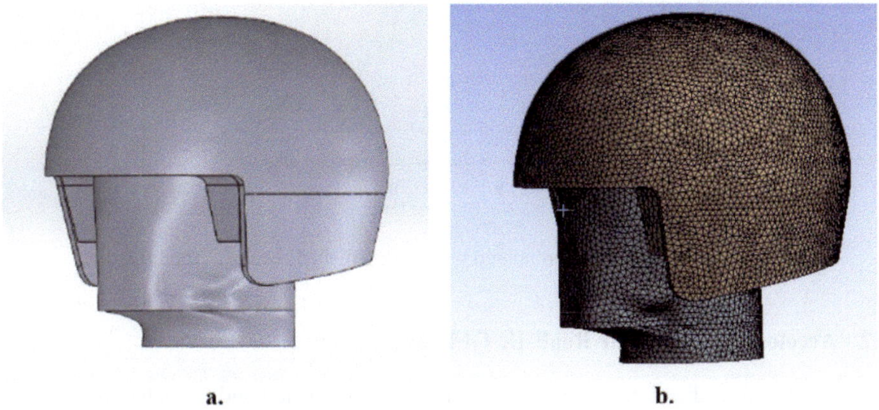

Fig. 2. Open Face Helmet Head Foam Model, a. CAD Model, b. Mesh

4 Results and Discussion

The results of the simple helmet head foam and realistic OFHHF found from numerical analysis are presented and discussed in this section. First, the validation of the developed numerical model is presented for already published research work. The same numerical model is used for realistic OFHHF, and results are discussed, and their outcomes are concluded.

4.1 Acceleration Plots for Simple Helmet Head Foam

The low-velocity analysis type was used for simulation, using Autodyn solver, as the velocity was less than 10 $\frac{m}{s}$ criterion. The acceleration plot of the head foam was then compared with the published research of Shuaeib et al. [34] as shown in Fig. 3.

The acceleration plot showed that the results from the Autodyn model are in good agreement with the published research work results. The impact interval of 7.9 ms in the current study is also almost like the impact interval in the literature, i.e. 8 ms [34]. The maximum acceleration in the Autodyn model was 245g as compared to 235g in Shuaeib et al. [34], which gave a percentage error of 4.1%. This comparison shows that the developed numerical technique models the mechanical behavior of the liner and shell accurately. From these results, the numerical model is validated and is further used for the study of realistic OFHHF.

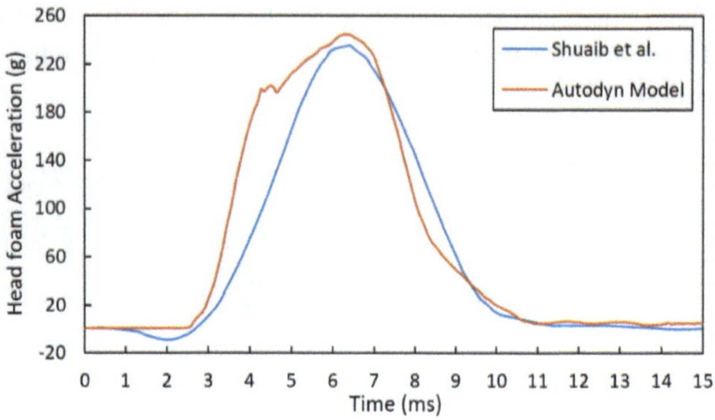

Fig. 3. Acceleration plot compared with the published results in the literature

4.2 Acceleration Plots for Realistic OFHHF

The possible four different locations of helmets are identified and modeled as shown in Fig. 4. The contact and body interaction were applied to mimic the adhesive bond present between the liner and shell during manufacturing. The body interaction between liner and shell was set as 'bonded' and the interaction of the helmet head foam model was set as 'frictionless. The maximum offset value for interaction was set to avoid slippage between bodies. To avoid penetration between the bonded bodies, the surface contact was implemented between the liner and shell as 'bonded'.

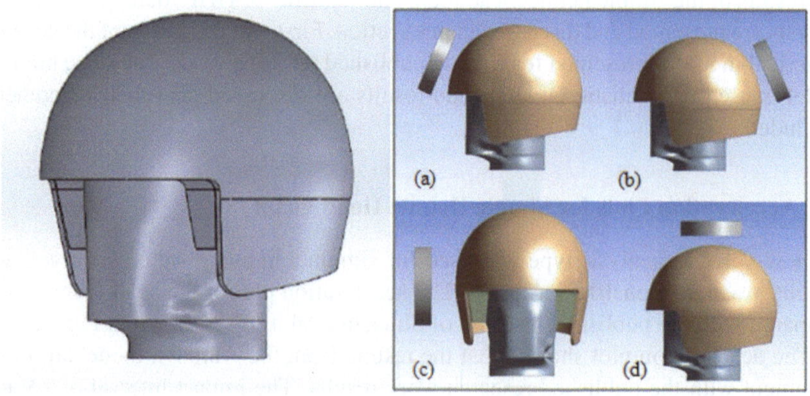

Fig. 4. Model of Head Foam and Open-Faced Helmet with impact locations, (a) Front, (b) Rear, (c) Side, and (d) Top

The tensile stiffness was set as $4 \times 10^5 \frac{N}{m}$. The open-faced helmet head foam model was simulated by impacting the flat anvil at four different locations named top, front, rear, and side according to ECE 22.05 test standard. The open-faced helmet head foam model was impacted at a velocity of 7.5 $\frac{m}{s}$ on the fixed anvil. Since the velocity was less than 10 $\frac{m}{s}$, the analysis setup was low velocity in the explicit section of the module.

This allowed the solver to automatically control the simulation setup. The end time for the analysis was 15 ms. To find the head foam acceleration, an acceleration probe was used on the head foam.

The dynamic analysis of the realistic OFHHF model was carried out at four different locations and the acceleration versus time graphs were obtained. Figure 5 shows the OFHHF acceleration at the top, front, rear, and side locations upon impact with a flat anvil.

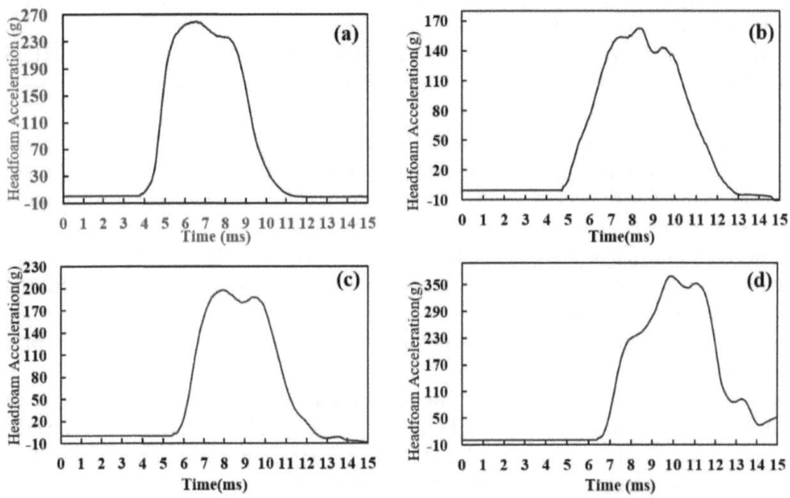

Fig. 5. Impact location for each location, (a) Top, (b) Front, (c) Rear, and (d) Side

As shown in Fig. 5, the maximum acceleration at the top, front, rear, and side locations were 260g, 162g, 197g, and 368g, respectively. The time duration at the front of the helmet was the largest. The plots showed that the impact occurs at different times due to variations in the distance between the anvil and the helmet. By comparing the maximum acceleration of each location to the threshold acceleration of ECE 22.05 standard, the result showed that the helmet impact on the top and side locations exceeded the allowable limit and thus is critical to human head foam.

5 Conclusion

This research work aimed to investigate the impact performance of a more realistic OFHHF combination. EPP and ABS were used as helmet shells and liner materials, respectively. An accurate material model was used for the EPP liner and ABS shell that made up the basic structure of the helmet. For validation purposes, a simple OFHHF model was reproduced from an already published work. Afterward, a more realistic OFHHF was modeled in SOLIDWORKS software. The SOLIDWORKS model was imported to the ANSYS Explicit Dynamics module. The impact of the assembled model upon the flat anvil was investigated using explicit dynamics (autodyn Solver).

The OFHHF model was analyzed using different impact locations, i.e. front, rear, side, and top at a velocity of 7.5 $\frac{m}{s}$. The maximum acceleration at the top, front, rear, and side locations were found to be 260g, 162g, 197g, and 368g, respectively. The results showed that the helmet impact on top and side locations has exceeded the allowable limit as per ECE 22.05 standard and thus is critical to human head injury

References

1. Global status report on road safety (2018). https://www.who.int/publications/i/item/9789241565684. Last accessed 3 Mar 2025
2. Vecino-Ortiz, A.I., Madhuram, N., Sarah, E., Deivis, N.G., Nino, P., Adnan, A.H.: Saving lives through road safety risk factor interventions: global and national estimates. Lancet. **400**(10247), 237–250 (2022)
3. Zhang, X., Bin, Y., Jinguo, W., Xin, L., Ronghua, Z.: Research progress on helmet liner materials and structural applications. Materials. **17**(11), 2649 (2024)
4. Khor, D., et al.: The impact of helmet use on outcomes after a motorcycle crash. Injury. **48**(5), 1093–1097 (2017)
5. Hering, A.M., Derler, S.: Motorcycle helmet drop tests using a Hybrid III dummy. In: International Research Council on the Biomechanics of Injury (IRCOBI) Conference, pp. 307–320, Montpellier, France (2000)
6. Høye, A.: Bicycle helmets–to wear or not to wear? A meta-analyses of the effects of bicycle helmets on injuries. Accid. Anal. Prev. **117**, 85–97 (2018)
7. Zhang, K.: The impact of the policy of wearing safety helmets while driving electric vehicles in Suzhou on road traffic injuries. Ph.D. Thesis, Suzhou University, Suzhou, China (2023)
8. Suphanut, K., Thiraphat, N., Hiroshi, H.: FE analysis of motorcycle helmet performance under severe accidents. Appl. Sci. **12**(11), 5676 (2022)
9. Nadifa, A., Tara, R., Pammla, P., Alexander, M.C.: Do motorcycle helmets reduce road traffic injuries, hospitalizations and mortalities in low and lower-middle income countries in Africa? A systematic review and meta-analysis. BMC Public Health. **22**(1), 1–20 (2022)
10. Thanapong, C., Panuwat, W., Kattreeya, C., Napat, L., Sajjakaj, J., Vatanavongs, R.: Factors affecting severity of motorcycle accidents on Thailand's arterial roads: multiple correspondence analysis and ordered logistics regression approaches. IATSS Res. **46**(1), 101–111 (2022)
11. Ishaq, M., Ahmed, A., Alam, I., Khan, M., Shah, S.N., Ali, M.: Patterns of motorbike accidents related head injuries in patients presenting to a tertiary care hospital of Peshawar. Pak. J. Neurol. Surg. **21**(2), 79–85 (2017)
12. Bachulis, B.L., William, S., Gorrell, G.W., Long, W.B.: Patterns of injury in helmeted and nonhelmeted motorcyclists. Am. J. Surg. **155**(5), 708–711 (1988)
13. Lin, M.R., Hwang, H.F., Kuo, N.W.: Crash severity, injury patterns, and helmet use in adolescent motorcycle riders. J. Trauma Inj., Infect. Crit. Care. **50**(1), 24–30 (2001)
14. Olsen, C.S., et al.: Motorcycle helmet effectiveness in reducing head, face and brain injuries by state and helmet law. Inj. Epidemiol. **3**(8), 27747545 (2016)
15. Global status report on road safety. https://iris.who.int/bitstream/handle/10665/44122/9789241563840_eng.pdf. Accessed 3 Mar 2025
16. Kostopoulos, V., Markopoulos, Y., Giannopoulos, G., Vlachos, D.: Finite element analysis of impact damage response of composite motorcycle safety helmet. Compos. Part B Eng. **33**, 99107 (2002)
17. Xiancheng, Y., Ingrid, L., Itziar, P.S., Atulit, D., Mazdak, G.: The protective performance of modern motorcycle helmets under oblique impacts. Ann. Biomed. Eng. **50**, 1674–1688 (2022)

18. Sun, X.H.: Research on personalized helmet design method based on head shape characteristics. Sci. Technol. Innov. **29**, 177–178 (2018)
19. Michelle, A., Galvanetto, U., Iannucci, L.: Numerical simulations of motorcycle helmet impact tests. Int. J. Crashworthiness. **12**(1), 1–7 (2007)
20. Liu, X.L.: Preparation of modified polyurethane foam/UHMWPE fabric sandwich material and study on low speed impact performance. Ph.D. Thesis, Jiangnan University, Wuxi, China (2023)
21. Bliven, E., et al.: A novel strategy for mitigation of oblique impacts in bicycle helmets. J. Forensic Biomech. **10**(1), 1–6 (2019)
22. Fábio, A.O.F., Ricardo, J.A.S., Mariusz, P., Gonçalo, M.: Helmet design based on the optimization of biocomposite energy-absorbing liners under multi-impact loading. Appl. Sci. **9**(4), 735 (2019)
23. Cernicchi, A., Ugo, G., Iannucci, L.: Virtual modelling of safety helmets: practical problems. Int. J. Crashworthiness. **13**(4), 451–467 (2008)
24. Recycling of polystyrene in helmets. https://www.greenmax-machine.com/the-easily-overlooked-polystyrene-in-helmets-can-be-successfully-recycled-by-the-polystyrene-compactor.html. Accessed 3 Mar 2025
25. Meng, S., Patrik, I., Nils, L.: Evaluation of full-face, open-face, and airbag-equipped helmets for facial impact protection. Accid. Anal. Prev. **191**, 107181 (2023)
26. Mustafa, H., Pang, T.Y., Perret-Ellena, T., Subic, A.: Finite element bicycle helmet models development. Procedia Technol. **20**, 91–97 (2015)
27. Gilchrist, A., Mills, N.: Modelling of the impact response of motorcycle helmets. Int. J. Impact Eng. **15**(3), 201–218 (1994)
28. Kostopoulos, V., Markopoulos, Y., Giannopoulos, G., Vlachos, D.: Finite element analysis of impact damage response of composite motorcycle safety helmet. Compos. Part B Eng. **33**, 99–107 (2002)
29. Caserta, G.D., Iannucci, L., Galvanetto, U.: Shock absorption performance of a motorbike helmet with honeycomb. Compos. Struct. **93**, 2748–2759 (2011)
30. Coelho, R., Sousa, R.A.D., Fernandes, F., Teixeira-Dias, F.: New composite liners for energy absorption purposes. Mater. Des. **43**, 384–392 (2014)
31. Blanco, D.H., Cernicchi, A., Galvanetto, U.: Design of an innovative optimized motorcycle helmet. Proc. Inst. Mech. Eng., Part P: J. Sports Eng. Technol. **228**(2), 95–110 (2014)
32. Teng, T.L., Liang, C.L., Nguyen, V.H.: Development and validation of finite element model of helmet impact test. J. Mater.: Des. Appl. **227**(1), 82–88 (2012)
33. Pinnoji, P., Mahajan, P., Bourdet, N., Deck, C., Willinger, R.: Impact dynamics of metal foam shells for motorcycle helmets. Int. J. Impact Eng. **37**, 274–284 (2010)
34. Shuaeib, F., Hamouda, A., Wong, S., Umar, R.R., Ahmed, M.M.: A new motorcycle helmet liner material: the finite element simulation and design of experiment optimization. Mater. Des. **28**, 182–195 (2007)
35. Khan, S., et al.: Observation of the starting and low speed behavior of small horizontal axis wind turbine. J. Wind Energy. **2014**(1), 527198 (2014)
36. Bhatti, U.N., Khan, S., Ali, S., Horoub, M.M.: A numerical modeling study of the effects of various joint boundary conditions on stiffness behavior of 6dof platform's top plate. In: 2018 9th International Conference on Mechanical and Aerospace Engineering (ICMAE), pp. 243–248. IEEE, New York (2018)
37. Ali, H., Ali, S., Khan, S., Siddiqui, M.U.: Experimental investigations of a single-slope solar still: energy and exergy analysis. Energies. **16**(22), 7659 (2023)

38. Uniform provisions concerning the approval of protective helmets and of their visors for drivers and passengers of motor cycles and mopeds. https://unece.org/sites/default/files/2025-05/R022r5e%20%282%29.pdf. Accessed 3 Mar 2025
39. Gibson, L.J., Ashby, M.F.: Cellular Solids - Structure and Properties, 2nd edn. Cambridge University Press, New York (1997)

Anatomical Image-Based 3D Modeling of Human Tibia and Its Biomechanical Analysis

Muhammad Rizwan[1], Muddassar Sharif[1], Fayyaz[2], Sikandar Khan[2,3](✉), Numan Khan[4], Ashfaq Khan[5], and Salem Bashmal[2,3]

[1] Department of Mechanical Engineering, Pakistan Institute of Engineering and Applied Science, Islamabad, Pakistan
mudasar@pieas.edu.pk

[2] Department of Mechanical Engineering, King Fahd University of Petroleum & Minerals, Dhahran 31261, Saudi Arabia
{g202202260,sikandarkhan,bashmal}@kfupm.edu.sa

[3] Interdisciplinary Research Center for Intelligent Manufacturing and Robotics, King Fahd University of Petroleum & Minerals, Dhahran 31261, Saudi Arabia

[4] Department of Engineering, University of Campania "L. Vanvitelli", Via Roma 29, 81031 Aversa, Italy
numan.khan@unicampania.it

[5] School of Engineering and Built Environment, Sheffield Hallam University, Sheffield, UK
ashfaq.khan@shu.ac.uk

Abstract. Over the years, biomechanics has created strategies for injury prevention and treatment. The tibia is one of the many bones that make up the human body and is a key load-bearing structure. Regular activities including walking, running, and kneeling exert a lot of stress on it. It is especially prone to injuries involving fractures, which are caused due to accidents or abnormal mechanical stress. The intricate geometry of human bones, notably the tibia, makes its modeling extremely challenging. Such intricate designs are often challenging for conventional CAD tools to create accurately. This research aims to investigate the structural integrity of a human tibia under usual physiological loading conditions by modeling it using a polygonal modeling approach based on anatomical MRI and CT scans images. The open-source digital graphics software Blender, which is good for creating complex biological structures, was utilized in the modeling process. At least two tibia anatomical images were required to develop an accurate polygonal model. The study used the open-source meshing software Gmsh to preprocess the model and create a high-quality finite element mesh as polygonal models cannot be used directly for finite element analysis (FEA). Following that, ANSYS Workbench was employed to import the meshed tibia model and carry out a thorough finite element analysis under static compressive and bending loads. Significant details about the tibia's mechanical behavior were found by the bending investigation. It was found that lateral bending produced more stress and deformation than frontal bending, indicating that lateral loading increases the possibility of structural damage. The findings emphasize the importance of considering loading direction when assessing the tibia's biomechanics. These results contribute to the design of improved fracture plates and tibial implants, offering feasible methods to reduce the risk of injuries and speed up patient recovery.

Y. Zhao (Ed.): ICAMM 2025, SPM 96, pp. 273–281, 2025.
https://doi.org/10.1007/978-3-032-09862-7_26

Keywords: Human Tibia · Finite Element Analysis · Biomechanical Analysis · Blender Software · Gmsh Software · MRI and CT scans

1 Introduction

People have used biomechanics to improve performance, avoid injuries, and aid in therapy across time [1]. Forces acting on living things create movements, which are very significant for living things because they are healthy stimuli for growth and development. At the same time, overloading tissues can cause injury. It is biomechanics that provides conceptual and mathematical tools to understand these forces and movements of living things [2]. The tibia or shin bone is the weight-bearing bone of the human body. It has a strong shaft of triangular cross-section, which expands at its proximal and distal ends. Its proximal end is broadened by lateral and medial condyles to help in load bearing. Its upper portion forms knee joint and its lower portion forms ankle joint [3]. Body load is transmitted to the tibia by the femur via viscoelastic material, called meniscus or cartilage. Tibia helps us in walking, running, kneeling, and a lot of other activities of daily life. It is the most used part of our body. Large bones like tibia are more susceptible to fracture [4]. Biological objects have very complex shapes. They have irregular cross-sections and don't have any symmetry. So, their modeling is a challenging task [5–7]. Reverse engineering is used for modeling these complex geometries [8–10]. With the advancement in computing technologies, CAD advancement is manifested in the field of design and analysis [11–13]. Nowadays, CAD is being widely used in biomedical engineering. CAD and solid freeform manufacturing make it possible to design and fabricate very complicated biological scaffolds with functional components. It has numerous applications in clinical medicine, customized medical implant design, and tissue engineering [14, 15]. Medical imaging techniques such as MRI and CT-scan are mostly used for modeling biological objects. Normally, CT scanning is used for hard tissues or bones.

In the past, different studies were carried out on a human tibia to understand its biomechanical behavior. Emon et al. [16] performed finite element analysis on a CT images-based 3D model of tibia under static load. The induced deformation, stresses, and strain were obtained during the normal gait cycle. Kim et al. [17] investigated the flexible bone plates that were used for the fixation of fractured tibia. By applying finite element analysis, it was observed that composite bone plates provide high healing rates for fractured bones. Suryarao et al. [18] studied the human tibia under static loading conditions and stress concentration regions were identified. Soni et al. [19] made a patient-specific human knee joint by using computed tomography (CT) data. This data was converted into a CAD model by image processing software. Stress distribution and total deformation of the knee were observed by assigning two different materials and results were compared. Kaveh et al. [20] investigated the three-point bending of sheep tibia. An experimental study of wet bones and dry bones was done to find out the mechanical properties of bone. It was seen that wet and dry bones had different mechanical strengths. By using a simple finite element model and considering two different bone shaft cross-sections, it was deduced that the elliptical cross-section had greater strength as compared to the circular cross-section. Martens et al. [21] examined multiple human femur bones under a four-point bending load and two different modes of fracture were explored.

Keeping in view the published literature, it was seen that the human bone's CAD modeling was done using medical imaging techniques, i.e. CT scan and MRI, etc. and reverse engineering techniques. These two techniques are time-consuming and expensive. The human bones have a complex shape and therefore, conventional CAD modeling software, i.e. SOLIDWORKS and CREO Parametric, etc., are not able to model them directly. This study focuses on modeling the human tibia bone using a different approach. This research uses a polygonal modeling method using open-source digital art software. This software requires at least two orthogonal reference images showing anatomical features of the tibia bone. The conversion of this non-CAD model to a usable finite element model was done using the open-source software Gmsh. The finite element mesh of the tibia bone was imported to ANSYS, where static structural analysis was carried out under static compressive and bending load. From bending analysis, it was found that lateral bending has greater deformation and stress values. The lateral bending can be more dangerous than the front bending of the tibia. This study provides a base for having an insight into the fractured tibia bone plate and implant design.

2 Materials and Methods

In this study, the human tibia is modeled using digital art software to perform finite element analysis. The methodology followed in this study is shown in Fig. 1 and is explained in the following sections.

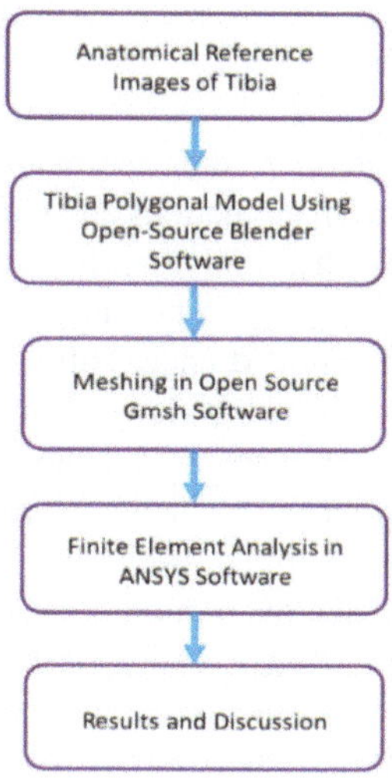

Fig. 1. Methodology Adopted for Current Research

2.1 Anatomical Reference Images of Tibia

To model the human tibia, its exact anatomical reference images were required. In this modeling technique, tibia is modeled by using its two anatomical images. The tibia model is dependent on reference images. The reference images used in this study were obtained from a site of an anatomy expert [22]. These images have appropriate information on the contours of the tibia and are taken with a similar camera on a similar scale. The front-view and right-side view images were used for the approximation of the tibia bone. The two reference images of the tibia are shown in Fig. 2.

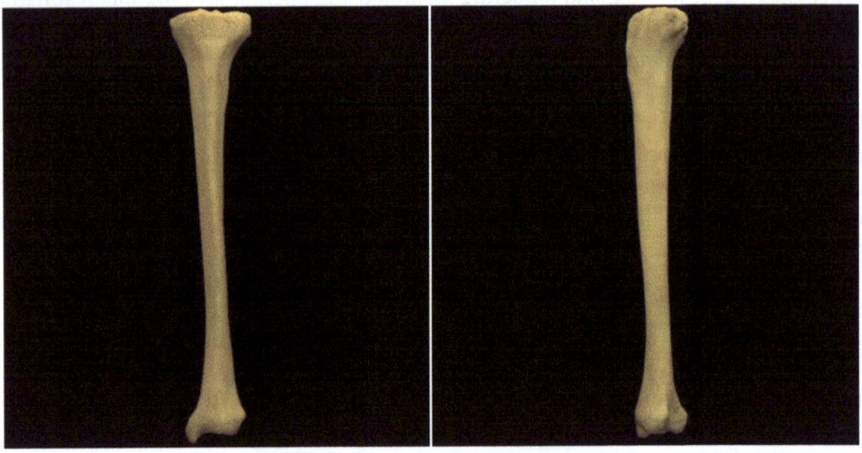

Fig. 2. Reference Images of Tibia [19]

2.2 CAD Modeling of Tibia

The modeling of tibia is done in an open-source software Blender. The Blender is digital art software, which is mostly used in animations and gaming industries. It uses vertices, lines, and faces to make a model. These polygons have no mathematical relation. These polygons can be easily moved in space to get any complex geometry. For modeling purposes, a split screen can be used in Blender software to access different views. To model the tibia, the front and right sides views were used, although the screen can be split into multiple views. The background images were opened in split screens. A cylindrical mesh from the mesh menu was added and this mesh is scaled and aligned with the contours of the bone shaft in both the views. The top and bottom faces of the cylindrical mesh were deleted. Then by extruding these cylinder's vertices and closely approximating the tibia geometry, we got a base mesh of tibia in edit mode of Blender software. The base mesh achieved was crude, and extra work was done on it to get the polygonal model of the tibia. In object mode, by applying smooth operations and a multiresolution modifier, we got a realistic model of the tibia. Also, an appropriate thickness was applied to the model by using a solidify modifier. The STL file of the model was exported from Blender. The final polygonal model of the tibia in Blender software is shown in Fig. 3.

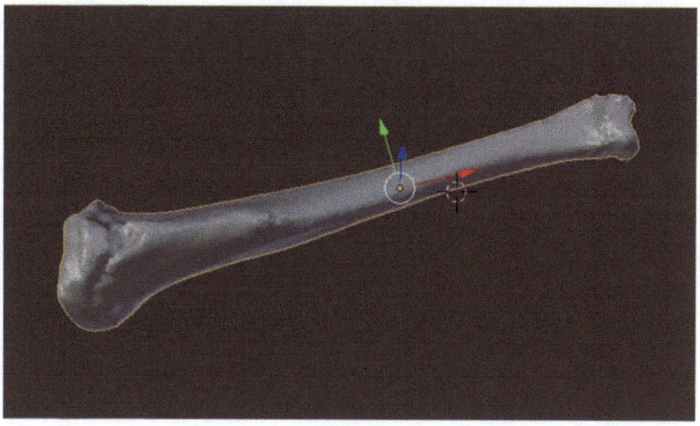

Fig. 3. Model of Tibia in Blender Software

2.3 Meshing

The model obtained from Blender software consists of only polygons. It doesn't have any mathematical identity. The obtained model cannot be used for finite element analysis. The STL file exported from Blender was opened in Gmsh, a finite element mesh generator. Finite element meshing of this polygonal model was done in this open-source software. Mesh was exported in INP format. The quality of this mesh was then improved in the ANSYS workbench. The final mesh has an average skewness of 0.1945 and is shown in Fig. 4.

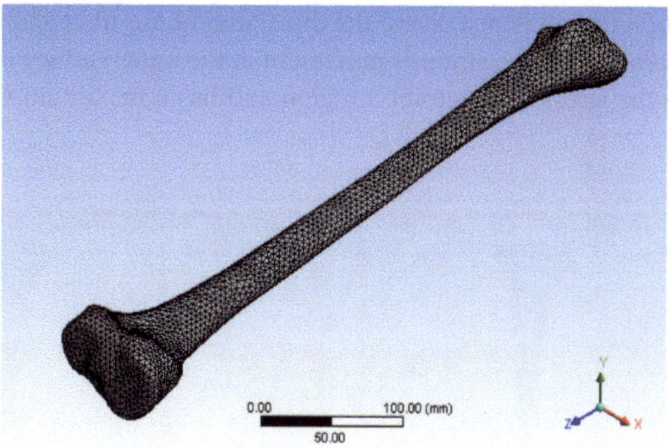

Fig. 4. Meshed Model of Tibia imported in ANSYS workbench

2.4 Material Properties and Analysis Settings

The static structural finite element analysis of the tibia was performed in ANSYS workbench. INP file exported from Gmsh was imported into ANSYS workbench. A good-quality tetrahedral mesh was obtained. To simplify the analysis, only cortical bone was considered and bone properties were assumed to be linear isotropic. The load and boundary conditions were applied according to the literature [18]. The material properties that were used for the analysis of the tibia is shown in Table 1.

Table 1. Material Properties of Tibia [19]

Property	Value
Elastic Modulus (MPa)	17500
Density (kg/m^3)	2000
Poisson's Ratio	0.3
Shear Modulus (MPa)	318.9
Tensile Strength (MPa)	135
Yield Strength (MPa)	125

The bottom end of the tibia was fixed and a pressure of 0.0286 MPa equivalent to half of the body weight of 70 kg was applied on the upper end of the tibia and von Mises stress and strain distribution were studied.

3 Results and Discussion

The contours of stress and strain distribution are shown in Fig. 5. It can be seen that the stress concentration region is just above the distal end of the tibia and has a maximum value of 0.755 MPa. Similarly, strain is maximum at the outer surface of the tibia at the same location as the stress concentration region and has a maximum value is 0.432×10^{-6}.

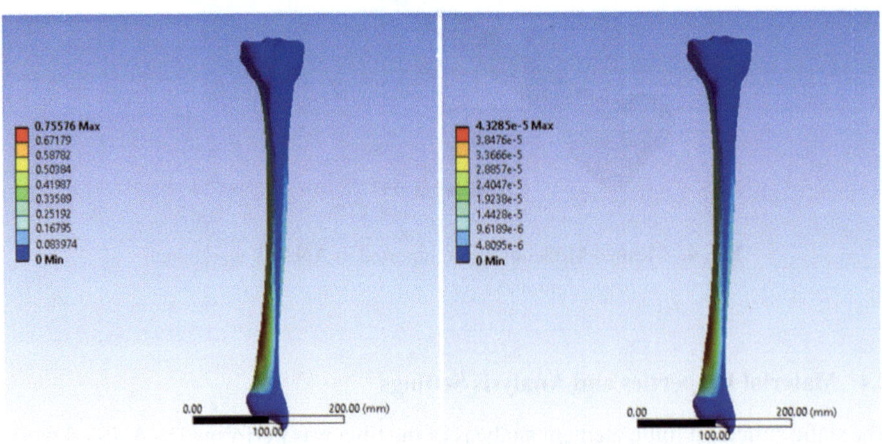

Fig. 5. Von Mises Stress (MPa) and Equivalent Strain

The human tibia experiences complex types of loading during life. Bending load was also chosen to study its behavior. Tibia was treated like a beam. It was fixed from both ends and a pressure of 1 MPa was applied at its center. The pressure was within the range of the critical bending moment. Two different locations of the tibia were selected to apply the load to see its behavior, i.e., Front side of the tibia and the Lateral side of the tibia. The details of these two locations are given as follows.

3.1 Case 1: Front Side of Tibia

Figure 6 shows the total deformation and von Mises stress distribution of first-case loading. Total deformation is maximum at the location where the load is applied. It has a maximum value of 0.052 mm. However, the maximum stress is located at the tensile region of bending opposite to the load applied.

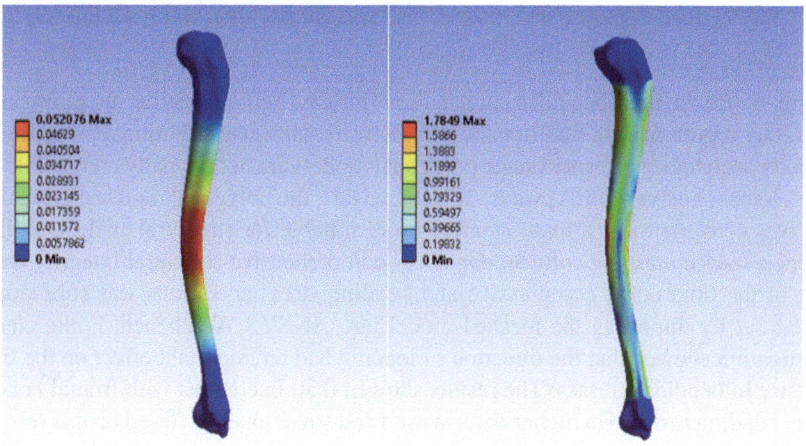

Fig. 6. Total Deformation (mm) and Von Mises Stress (MPa) for Case 1 Loading

3.2 Case 2: Lateral Side of Tibia

The deformation and stress distribution resulting in the case of lateral bending load is shown in Fig. 7. Similar contours were obtained in the second case. The maximum values of total deformation and stress are 0.067 mm and 2.39 MPa, respectively.

4 Conclusion

This study efficiently demonstrated a way to use an effective image-based polygonal modeling approach for modeling the human tibia in three dimensions. Because of its highly intricate geometry and crucial load-bearing role, the human tibia presents significant challenges for traditional CAD modeling methods. To overcome these limitations, an accurate polygonal model was created using Blender software, an open-source

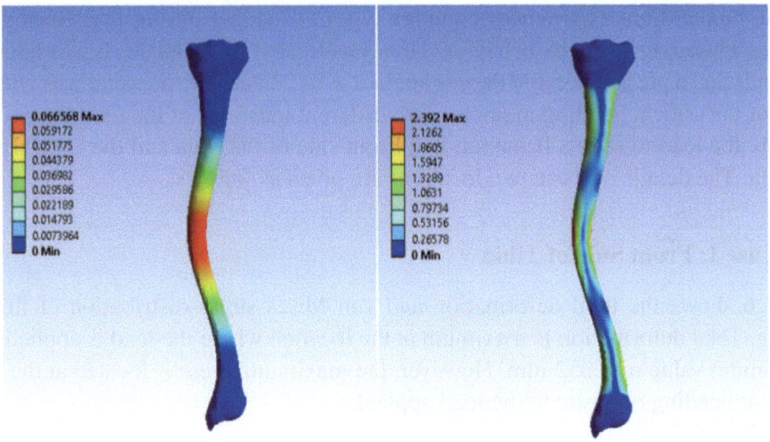

Fig. 7. Total Deformation (mm) and Von Mises Stress (MPa) for Case 2 Loading

graphics software that specializes in handling intricate geometries that are often hard to accurately represent with traditional CAD software. However, additional preprocessing methods need to be performed since polygonal models are not typically compatible with finite element analysis (FEA) ways. To achieve this, the polygonal representation of the tibia was converted into a finite element mesh suitable for structural analysis utilizing the open-source meshing software Gmsh. A comprehensive structural integrity assessment of the tibia under compressive and bending stress conditions was subsequently carried out by importing the meshed model into ANSYS Workbench. Finite element investigation showed that the direction of loading had an important effect on the tibia's response to bending stresses. The results showed that, in contrast with frontal bending, lateral bending resulted in higher deformation and stress extents. Based on this research, lateral bending increases the tibia's risk of structural failure and needs to be considered into account during biomechanical investigations and avoiding injuries. In addition, findings establish a basis for the design of appropriate orthopedic interventions, including implants and tibial bone plates, that are tailored to suit certain loading conditions while improving patient outcomes. The present research shows a cost-effective approach to studying the structural behavior of biological structures by combining Blender software for polygonal modeling and Gmsh for finite element meshing.

References

1. Gordon, D., Robertson, E., Graham, E., Hamill, C.J., Kamen, G., Whittlesey, S.: Research Methods in Biomechanics, 2nd edn. Human Kinetics, Champaign (2013)
2. Duane, K.: Fundamentals of Biomechanics, 2nd edn. Springer, New York, USA (2007)
3. Angeli, L.C.E., Juszczyk, J.M., Juszczyk, M.M.: Shape and function of the diaphysis of the human tibia. J. Biomech. **46**(11), 1882–1892 (2013)
4. Drake, R.L., Vogl, W., Mitchell, A.W.M.: Gray's Anatomy for Students E-Book. Elsevier Health Sciences, Oxford (2009)

5. Balaşa, M., Mihai, S., Filip, V., Negrea, A., Tomescu, G.: Modelling the tibial bone using CAD techniques, starting from the 3D scan model. In: International Conference of Mechatronics and Cyber-Mixmechatronics, pp. 156–165. Springer, New York (2018)
6. Kumar, K.C.N., Chavadaki, S., Chaudhry, V., Pratap, D.S., Amir, S.: Development of an accurate CAD model of femur bone. Mater. Today Proc. **26**(2), 356–363 (2020)
7. Tarniţă, D., Popa, D., Tarniţă, D.N., Grecu, D.: CAD method for three-dimensional model of the tibia bone and study of stresses using the finite element method. Romanian J. Morphol. Embryol. **47**(2), 181–186 (2006)
8. Majstorovic, V., Trajanovic, M., Vitkovic, N., Stojkovic, M.: Reverse engineering of human bones by using method of anatomical features. CIRP Ann. **62**(1), 167–170 (2013)
9. Galantucci, L.M., et al.: Reverse engineering techniques applied to a human skull, for CAD 3D reconstruction and physical replication by rapid prototyping. J. Med. Eng. Technol. **30**(2), 102–111 (2006)
10. Popov, I., Onuh, S.O.: Reverse engineering of pelvic bone for hip joint replacement. J. Med. Eng. Technol. **33**(6), 454–459 (2009)
11. Rehmat, A., Fayyaz, Bashmal, S., Sharif, M., Khan, S.: Numerical modeling of the shape optimization for a commercial car by decreasing drag and increasing stability. Arab. J. Sci. Eng. **48**(9), 12427–12437 (2023)
12. Khan, S., Khan, A., Irfan, M., Hussain, S.: Aerodynamic analysis and dynamic modeling of small horizontal axis wind turbine. In: 2012 International Conference of Robotics and Artificial Intelligence, pp. 117–124. IEEE, New York (2012)
13. Khan, S.: Forward kinematic analysis of a pick-and-place robot for industrial applications. In: 2023 International Conference on Engineering and Emerging Technologies (ICEET), pp. 1–6. IEEE, New York (2023)
14. Sun, W., Starly, B., Nam, J., Darling, A.: Bio-CAD modeling and its applications in computer-aided tissue engineering. Comput. Aided Des. **37**(11), 1097–1114 (2005)
15. Sun, W., Darling, A., Starly, B., Nam, J.: Computer-aided tissue engineering: overview, scope and challenges. Biotechnol. Appl. Biochem. **39**(1), 29–47 (2004)
16. Barua, E., Das, S., Deoghare, A.B.: Development of computational tibia model to investigate stress shielding effect at healing stages. Mater. Today Proc. **5**(5), 13267–13275 (2018)
17. Kim, H., Kim, S., Chang, S.: Bio-mechanical analysis of a fractured tibia with composite bone plates according to the diaphyseal oblique fracture angle. Compos. Part B Eng. **42**(4), 666–674 (2011)
18. Suryarao, N., Sreedevi, K., Ahamad, T.: Analytical study of the tibia bone under static load using finite element method. Int. J. Res. Appl. Sci. Eng. Technol. (IJRASET). **5**(8), 1145–1149 (2017)
19. Soni, A., Modi, Y.K., Agrawal, S.: Computed tomography-based 3D modeling and analysis of human knee joint. Mater. Today Proc. **5**(11), 24194–24201 (2018)
20. Saffar, K.P., JamilPour, N., Rajaai, S.M.: How does the bone shaft geometry affect its bending properties. Am. J. Appl. Sci. **6**(3), 463–470 (2009)
21. Martens, M., Audekercke, R.V., Meester, P.D., Mulier, J.C.: Mechanical behaviour of femoral bones in bending loading. J. Biomech. **19**(6), 443–454 (1986)
22. Anatomy and Physiology. https://www.anatomyexpert.com/lander. Accessed 3 Mar 2025

Synthesis and Properties of Specialty Materials

Research on Sound Absorption of Acoustic/Thermal Materials for Civil Aircraft

Haiquan Wu[1(✉)] and Chenyi Zhao[2]

[1] Shanghai Aircraft Design and Research Institute, State Key Laboratory of Airliner Integration Technology and Flight Simulation, Shanghai, China
wuhaiquan@comac.cc

[2] Shanghai Aircraft Design and Research Institute, Shanghai, China

Abstract. Noise in civil aircraft is an important index that affects passenger experience. Glass fiber porous materials are commonly used in aircraft to reduce noise. Two acoustic simulation models, JCA and impedance tube, are presented to study the sound absorption of porous materials. Through these two acoustic models, the effects of different characteristic parameters of porous materials on acoustic sound absorption and sound pressure changes can be effectively studied. The model provides a method for selecting better aerospace materials and reducing aircraft cabin noise.

Keywords: glass fiber · characteristic parameters · sound absorption · sound pressure

1 Introduction

Noise is one of the important factors affecting passenger comfort in civil aircraft. At present, the commonly used thermal/acoustic insulation materials on the aircraft are glass fiber, foam, etc. These thermal/acoustic materials are porous materials. Therefore, in order to select suitable thermal/acoustic insulation materials for aircraft, the sound absorption property of porous materials have great significance. The acoustic equivalent model of porous sound-absorbing materials can be divided into empirical models and phenomenological models. The Delany-Bazley model is commonly used empirical model, which can not provide accurate prediction. The phenomenological model considers the propagation of sound waves in the pores and cavities of materials [1, 2]. The Johnson-Champoux-Allard(JCA) phenomenological model has been applied by many scholars. JCA phenomenological model has 5 characteristic parameters, including resistivity, porosity, tortuosity, thermal characteristic length and viscous characteristic length [3, 4].

Y. Zhao (Ed.): ICAMM 2025, SPM 96, pp. 285–295, 2025.
https://doi.org/10.1007/978-3-032-09862-7_27

2 Thermal/Acoustic Insulation Materials

2.1 Porous Sound-Absorbing Materials in Aircraft

Porous sound-absorbing materials achieve their function by harnessing thermal and viscous effects within their pores [5]. These materials contain numerous interconnected micro-pores that convert sound waves into heat energy via friction, thereby dissipating the energy [6].

Porous sound-absorbing materials primarily consist of fiber and foam types [7], with fiber materials further categorized into inorganic, organic, and metal fibers [8]. Inorganic fibers are extensively employed due to their resistance to acids and alkalis, robust flame retardancy, and effective sound absorption [9].Currently, in compliance with civil aviation regulation emphasizing fire prevention, mainstream civil aircraft utilize glass fiber as the primary material for thermal and acoustic insulation, is shown in Fig. 1.

Fig. 1. Glass fiber.

2.2 JCA Theoretical Model

The JCA model describes the viscous and thermally reflective characteristics of acoustic energy in porous materials with arbitrarily shaped pores. The model incorporates two primary physical parameters: effective density $\rho e(w)$ and bulk modulus $k(w)$, enabling the calculation of characteristic impedance and propagation constant. The expression for $\rho e(w)$ as in Eqs. (1) and Eqs. (2).The expression for $k(w)$ as in Eqs. (3) and Eqs. (4).

$$\rho e(w) = \alpha\infty\rho 0\left(1 - \frac{j\sigma\Phi G(w)}{\rho 0\alpha\infty w}\right) \tag{1}$$

$$G(w) = \left(1 + \frac{4j\alpha\infty^2\eta\rho 0w}{\sigma^2\Phi^2\Lambda^2}\right)^{\frac{1}{2}} \tag{2}$$

$$K(w) = \frac{\gamma P0}{\gamma - (\gamma - 1)H(w)} \tag{3}$$

$$H(w) = \frac{1}{1 + \frac{8\eta}{j\Lambda'^2\rho 0\,\mathrm{Pr}\,w}\left(1 + \frac{j\Lambda'^2\rho 0\,\mathrm{Pr}\,w}{16\eta}\right)} \tag{4}$$

$\rho 0$ is the equilibrium pressure, γ is the specific heat capacity of air, Pr is the Prandtl constant, w is the angular frequency, and f is the frequency,. The five parameters in (1)–(4) (resistivity, porosity, tortuosity, thermal characteristic length and viscous characteristic length) are important parameters that affect the sound absorption coefficient, and this model is widely used in the field of acoustics research [10].

Airflow resistance(R): R of porous materials denotes the resistance encountered by air particles as they pass through the pores and gaps in the material, as in Eqs. (5). At low, R, sound waves pass through the material without being attenuated; at high, sound waves are reflected by the material or enter the structure [11]. R is defined as the ratio of the pressure difference across the specimen to the volumetric velocity of the airflow passing through it (in Pa·s/m^3 for R).

$$R = \frac{\Delta p}{qv} \tag{5}$$

Specific airflow resistance (Rs) is the ratio of the pressure difference on both sides of the specimen to the average airflow velocity passing through the specimen, which is the product of flow resistance and area, as in Eqs. (6).

$$Rs = R \cdot A \tag{6}$$

If the porous material is isotropic, the airflow resistivity (σ) is the Rs divided by the thickness of the specimen, specifically representing the flow resistance per unit thickness, as in Eqs. (7).

$$\sigma = \frac{Rs}{d} \tag{7}$$

Porosity(φ): φ is the proportion of the liquid phase volume (V_a) within porous materials relative to the total volume (V_m), as in Eqs. (8). Only the volume of open pores is considered in porous materials, with the closed pores forming part of the solid network or framework. In the case of rigid porous materials, it is sufficient to consider only the open-pore ratio, because the air within closed pores does not impact the transmission of sound waves, thereby not affecting the sound absorption capabilities of the material. Hence, the open-pore ratio is the only significant factor to be taken into account [12].

$$\varphi = \frac{Va}{Vm} \tag{8}$$

Tortuosity ($\alpha\infty$): $\alpha\infty$ is only related to the skeleton of porous materials and represents the appearance of fluid pathways within porous media, as in Eqs. (9).

$$\alpha\infty = \frac{\int VL|v|^2 dV}{\left|\int VLvdV\right|^2} \tag{9}$$

In (9), v is the microscopic flow velocity of the fluid, VL is the homogenized volume of the material.

Viscous characteristic length (Λ), Λ introduced by Johnson and Coll, primarily represents the impact of liquid viscosity and inertia on the sound absorption properties of porous materials within the high-frequency range, as in Eqs. (10).

$$\Lambda = \frac{\int Vt|v|^2 dV}{\int St|v|^2 dS} \tag{10}$$

In Eqs. (10), Vt is the fluid volume contained in the air, and St is the contact surface between the fluid in the hole and the skeleton. According to (10), Λ related to the geometric characteristics of the pores in the medium and is independent of other factors.

Thermal characteristic length (Λ'), Λ' introduced by Champoux and Allard, primarily represents the extent of heat exchange between the liquid phase within the voids of porous materials and its framework, as in Eqs. (11) [13].

$$\Lambda' = \frac{\int Vtdv}{\int StdS} \tag{11}$$

3 Establishment of Simulation Model

3.1 Verify the Correctness of Model

This paper develops a JCA sound absorption model using COMSOL software. The model employs a two-dimensional pressure acoustic framework. Initially, the model's accuracy is validated, with parameter choices outlined in Table 1 [14].

Table 1. Parameter Selection

Characteristic Parameters	Reference Value	Unit
Resistivity	29000	$Pa \cdot s/m^2$
Porosity	0.9792	1
Tortuosity	1	1
Thermal characteristic length	0.000174	m
Viscous characteristic length	0.000087	m

Figure 2 show the acoustic absorptivity results. The acoustic absorptivity coefficient in Fig. 2 is basically the same as the black line in section 3 of chapter 4 in [15]. Reaching a maximum value of 0.99 at 3000 Hz and 0.94 at 6000 Hz. The acoustic absorptivity shows a trend of first increasing, then decreasing, and then slowly increasing. Through comparison, the correctness of the model is verified.

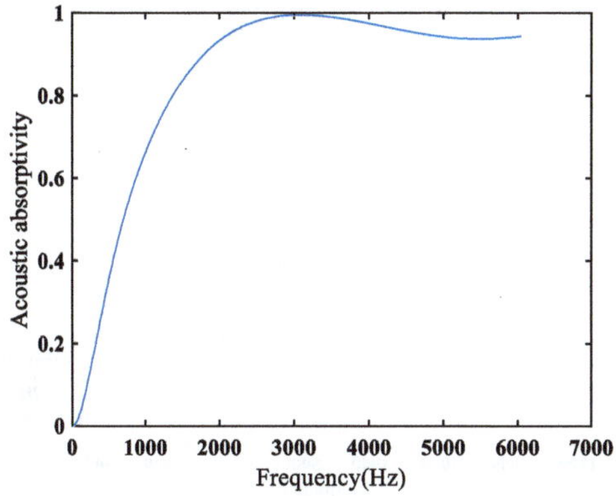

Fig. 2. Verification Model.

3.2 JCA Model of Glass Fiber

Porous materials have the characteristics of low density, light weight, good shock absorption, and higher acoustic absorptivity than traditional synthetic materials. In the aviation industry, fire prevention needs also need to be considered.

At present, the most commonly used thermal/acoustic insulation materials for civil aircraft is glass fiber. Through measurement and investigation, several characteristic parameters of fiberglass are shown in Table 2. Result is shown in Fig. 3.

Table 2. Characteristic Parameters of Fiberglass

Characteristic Parameters	Reference Value	Unit
resistivity	10000	Pa·s/m^2
porosity	0.99	1
tortuosity	1.1	1
thermal characteristic length	0.000384	m
viscous characteristic length	0.000192	m

From the Fig. 3, it can be seen that the thickness of glass fiber has a significant impact on its acoustic absorptivity . As the thickness increases, the maximum acoustic absorptivity coefficient shifts from high frequency to low frequency, and the extreme value of the maximum coefficient increases. Aircraft engine noise is an important indicator that affects passenger experience, with a noise frequency of approximately 5000 Hz.

Therefore, laying 3-inch glass fiber in the cabin near the aircraft engine can have a good effect.

Taking 3-inch glass fiber as an example, the background sound pressure and sound pressure level are shown in Fig. 4, the scattered sound pressure and sound pressure level are shown in Fig. 5, and the total sound pressure and sound pressure level are shown in Fig. 6.

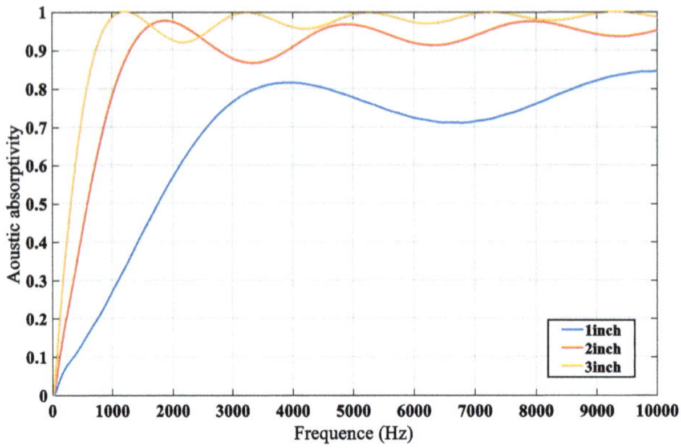

Fig. 3. Acoustic Absorptivity of Glass Fiber.

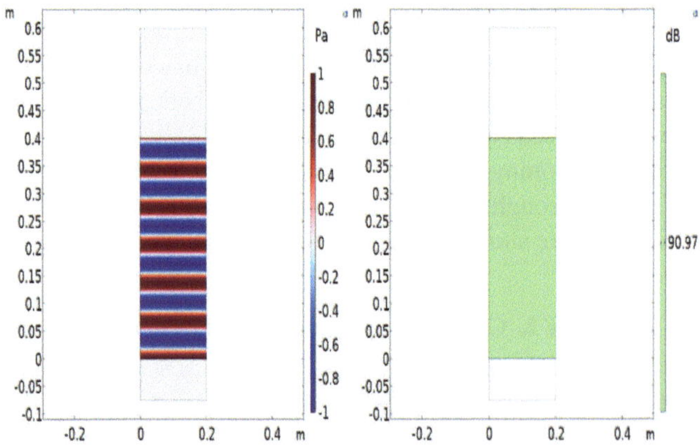

Fig. 4. Background Sound Pressure and Pressure Level.

By analyzing the sound pressure level spectrum, we can gain a deeper understanding of the characteristics of noise energy distribution and noise transmission paths. This provides strong basis and guidance for noise control and reduction. In practical applications, appropriate noise reduction measures should be selected according to the specific situation to ensure the noise reduction effect while considering both economy and feasibility.

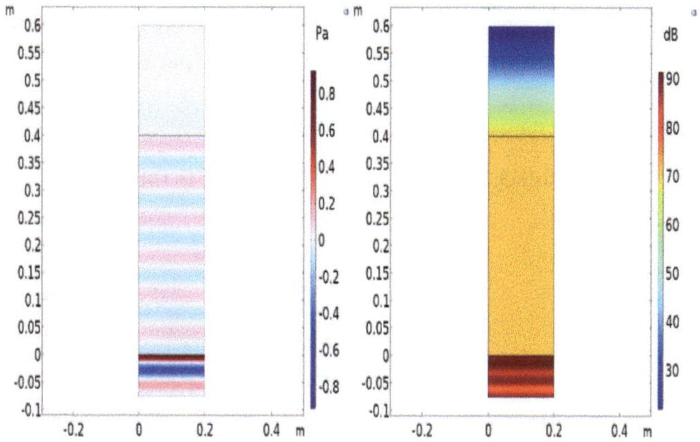

Fig. 5. Scattered Sound Pressure and Pressure Level.

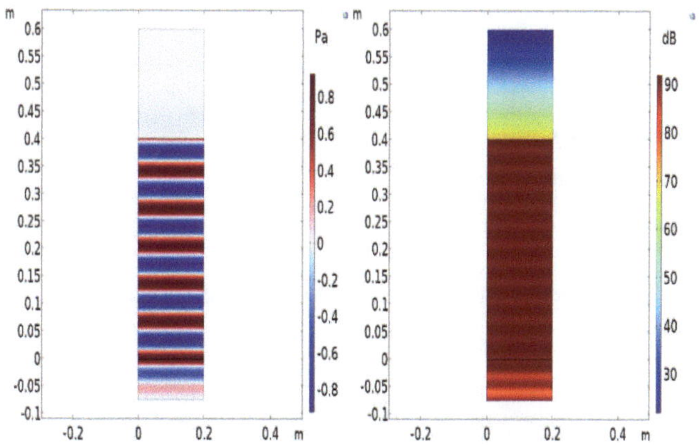

Fig. 6. Total Sound Pressure and Pressure Level.

4 Impedance Tube Simulation Testing Method

Impedance tube is used for studying the acoustic propagation characteristics of noise control materials, measuring the acoustic parameters of small specimens under vertical incidence conditions, including absorption coefficient, reflection coefficient, and characteristic impedance [16]. The entire system consists of standing wave tubes, high-end microphones, data acquisition cards, computers, etc. The testing software automatically calculates the sound absorption coefficient and can generate a coordinate curve between the sound absorption coefficient and frequency [17]. The frequency and amplitude of the instrument output signal can be freely set within the specified range. Due to the fact that only a small amount of sound-absorbing materials are needed as specimens, parameter research and design of sound-absorbing materials are particularly suitable. Because this article proposes a COMSOL three-dimensional impedance tube testing and simulation method [18].

By comparing with the same data of the first model, the correctness of the impedance tube simulation model is confirmed. Several characteristic parameters of impedance tube are shown in Table 3.

Table 3. Parameters of IMPEDAMCE Tube

Parameters	Reference Value	Unit
Impedance tube diameter	0.1	m
Impedance tube length	0.52	m
Sensor distance	0.05	m
The distance between two sensors	0.08	m
Maximum frequency	1600	Hz
Material thickness	0.254	m
Resistivity	10000	$Pa \cdot s/m^2$
Porosity	0.99	1
Tortuosity	1.1	1
Thermal characteristic length	0.000384	m
Viscous characteristic length	0.000192	m

The mesh of the impedance tube simulation model is shown in Fig. 7.

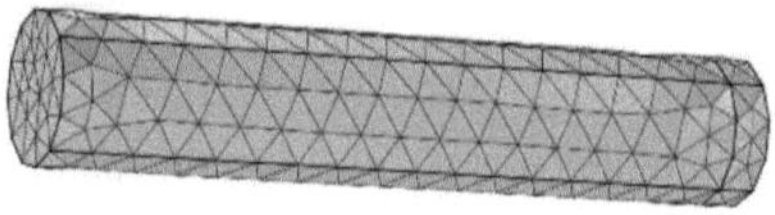

Fig. 7. Mesh of a simulation model.

The background sound pressure and sound pressure level are shown in Fig. 8, the scattered sound pressure and sound pressure level are shown in Fig. 9, and the total sound pressure and sound pressure level are shown in Fig. 10.

The acoustic absorptivity is shown in Fig. 11.

As can be seen from Figs. 7, 8, 9, 10 and 11, the acoustic performance simulation of the impedance tube can obtain the sound absorption coefficient and acoustic pressure field of the porous material. In this model, background sound pressure is 0, the scattered sound pressure is equal to the total sound pressure, and the total sound pressure is equal to the sum of scattered sound pressure and background sound pressure.

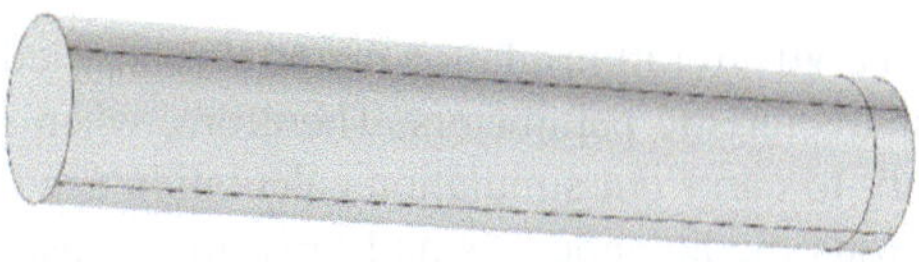

Fig. 8. Background Sound Pressure and Pressure Level of Impedance Tube.

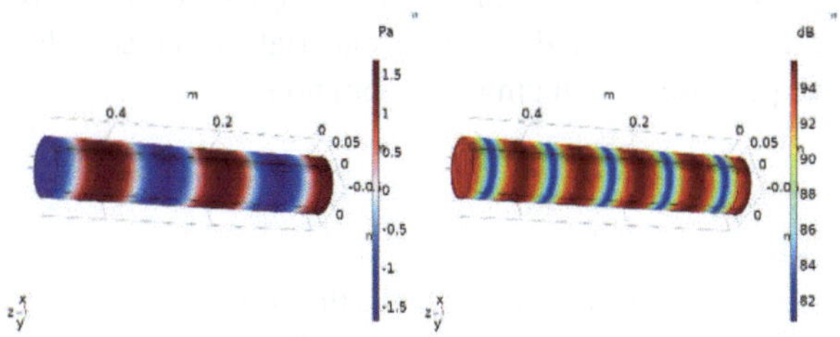

Fig. 9. Scattered Sound Pressure and Pressure Level of Impedance Tube.

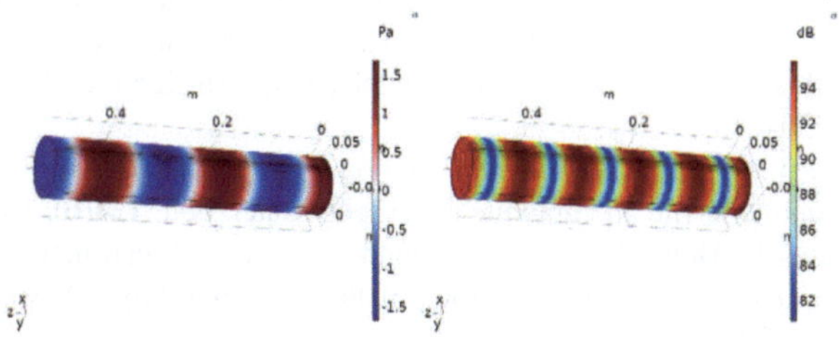

Fig. 10. Total Sound Pressure and Pressure Level of Impedance Tube.

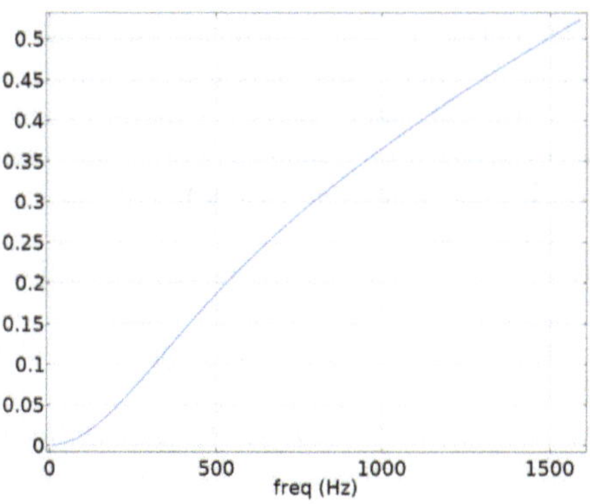

Fig. 11. Acoustic absorptivity.

5 Conclusion

The sound absorption of civil aircraft insulation material is affected not only by its thickness, but also by its characteristic parameters. Therefore, when selecting new porous materials, it is necessary to carry out simulation calculation according to their characteristic parameters. Porous media acoustics and pressure acoustics frequency domain modules established by COMSOL software, and the corresponding sound absorption coefficient can be obtained by inputting different characteristic parameters. The noise spectrum of different areas in the cabin of civil aircraft is different, and the sub-regional design is particularly important, especially near the engine, the noise spectrum is higher, and it is necessary to select the suitable porous materials for noise reduction processing, so as to achieve the purpose of reducing the cabin noise.

References

1. Champoux, Y., Allard, J.F.: J. Appl. Phys. **70**(4), 1975 (1991)
2. Zhongyun, L., Wang inherits.: Porous sound-absorbing material sound-absorbing performance optimization and engineering application. J. Measur. Test. Technol. **03**, 85 (2024). https://doi.org/10.15988/j.carolcarrollnki.1004-6941.2024.3.025
3. Johnson, D.L., Koplik, J., Dashen, R., et al.: J. Fluid Mech. **176**(1), 379 (1987)
4. Leilei, P., Shuo, F., Yuxuan, W., et al.: Acoustic insulation research status and progress of the functional textile materials. J. Mod. Text. Technol. **31**(6), 216–225 (2023). https://doi.org/10.19398/j.att.202304013
5. Yang, Y., Wang, J.: Synthesis and characterization of a microwave absorbing material based on magnetoplumbite ferrite and graphite nanosheet. Mater. Lett. **124**(6), 151–154 (2014)
6. Huang, D., Kang, F., Dong, C., et al.: A second-order cross fractal meta-material structure used in low-frequency microwave absorbing materials. Appl. Phys. A Mater. Sci. Process. **115**(2), 627–635 (2014)
7. Ran, T., Ying, H., Qing, R.: Composites of fuselage acoustic analysis of curved plate and sound insulation cotton optimization design. J. Civ. Aircr. Des. Res. **4**, 82–87 (2022). https://doi.org/10.19416/j.carolcarrollnki.1674-9804.2022.04.014
8. Bai, P.F., Ber, L.Y., He, S., Gao, L., Liu, L.: Common sound absorbing materials and sound absorbing mechanism. Shanxi Chem. Ind. **38**(03), 40–42. (in Chinese) (2018)
9. Zhang, W.T.: Study on Structure and Properties of High-Efficiency Sound-Absorbing Fiber Materials. South China University of Technology (2017) (in Chinese)
10. Chan, P., Liu, Q., Xu, B.: Acoustic properties of porous materials. Sichuan Build. Mater. **49**(12), 13–14 (2023) (in Chinese)
11. C522-03, Standard test method for airflow resistance of acoustical materials (2009)
12. Hailin, C., Li, D., Dan, L.: The sound absorption performance analysis and optimization of the fibrous porous materials. J. Electroacoust. Technol. **12**, 46–48 (2022). https://doi.org/10.16311/j.audioe.2022.12.004
13. Party thought is. Flexible fiber porous sound absorption performance research. The Northern Industrial University (2024). https://doi.org/10.26926/dcnki.Gbfgu.2024.000213.
14. Zhenmao, Z.: Sound absorption performance analysis of porous fiber materials in cruise liner based on Comsol. Installation. **06**, 84–87 (2024)
15. Yuecan, L.: Research on the Sound Absorption Performance of Porous Materials. Yanshan University, Hebei (2021)

16. Huading, L., Hequn, M.: Impedance tube not the standard size samples of normal incidence absorption coefficient measurement. Acta. **05**(13), 1012–1020 (2023). https://doi.org/10.15949/j.carolcarrollnki.0371-0025.2023.05.010
17. Yumi, H., Biao, Z., Chao, X., et al.: Based on impedance tube of the air conditioning with soundproof cotton performance study. J. Electron. Sci. Technol. **S1**, 181–184 (2023). https://doi.org/10.19784/j.carolcarrollnkiissn1672-0172.2023.99.040
18. Ying, Z., Xinxin, G., Zheng, H.: Research on impedance tube calibration based on acoustic simulation. Metrol. Sci. Technol. **67**(05), 58–63. (in Chinese) (2023)

Experimental Analysis on the Effect of Catalyst Concentration and Fuel Flow Rate Toward Formation of Carbon Nanotubes in Quasi-Pyrolysis Chamber Using Methane Diffusion Flame

Muhammad Amirrul Amin Moen[1,2], Dhevanessh R. Chandraguptha[1], Mohd Fairus Mohd Yasin[1,2], and Norikhwan Hamzah[1,2](✉)

[1] Faculty of Mechanical Engineering, Universiti Teknologi Malaysia, 81310 UTM Johor Bahru, Johor, Malaysia
norikhwan@utm.my
[2] High Speed Reacting Flow Research Laboratory (HiREF), Faculty of Mechanical Engineering, Universiti Teknologi Malaysia, 81310 UTM Johor Bahru, Johor, Malaysia

Abstract. Utilization of a quasi-pyrolysis chamber (QPC) for synthesizing carbon nanotubes (CNT) in a diffusion flame promises a low-cost and scalable process. However, the flame environment is complex and requires an in-depth understanding of the interrelation of flame and catalyst parameters in optimizing CNT growth in flame using QPC. Hence, this study investigates the effects of vapor catalyst concentration and carrier gas flowrate parameters on the growth rate and morphology of the synthesized CNTs in a QPC via methane diffusion flame. Methane fuel is used as a catalyst carrier gas, with 100% of the fuel flowing through the catalyst nebulizer to the burner outlet. It was observed that a catalyst concentration of 0.05 M obtained the maximum yield of CNTs with minimized bundling and agglomeration. Greater catalyst concentrations enhanced the CNT yield by 20%, but with notable structural irregularities. Lower catalyst concentrations of 0.005 M caused sparse growth due to reduced catalytic activity. In the case of methane flow rates, 0.4 slpm shows the most consistent growth of CNT. A higher flow rate of 0.6 slpm decreased CNT yield by 90%, disrupting flame stability, while a lower flow rate of 0.2 slpm resulted in minimal catalyst carryover into the flame and poor CNT yield. Moreover, modification of the methane flow rate changed the flame structure, extending the flame at excessive flow rates and decreasing temperatures at the chamber inlet, where the majority of CNT growth takes place. The results underscore the necessity to optimize catalyst and methane parameters to achieve high-yield, economical CNT synthesis in advancing flame synthesis as a viable industrial process.

Keywords: Carbon nanotubes · quasi-pyrolysis chamber · flame synthesis · vapor catalyst

Y. Zhao (Ed.): ICAMM 2025, SPM 96, pp. 296–305, 2025.
https://doi.org/10.1007/978-3-032-09862-7_28

1 Introduction

Synthesis process using flame-based techniques has the potential to completely transform the conventional CNT synthesis process, reducing production cost while minimizing the carbon footprint associated with its manufacturing process [1]. Theoretically, the autothermal condition and the constant supply of carbon sources inside a flame environment enable CNT growth faster than CVD, greatly lowering the overall carbon footprint and production costs [2]. For the past three decades, researchers have been trying to develop an overall understanding of the complex flame environment for flame-based synthesis methods [3]. Many researchers are currently exploring novel approaches by combining modeling and experimental effort to capitalize on the advantages promised by flame synthesis process [4]. The development of a flame-based, industrial-scale synthesis method for carbon nanotubes has, however, remained rather limited. The challenge of creating a stable and well-controlled synthesis process has been shown to be a significant barrier to the flame synthesis method's widespread use [5].

Quasi-pyrolysis chamber (QPC) usage is one of the current strategies being used to enhance CNT growth in flame [6]. The QPC is a stainless-steel tube, placed with a certain distance on top of a diffusion flame burner to separate the CNT growth region from away from the soot and highly oxidizing area of the flame sheet. With wire mesh at the inlet of the QPC facing the burner outlet to quench flame formation, the QPC capable of producing a high-temperature, high-carbon area within it that is ideal for CNT growth using vapor type catalyst [7]. Nevertheless, recent application of QPC in CNT synthesis means that there is still much to learn about how crucial parameters related to fuel, catalyst, and flame affect the CNT growth during the flame synthesis process in a QPC.

Previous works show that CNT inception and growth in flame environment occurred on top of catalyst nanoparticles with availability of carbon precursor in high temperature environment around 600 to 900 °C [8]. Even though it has been shown that utilization of QPC with diffusion flame capable of producing high temperature region within the range of CNT growth temperature, the effect of catalyst parameters such as catalyst solution concentration and carrier gas flow rate toward CNT morphology and growth rate has yet to be explored. In CVD, it has been shown that catalyst concentration in vapor type catalyst plays a vital role in determining CNT growth efficiency [9]. It also has been shown that precise control of catalyst solution concentration and the catalyst carrier gas flow rate enable manipulation of CNT yield and morphology [10]. Furthermore, the type of carrier gas also has shown significant affects in morphology of the synthesized CNT in CVD [11].

This study experimentally assesses the impact of catalyst concentrations and the flow rate of the catalyst vapor carrier gas on the morphology and yield of produced CNTs. The synthesis process is conducted using a nickel-based catalyst in an ethanol solution at varying concentrations, with methane gas serving as the carrier gas and fuel, utilizing QPC at a set height above the burner (HAB). The results offer preliminary insights into the relationship between catalyst concentration and carrier gas flow rate in the synthesis of CNTs, contributing to both fundamental knowledge of the process and practical guidance for future industrial-scale developments.

2 Experimental Details

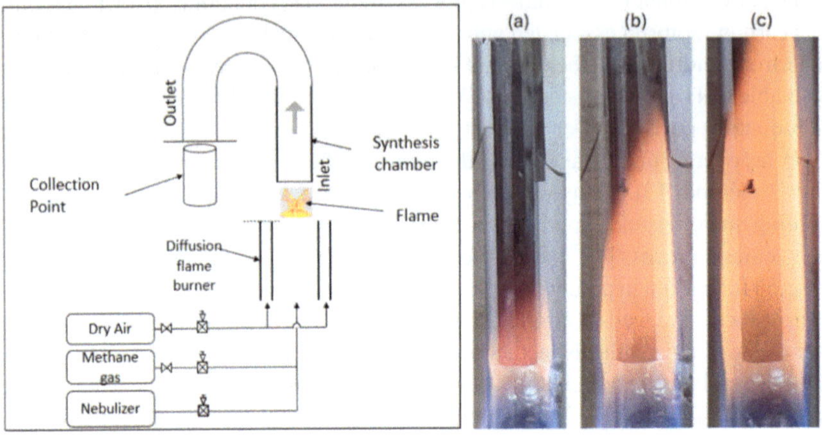

Fig. 1. Schematic diagram and actual images of flame shape surrounding the QPC at methane flow rate of (a) 0.2 slpm (b) 0.4 slpm and (c) 0.6 slpm.

Figure 1 shows a schematic diagram of a co-flow diffusion flame with a QPC positioned atop the burner. Methane gas (99.5% purity, Alpha Gas Solution Sdn. Bhd.) serves as fuel, whereas dry air is used as oxidizer. The burner design enables independent delivery of fuel and oxidizer through concentric rings with fuel and oxidizer through the inner and the outer ring respectively. The CNT synthesis takes place within the QPC, a stainless-steel chamber with an external diameter of 9.5 mm, positioned at a HAB of 20 mm. To ensure quasi-pyrolysis condition within the chamber, a wire mesh with size #20 is installed at the QPC's inlet which quenches the flame and allow carbon precursor to enter the chamber. More importantly, the wire mesh also used to redistribute the flame flow field and create engulfing flame sheet surrounding the QPC creating ideal growth temperature for CNT within the chamber. Methane flow rates are adjusted from 0.2 to 0.6 slpm with fixed oxidizer flow rate of 3.7 slpm. The methane gas flow is directed to go through catalyst nebulizer before entering the burner to carry-over the floating catalyst mist produced by the nebulizer into the QPC.

Catalyst solution is prepared by dissolving nickel (II) nitrate hexahydrate ($Ni(NO_3)_2{\cdot}6H_2O$) in ethanol according to concentration required. Ethanol is selected as diluent for the catalyst solution to ensure that fast drying of the vapor inside the flame and minimize flame temperature reduction. Furthermore, vaporization of ethanol inside the flame will contribute as fuel, further enhance the temperature inside the QPC. Utilization of water as diluent is not suitable since with will absorb heat from the flame to vaporize the water, reducing overall flame temperature. In the work, concentration of catalyst solution is varied from 0.005 to 0.5 M. The catalyst was dissolved in 50 ml of ethanol until fully dissolved and was then transferred to the nebulizer. The nebulizer (Omron; model NE-U780) capable of producing vapor with average particle size of 4.61 μm and fixed nebulizing rate of 3 ml/min. The methane gas then carry-over the nebulized catalyst solution to the outlet of the burner. At the outlet of the burner where the methane diffusion flame is established, the catalyst solution then exposed to higher temperature which quickly vaporizes ethanol, producing nickel nanoparticles as CNT growth catalyst.

As the flame established surrounding the QPC, high temperature ensures that catalyst nanoparticles are formed and become growth sites for CNT inside the QPC. Due to buoyancy effect of the gas inside the QPC, the floating synthesized CNT then carried over to the outlet of the QPC. Bottle filled cold water is used to collect the CNT as the exhaust gas from the QPC carrying synthesized CNT flow through the surface of the water. The synthesis process is performed for around 1 h until the catalyst solution is exhausted. To collect the synthesized CNT, the water inside the bottle is vaporize slowly using hot plate at temperature around 70 °C until all water is vaporize. The weight of the collected CNT then is measured using analytical balance (ATX-R, Shimadzu), and the CNT are analyzed as is. The synthesized CNTs were characterized using Field Emission Scanning Electron Microscopy (FESEM) and Raman Spectroscopy.

3 Results and Discussion

3.1 Flame Shapes at Varied Fuel Flow Rate

The influence of methane flow rate on the flame structure surrounding the QPC was studied at 0.2 slpm, 0.4 slpm, and 0.6 slpm while maintaining HAB at 20 mm as shown in Fig. 1(a) to (c). The lower range of the fuel flow rate is set at 0.2 slpm since at this flow rate, the flame is started to be too lean to produce high temperature suitable for the growth in QPC. While the higher range of 0.6 slpm is limited by the flame length produced which is beyond the length of the QPC. At 0.2 slpm, the flame is relatively short covering only region near to the inlet of the QPC as shown in Fig. 1(a). Although the established flame is strongly anchored to the burner outlet, the limited supply of fuel, which is half of oxidizer, producing globally lean diffusion flame indicated by the blue flame. Furthermore, as the flame outside of the QPC covers only the region near to the inlet, most of the region downstream of the QPC is not suitable for CNT growth because of the low temperature. With methane flow rate doubled to 0.4 slpm, the formed flame height is relatively longer and covering most of the QPC as shown in Fig. 1(b). Higher methane supply producing combustion process at richer global equivalence ratio with soot formation producing yellowish flame. At 0.6 slpm methane flow rate, the velocity of the fuel at the outlet of the burner further increased causing the flame to be lengthened further beyond the QPC, reducing the efficiency of fuel utilization to produce CNT. With much higher methane supply, the combustion occurred at a very rich global equivalence ratio producing characteristically long lazy flame with abundance of soot. The dimmed glow of the QPC tube observed at the region near the inlet indicates that the temperature inside the QPC at the region is relatively low compared at 0.4 slpm flow rate.

3.2 CNT Morphologies

Figure 2(a) to (c) shows FESEM analysis images of CNT synthesized at varied catalyst concentrations. General observations of the morphological conditions of the CNT

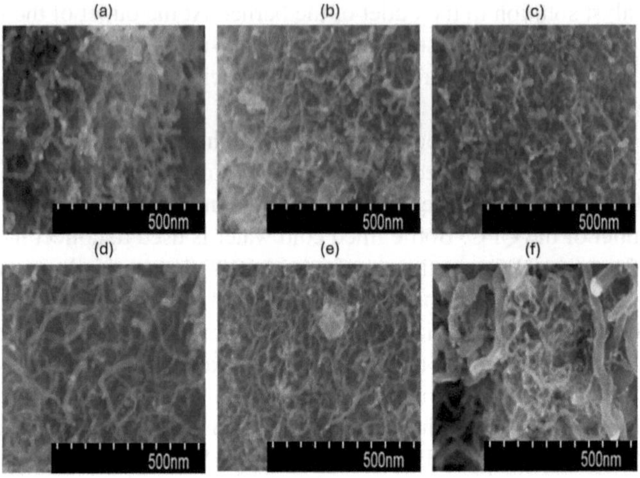

Fig. 2. FESEM images of CNT at outlet of the synthesis chamber with catalyst concentration of (a) 0.005 M (b) 0.05 M (c) 0.5 M and fuel flow rate of (d) 0.2 slpm (e) 0.4 slpm and (f) 0.6 slpm.

produced do not indicate any significant difference between the samples. The CNTs are mostly curled and entangled with catalyst nanoparticles disperse in between the CNT strands. This is a typical morphology for CNT synthesized in flame environments where the carbon precursor concentrations and temperature are not uniform throughout the growth region [12].

Visual observation of the FESEM images of the synthesized CNTs at various fuel flow rate shows different general morphologies as shown in Fig. 2(d) to (f). In previous analysis of CNT synthesized in varied catalyst concentration, the condition with the QPC is similar for all tested concentration, hence producing similar morphology of CNT without any significant difference. On the other hand, different fuel flow rate changes two important factors in synthesis process; first, it changes the composition of the flame as per discussed previously which changes the temperature distribution, carbon precursor composition and concentration. At fuel flow rate of 0.2 slpm, even though the total amount of CNT collected is minimal due to inadequate carbon precursor supply compared to fuel flow rate at 0.4 slpm, the morphology is visibly similar. However, at higher flow rate of 0.6 slpm, the samples shows that the CNTs are entangled not only with other strands of CNTs, but also with other various carbonaceous materials as seen in Fig. 2(f). With very high amount of carbon source and lower temperature at the inlet of the QPC, amorphous carbon formation and other carbon structures are produced which then collected together with CNTs at the outlet of the QPC. Secondly, since the fuel is used as carrier, with higher flow rate of fuel, the velocity of the catalyst vapor going through the QPC also increases, reducing the time for CNT growth inside the QPC. As a result, at a higher flow rate of 0.6 slpm, the synthesized CNTs are generally observed to be much shorter than the one produced at lower flow rate.

Analysis of the synthesized CNTs diameter from the FESEM images in Fig. 3(a) reveals that lower catalyst concentrations produce smaller and more uniform CNTs, while higher concentrations produce larger diameters. At 0.005 M, CNT diameters are

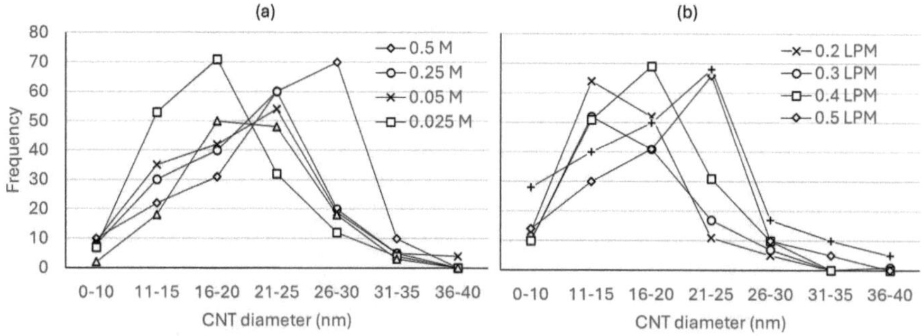

Fig. 3. CNT diameters at (a) varying catalyst concentrations and (b) fuel flow rate.

mostly within 10–20 nm, with negligible bundling and high structural uniformity. Lower catalyst concentration enables smaller catalyst nanoparticles formation producing sparse but minimally defective CNT growth which is confirmed by the respective FESEM images. Increment of catalyst concentration to 0.025 M, the CNT diameter distribution widens, with a peak in the range 15–20 nm, demonstrating increased catalyst nanoparticles average sizes and denser CNT growth. The catalyst concentration of 0.05 M attains the most uniform diameter distribution, with a peak in the range 20–25 nm and uniform morphology. At 0.25 M, the diameter distribution is shifted to a higher range of 20 to 30 nm, demonstrating increased bundling and clustering from larger catalyst nanoparticles size. The effect becomes more pronounced at 0.5 M, where CNT diameters increase further to 30–35 nm, along with severe agglomeration and irregular morphology, observed through FESEM images. In general, lower catalyst concentrations (0.005 M, 0.025 M) support smaller and uniform CNTs, while higher concentrations (0.25 M, 0.5 M) induce enhanced clustering and defects. The 0.05 M catalyst concentration is an optimal trade-off, forming densely packed CNTs, verifying the determining role of catalyst concentration in CNT morphology.

Interestingly, variation in fuel flow rate has slightly less significant effect toward the distribution of CNT diameter compared to the catalyst concentration as shown in Fig. 3(b). Overall, smaller and uniform CNTs being achieved at low fuel flow rates (0.2–0.3 LPM) and larger diameters at high fuel flow rates (0.5–0.6 LPM). In previous works it has been shown that CNT diameters are highly correlated to the size of catalyst nanoparticles [13–16]. In this case, since the concentration of catalyst used is fixed at 0.5 M, the distribution of CNT diameter is quite narrow reflecting the narrow distribution of catalyst nanoparticles produced using similar concentrations.

3.3 Raman Spectroscopy Analysis.

Raman spectra of CNTs synthesized at 0.005 M, 0.05 M, and 0.5 M catalyst concentrations reveal notable structural differences, as illustrated in Fig. 4(a). The I_D/I_G ratio increases with rising catalyst concentration. Since all samples were synthesized under identical fuel flow rates, the observed variation in I_D/I_G ratio can be attributed solely to

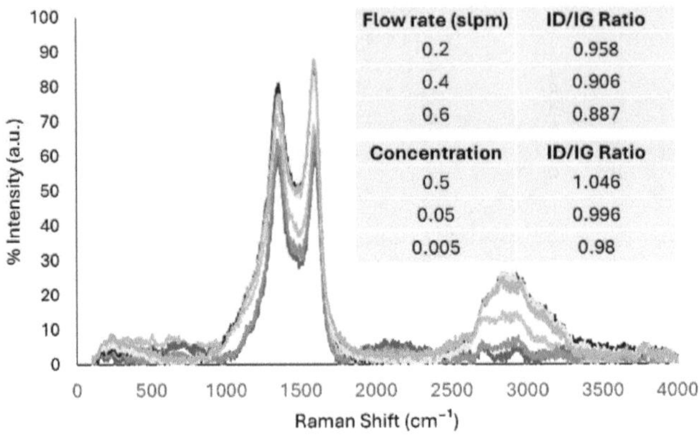

Fig. 4. Raman spectra for CNTs synthesized using different catalyst concentration with constant methane flow rate of 0.4 slpm and using different fuel flow rate with constant catalyst concentration of 0.05 M.

changes in catalyst concentration. As catalyst vapor enters the QPC, catalyst nanoparticles form upon ethanol vaporization. At lower concentrations, smaller catalyst nanoparticles are produced, which generally yield CNTs with narrower diameters and fewer structural defects, resulting in a lower I_D/I_G ratio. Conversely, higher catalyst concentrations tend to generate larger nanoparticles, promoting the growth of thicker and more defective CNTs with higher I_D/I_G ratio. Nevertheless, due to the heterogeneous environment within the QPC—characterized by variations in carbon precursor availability and temperature, the synthesized CNTs are not pristine and still exhibit relatively high I_D/I_G ratio due to the presence of defects. It is interesting to observe the reduction in I_D/I_G ratio of the samples with reduction in fuel flow rate which indicates increment of CNT quality as shown in Fig. 4(b). The fuel flow rate basically changes the CNT growth region, the concentration of carbon precursor available, and the CNT growth time within the QPC.

3.4 Effect of Varying Catalyst Concentrations on CNT Yield

Figure 5 shows the change in CNT mass collected relative to total mass of CNT collected with 0.05 M catalyst concentration which is taken as the baseline. With higher catalyst concentration, the CNT yield increases, but not linearly. Comparing the two lowest catalyst concentration, the mass of CNT yield is very low. Even with five times increment in concentration of 0.005 M to 0.025 M, the increment in mass of CNT yield is only increase around 20%. Interestingly, as the catalyst concentration is doubled to 0.05 M, the CNT mass yield is increased nearly 80% compared to the lower concentration. But, with further increment in the concentration of the catalyst to 0.25 M and 0.5 M, which represent 5 and 10 times of catalyst concentration, the measured of the CNT mass collected CNTs relatively stagnant, with minimal increment about 15 to 35%. This clearly shows that there is an ideal concentration that capable of producing higher CNT yield for certain fuel flow rates. This is important since in large manufacturing process, catalysts may become key parameters that significantly contribute toward overall cost of CNT production. Hence, optimization of catalyst concentration does require effective use of catalyst. Nevertheless, since the samples are weighted as-is once collected at the outlet of the QPC, there is a possibility the weight measured includes all the weight of the catalyst nanoparticles and all others amorphous carbon formation and soot. Furthermore, the mass of the catalyst uses increases as the catalyst concentration increases with the weight of the catalyst used for the 0.5 M is about 100 times the weight of catalyst used in 0.005 M. This may contribute toward higher mass collected at the outlet of the QPC for catalyst solution with higher concentration.

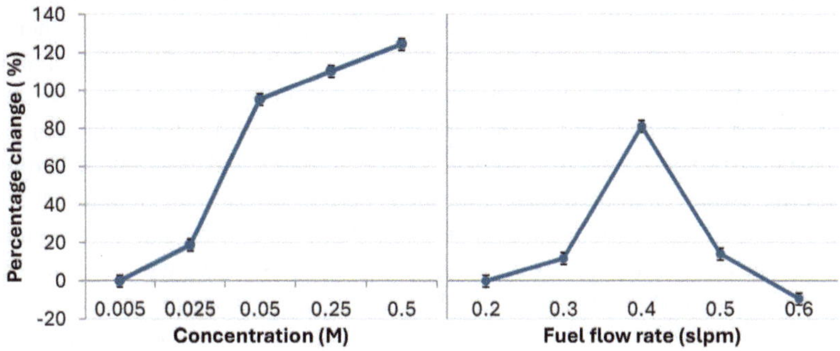

Fig. 5. Percentage of change in CNT yield at various catalyst concentration and fuel flow rate.

3.5 Effect of Varying Fuel Flow Rate on CNT Yield

Figure 5 shows the mass of the CNT yield for fuel flow rate from 0.2 to 0.6 slpm with constant 0.05 M catalyst concentration at 20 mm HAB. At low fuel flow rate, the yield of CNT is very low around 0 to 10%. Insufficient supply of methane at these flow rates maybe impedes CNT inception and growth inside the QPC. But when the fuel flow rate increases to 0.4 slpm, the CNT yield increases up to 80%. With further increment of fuel flow rate up to 0.5 and 0.6 slpm, the CNT yield remains low nearly similar to the yield produced by the low fuel flow rate. Higher flow rates introduce inefficiencies through turbulence and shorter residence times, while the fixed HAB leading to non-optimal precursor decomposition and nucleation. This result shows a clear needs to ensure that the fuel flow rate used is optimized to the location of the QPC. In this analysis, QPC is fixed at HAB 20 mm that may contribute toward high yield produced by the fuel flow rate of 0.4 slpm. With higher or lower HAB of QPC, the lower or higher fuel flow rate may produce much more CNT since the physical length of the flame changes, affecting the distribution of heat and also carbon precursor inside the QPC. The results show that optimization of flame flow rate is required for CNT synthesis setup using QPC. As fuel as one of the main contributors toward the overall cost of CNT production, and with increasing concern of carbon footprint in the synthesis process, this result shows the need to ensure that optimization of fuel flow rate is critical and beneficial to ensure efficient CNT synthesis process using QPC.

4 Conclusion

This study provides an experimental analysis of the effects of vapor catalyst concentration and methane gas flow rate on CNT growth within a QPC using methane diffusion flame. The catalyst concentrations were varied between 0.005 M to 0.5 M and methane flow rates between 0.2 slpm and 0.6 slpm. The results show that both catalyst concentration and methane flow rate significantly influence CNT yield and morphology. The optimal condition was identified at a catalyst concentration of 0.05 M and a methane flow rate of 0.4 slpm, which provided the best trade-off between high yield, uniform diameter distribution, and reduced defect density. Methane flow rate also played a vital role in flame structure, temperature profile, and residence time within the QPC. At 0.4 slpm, the flame exhibited optimal height and coverage, delivering sufficient thermal energy and carbon precursor availability along the entire length of the QPC. Flow rates below this value produced short, lean flames with inadequate carbon supply, while higher rates (0.6 slpm) extended the flame beyond the QPC length, reduced residence time, and introduced instability. These effects contributed to decreased CNT yield due to diminished catalyst effectiveness and increased amorphous carbon formation. Yield measurements further confirmed that while total CNT mass may increase with catalyst concentration, the effectiveness of catalyst utilization which evaluated as CNT mass per gram of catalyst, peaked at 0.05 M. This finding emphasizes the importance of optimizing catalyst usage for cost-efficiency, particularly in industrial-scale applications. Overall, the study establishes the importance of precise control over both vapor catalyst concentration and fuel flow rate in achieving high-performance CNT synthesis via flame-based methods. These findings not only contribute valuable experimental validation to flame synthesis research but also underscore its viability as a sustainable and economically attractive method for large-scale CNT production.

Acknowledgements. The author would like to acknowledge the financial support from Ministry of Education Malaysia under Fundamental Research Grant Scheme (FRGS) (FRGS/1/2022/TK09/UTM/02/7) and Universiti Teknologi Malaysia for the funding under UTM Fundamental Research grant (UTMFR) (Q.J130000.3824.23H99).

References

1. Merchan-Merchan, W., Saveliev, A.V., Kennedy, L., Jimenez, W.C.: Combustion Synthesis of Carbon Nanotubes and Related Nanostructures. Elsevier Ltd. (2010)
2. Selamat, M.S.R., Zainal, M.T., Yasin, M.F.M., Hamzah, N., Sidik, N.A.C.: Modelling of the flame synthesis of single-walled carbon nanotubes in non-premixed flames with aerosol catalyst. J. Adv. Res. Numer. Heat Transf. **13**(1), 39–51 (2023)
3. Li, L., et al.: The influences of oxygen concentration and external heating on carbon nanotube growth in diffusion flame. CFD Lett. **13**(12), 45–62 (2021)
4. Gore, J.P., Sane, A.: Flame Synthesis of Carbon Nanotubes. IntechOpen. [Online]. Available: www.intechopen.com
5. bin Moen, M.A.A., De, V.T.Y., Yasin, M.F.M., Hamzah, N.: Analysis of fuel concentration effect toward carbon nanotubes growth in methane diffusion flame using CFD simulations. In: Communications in Computer and Information Science, pp. 313–325. Springer Science and Business Media Deutschland GmbH (2024)

6. De, V.T.Y., Mhd Louai, K., Mohd Yasin, M.F., Hamzah, N.: Numerical investigation on the effect of methane diffusion flame parameters toward the carbon nanotubes growth inside quasi-pyrolysis chamber. Mater. Today Proc. (2023)
7. Hidayat, N., Zainal, M.T., Yasin, M.F.M., Hamzah, N.: Numerical study on the influence of the quasi-pyrolysis chamber on the growth region of carbon nanotube in counter flow diffusion flame. CFD Lett. **14**(3), 11–21 (2022)
8. Zainal, M.T., Yasin, M.F.M., Wahid, M.A., Sies, M.S.M.: Design and simulation of a quasi-pyrolysis chamber for carbon nanotube flame synthesis. CFD Lett. **12**(2), 17–31 (2020)
9. Azmi, N.A.N., Zainal, M.T., Yasin, M.F.M., Hamzah, N.: Effect of flame temperature and flow rate on CNT growth region in premixed flat flame burner. CFD Lett. **14**(4), 1–14 (2022)
10. Zainal, M.T., Hamzah, N., Yasin, M.F.M., Wahid, M.A., Sies, M.S.M.: Parametric study of burner structure on carbon nanotube growth using flame synthesis technique. CFD Lett. **12**(5), 1–15 (2020)
11. Hamzah, N., Yasin, M.F.M., Saat, A., Yusop, M.Z.M., Subha, N.A.M.: Rapid synthesis of carbon nanotubes using flame: a review of the process parameters. Carbon Lett. **22**(1), 1–14 (2017)
12. Kumar, M., Ando, Y.: Chemical vapor deposition of carbon nanotubes: a review on growth mechanism and mass production. J. Nanosci. Nanotechnol. **10**(6), 3739–3758 (2010)
13. Cuenya, B.R.: Synthesis and catalytic properties of metal nanoparticles: size, shape, support, composition, and oxidation state effects. Thin Solid Films. **518**(12), 3127–3150 (2010)
14. Shah, N., Kim, H., Lee, C.H., Jeong, G.H.: Flame synthesis of carbon nanotubes and their applications: a review. J. Mater. Res. Technol. **17**, 4018–4034 (2022)
15. Yu, M.F., Files, B.S., Arepalli, S., Ruoff, R.S.: Tensile loading of ropes of single wall carbon nanotubes and their mechanical properties. Phys. Rev. Lett. **84**(24), 5552–5555 (2000)
16. Hernandez, Y., et al.: High-yield production of graphene by liquid-phase exfoliation of graphite. Nat. Nanotechnol. **3**(9), 563–568 (2008)

Fluid Flow and Thermodynamic System Optimization

Flow Exchange Process and Corresponding Heat Transfer Characteristics Between Latticework Subchannels

Minlong Li[1,2], Huishe Wang[1,2(✉)], Ke Yang[1,2], and Rongguo Yu[2]

[1] University of Chinese Academy of Sciences, Haidian District, Beijing 100190, China
{liminlong,wanghs,yangke}@iet.cn
[2] Institute of Engineering Thermophysics, Chinese Academy of Sciences, Beijing, China
yurongguo@iet.cn

Abstract. To establish a one-dimensional calculation model of a latticework of turbine blades, the flow exchange process and corresponding heat transfer characteristics between latticework subchannels are studied by numerical simulation, and the influence of latticework geometric characteristics and inlet Reynolds number on the flow exchange process was summarized. It is found that the blocking effect of the transverse vortex at the inlet of the subchannel and the stagnation effect of the subchannel with a small length-diameter ratio lead to the flow loss at the inlet of the subchannel. When the length-diameter ratio of the subchannel increases, the flow loss at the inlet of the subchannel decreases. The throttling effect of the turning channel leads to the flow exchange between the subchannels, and only the flow exchange is performed on the three adjacent interfaces before and after the turning channel. The flow exchange value of the subchannel is 30% ~ 80% of its average flow. As it approaches the turning channel, the flow exchange ratio of each interface is 1:3:6 in turn. The inlet Reynolds number does not affect the flow exchange process between subchannels. The increase in rib angle will enhance the flow exchange between subchannels. When the subchannel height ratio increases, the flow exchange process between the subchannels first increases and then decreases. The heat transfer intensity and flow resistance coefficient in the subchannel are proportional to the change in flow rate. Finally, the flow exchange calculation model of the latticework subchannel is established, which provides the basis for building the one-dimensional calculation model of the latticework duct.

Keywords: Turbine Blade · Latticework Duct · Flow Exchange · Heat Transfer Characteristics · Prediction Model

1 Introduction

Increasing the turbine inlet temperature of the gas turbine is an important means to improve the power and efficiency of the gas turbine. In order to ensure the normal operation of turbine blades at high temperatures, it is necessary to vigorously develop turbine blade cooling technology. The internal cooling of the turbine blade is an important

Y. Zhao (Ed.): ICAMM 2025, SPM 96, pp. 309–333, 2025.
https://doi.org/10.1007/978-3-032-09862-7_29

part of the blade cooling system. The commonly used internal cooling structures include turbulence rib [1] (see Fig. 1 a), pin fin [2], dimple [3], and lamella [4]. As a typical internal cooling structure (see Fig. 1b), the latticeworks make the blade have better structural strength while efficient and uniform heat transfer [5–8].

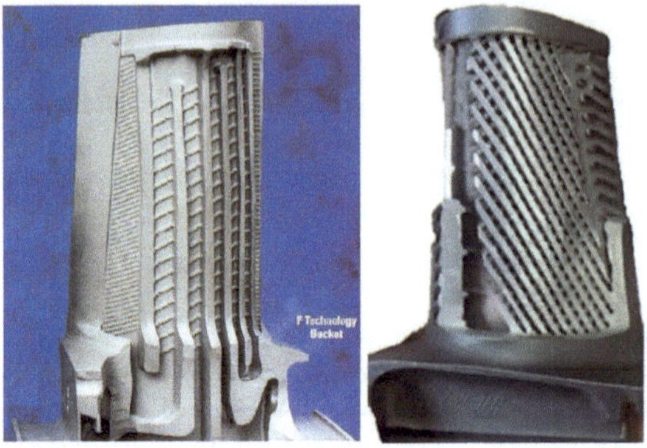

(a) Turbulence rib [9] (b) Latticework [10]

Fig. 1. Internal cooling structure of turbine blade.

Compared with the turbulence rib structure, the latticework cooling structure has a larger number of subchannels and stronger interference between subchannels. Because there are many interfaces between the top and bottom subchannels of the latticeworks, there is a flow exchange process between the channels.

In the study of the latticework cooling structure, Acharya et al. Found [11] that the rotation number had no significant effect on the Nusselt number and flow resistance coefficient of the latticework. It is considered that the bending effect of the latticework turning channel is the main factor for the existence of a higher Nusselt number in the subchannel. Gillespie et al. Studied [12] the heat transfer distribution characteristics of wedge-shaped latticeworks by transient liquid crystal measurement technology. It was found that the heat transfer on the side wall of wedge-shaped latticeworks was significantly enhanced, but the heat transfer uniformity of wedge-shaped latticeworks decreased.

With the development of experimental technology, the local flow characteristics in the latticework channel gradually appear. Tariq et al. Studied [13] the heat transfer distribution characteristics in latticework channels through liquid crystal thermal imaging technology. It is found that in addition to the inlet of the subchannel, the subchannel near the side wall of the latticework has the strongest heat transfer capacity. It is found that there is a leakage flow in front of the turning channel of the latticework channel. Tsuru et al. Used [14] magnetic resonance imaging technology to summarize the flow structure in the latticework channel. The experimental measurements show that there are mainly longitudinal vortex flow, leakage flow, rotating flow, and impact flow structures in the latticework subchannel. When comparing the velocity distribution of the subchannel section with the rib angle of 45°, it is found that the influence between the top and bottom subchannels of the latticework is not only distributed at both ends of the subchannel but also at the interface of the subchannel.

Numerical simulation has become the main technique to study the flow and heat transfer characteristics of latticework subchannels. Hagari et al. Reproduced [15, 16] the latticework experiment in the open literature by numerical simulation method. The calculation results are similar to the experimental results, which proves the reliability of the numerical simulation method in the study of latticeworks. It is well known that more flow field details can be obtained by numerical simulation. Du et al. Studied [17] the process of separating the latticework flow structure by numerical simulation. It was found that before the inlet of the latticework turning channel, a large amount of flow in the subchannel was leaked to the bottom subchannel, and only a small amount of cooling air flowed downstream through the turning channel. When Bu et al. Studied [18] the composite cooling structure of latticeworks and pin fin based on numerical simulation, they found that there was a large amount of flow exchange phenomenon at the turning channel of latticeworks, and the leakage flow formed a low-speed area in the downstream subchannel, which affected the flow structure in the downstream subchannel. Du et al. Studied [19] the flow and heat transfer characteristics of the composite cooling structure of the latticework side wall and impact hole by numerical simulation. It was found that when the impact hole moved from the steering area to the impact area, the rotating flow at the end of the latticework subchannel was enhanced, and the flow structure in the latticework subchannel was insensitive to the shape of the impact hole. It is considered that the leakage flow at the turning channel leads to an increase in the flow loss downstream of the subchannel. Based on numerical calculations, Khalatov and Nam [20] found that the rotational momentum exchange of cooling air at the interface between the top and bottom subchannels of the latticeworks is the main reason for maintaining the longitudinal vortex rotation in the latticework subchannels. Luo et al. Used [21] numerical simulation to study the effect of latticework end wall shape on the flow and heat transfer characteristics of subchannels. It was found that near the side wall surface of the wavy end wall shape, the rotational flow velocity in the latticework subchannel increased, resulting in a decrease in the leakage flow between the subchannels, thereby enhancing the jet effect at the side wall surface of the subchannel and improving the heat transfer performance in subchannel.

At present, the research on the flow exchange between the latticework subchannels is mainly based on the description of the flow structure characteristics. There are few summaries of the flow exchange relationship between the entire latticework subchannels, and the influence of the geometric parameters of the latticework subchannels on the flow exchange process of the latticework subchannels is insufficient. In this paper, the flow exchange process and corresponding flow and heat transfer characteristics between latticework subchannels are studied under the influence of rib angle, rib width to rib spacing ratio, subchannel height ratio, and inlet Reynolds number. The flow exchange law between latticework subchannels is summarized, which provides experience for building a one-dimensional calculation model of latticework.

2 Geometry

The geometric model in [22] was selected for the latticework model, as shown in Fig. 2. The rib pitch is $S = 22.54$ mm, the height of the inlet channel is $H = 48$ mm (the height ratio of the subchannel is $h/S = 1.1$), the inlet width is $W = 134.17$ mm, the length of the latticework test section is $L = 286.9$ mm, and the length of the inlet and outlet extension section is 200 mm.

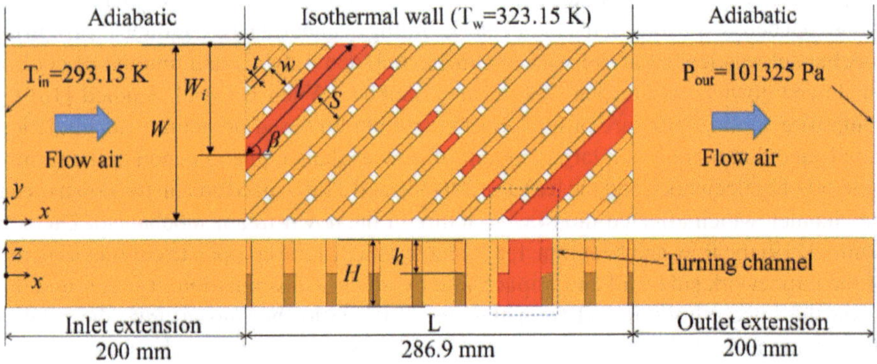

Fig. 2. Geometric model of latticework cooling channel.

The range of geometric factors is set by combining the structural characteristics of the latticework of the real turbine blade. The geometric influence factors of rib angle β are set to 15°, 30°, and 45°, the ratio of rib width to rib spacing t/w are set to 0.3, 0.5, and 1.0, and the ratio of subchannel height h/S are set to 0.5, 0.8, and 1.1.

3 Computational Details

3.1 Data Transform

The equivalent diameter is defined as:

$$D_h = \frac{4WH}{2(W + H)} \tag{1}$$

where W is the channel width, H is the channel height.

Reynolds number is defined as:

$$\mathrm{Re} = \frac{\rho \cdot u_{in} \cdot D_h}{\mu} \tag{2}$$

where ρ is the inlet cooling air density, u_{in} is the average inlet cooling air velocity, and μ is the inlet cooling air dynamic viscosity.

The heat transfer capacity is measured by the Nusselt number, and the Dittus-Boelter empirical formula [23] is selected to dimensionless the Nusselt number, which is defined as:

$$Nu = \frac{h_f \cdot D_h}{\lambda} \tag{3}$$

$$Nu_0 = 0.023\mathrm{Re}^{0.8} \cdot \mathrm{Pr}^{0.4} \tag{4}$$

where h_f is the convective heat transfer coefficient, λ is the thermal conductivity, and Pr is the Prandtl number.

The convective heat transfer coefficient is calculated by the heat balance method. The heat transfer of the cooling structure can be written as:

$$Q = q \cdot c_p \cdot (T_{out} - T_{in}) \tag{5}$$

$$Q = h \cdot A \cdot \Delta T \tag{6}$$

where Q is the heat transfer, q is the mass flow rate, c_p is the constant pressure specific heat capacity, T_{in} and T_{out} are respectively the average temperature of the inlet and outlet, A is the heat transfer area of the cooling channel, ΔT is the temperature difference between the latticework wall and the inlet cooling air.

The flow resistance coefficient is used as the evaluation index of the flow characteristics of the latticework cooling structure. The Karman-Nikuradse empirical formula [23] is used to dimensionless the flow resistance coefficient, which is defined as:

$$f = \frac{\Delta P_t}{4(L/D_h) \cdot \frac{1}{2}\rho {u_{in}}^2} \tag{7}$$

$$f_0 = 2(2.236 \ln Re - 4.639)^{-2} \tag{8}$$

where ΔP_t is the total pressure difference between the inlet and outlet of the cooling channel, L is the flow length of the latticework.

Root Mean Square Error (RMSE) and standard deviation (s_y) are used as evaluation indexes for turbulence model adaptability, which are defined as:

$$\mathrm{RMSE} = \sqrt{\frac{\sum (y - \hat{y})^2}{n}} \tag{9}$$

$$s_y = \sqrt{\frac{\sum (y - \overline{y})^2}{n}} \tag{10}$$

where n is the number of samples, y is the true value of the sample, $\hat{y}$ is the predicted value of the sample, $\bar{y}$ is the average of samples.

3.2 Boundary Condition

Based on the ANSYS Fluent platform, the latticework model is numerically simulated. The cooling air is set as an ideal gas, and the inlet mass flow rate is set according to the inlet Reynolds number (the inlet Reynolds numbers are 21000, 44000, 60000, and 80000). The inlet temperature of the cooling air is 293.15 K, and the inlet and outlet extension sections are set as adiabatic walls. The wall surface of the latticework heat exchange section adopts an isothermal wall. The wall temperature is $T_w = 323.15$ K, and the outlet pressure is set to $P_{out} = 101325$ Pa. The Simple algorithm is used to correlate the pressure and velocity. While monitoring the convergence of the residual error in the calculation process, the change value of the monitoring outlet mass flow is less than 0.01% of the inlet mass flow, which ensures the convergence of the calculation.

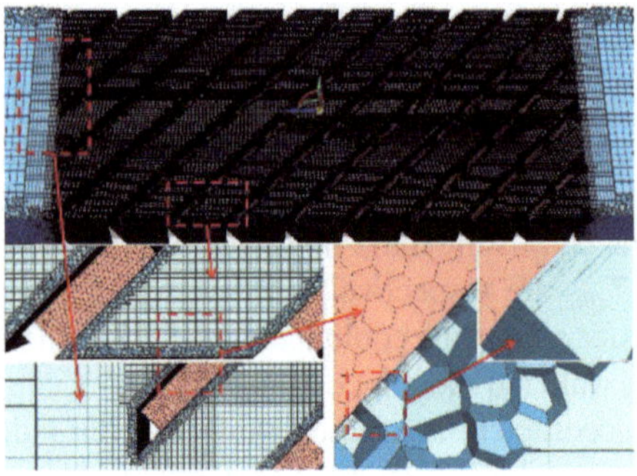

Fig. 3. Diagram of latticework mesh division.

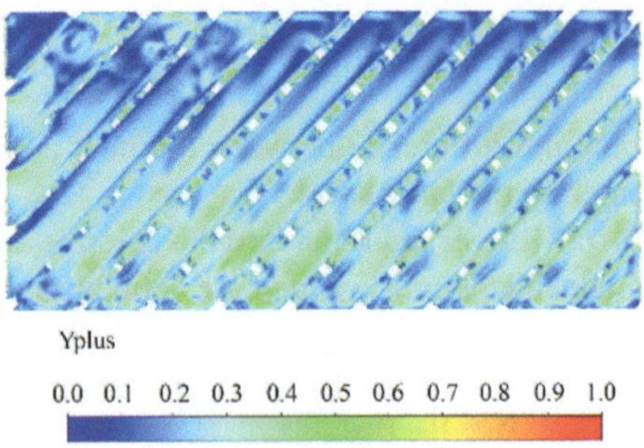

Fig. 4. Yplus cloud diagram of latticework wall.

The latticework calculation model is meshed by Fluent Meshing. The body mesh is set as Poly-Hexcore (see Fig. 3), and the mesh of the latticework heat exchange section is locally encrypted. The uniform method is used to divide the boundary layer mesh. The height of the first layer mesh is 0.005 mm, and the number of boundary layers is set to 10. The growth rate of the boundary layer mesh is 1.2, which ensures that Yplus ≤ 1, as shown in Fig. 4.

In order to ensure the accuracy of the calculation results, the latticework calculation models with the number of meshes of 4.13 million, 7.8 million, 8.73 million, 9.83 million, 13.83 million, 15.19 million, and 20.58 million were selected to verify the independence of the number of meshes. As shown in Fig. 5, when the number of mesh exceeds 14 million, the relative error of the calculation results is less than 0.2%. Based on the comprehensive study of cost and computational efficiency, the number of 14 million grids is used as the reference value for the meshing of the latticework calculation model.

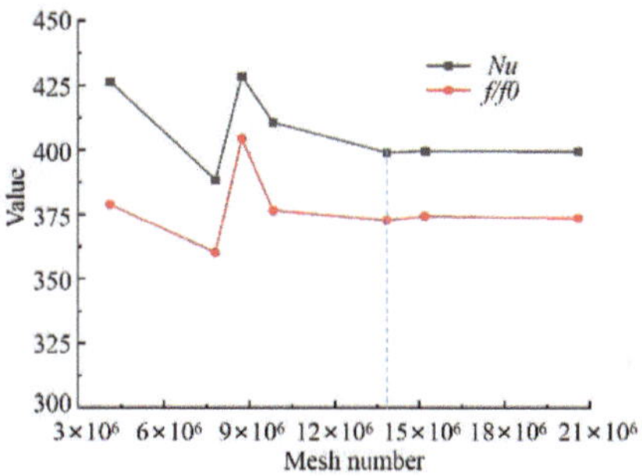

Fig. 5. Diagram of mesh number independence verification.

3.3 Turbulence Model Validation

The experimental data of the latticework inlet Reynolds number of 44000 in [22] were selected to verify the turbulence model. The turbulence models validated are the k-ω model, k-ω SST model, k-ε model, and k-ε Realizable model. As shown in Fig. 6, the applicability of each turbulence model is judged by comparing the prediction accuracy and distribution characteristics of the relative Nusselt number of the subchannel wall. The root mean square error and standard deviation are used to evaluate the data prediction accuracy and data distribution characteristics respectively, as shown in Table 1.

Meanwhile, the k-ω SST model and the k-ε Realizable model are compared and verified by the flow resistance coefficient of the latticework subchannel. As shown in Table 2, the predicted results of the k-ω SST turbulence model are closer to the experimental values [22]. Therefore, the k-ω SST model is chosen as the turbulence model for numerical calculation.

According to the governing equations of the turbulence model, the k-ω SST model is more accurate, more stable, and more applicable than the k-ε Realizable model when calculating separate and rotational flows. The general governing equation of the k-ω SST turbulence model is as follows:

$$\frac{\partial}{\partial t}(\rho k) + \frac{\partial}{\partial x_j}(\rho k \overline{u}_j) = \frac{\partial}{\partial x_j}\left(\Gamma_k \frac{\partial k}{\partial xj}\right) + F_k - Y_k \tag{11}$$

$$\frac{\partial}{\partial t}(\rho \omega) + \frac{\partial}{\partial x_j}(\rho \omega \overline{u}_j) = \frac{\partial}{\partial x_j}\left(\Gamma_\omega \frac{\partial \omega}{\partial x_j}\right) + F_\omega - Y_\omega + D_\omega \tag{12}$$

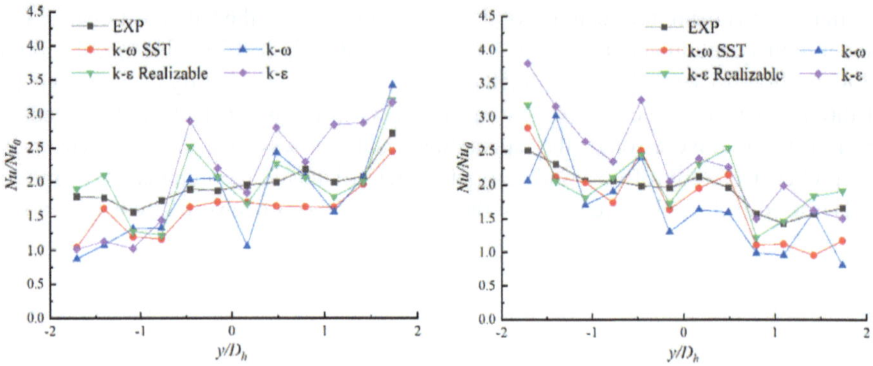

(a) Trailing surface (b) Leading surface

Fig. 6. The relative Nusselt number diagram of subchannel.

Table 1. Error analysis of relative Nusselt number.

Index	RMSE		s_y	
Turbulence model	a	b	a	b
k-ω SST	0.391	0.367	0.357	0.574
k-ε Realizable	0.332	0.351	0.507	0.5
k-ω	>0.5	>0.5	/	/
k-ε	>0.5	>0.5	/	/

Table 2. Predicted value of flow resistance coefficient.

Factor	Contrast parameter		
Re	EXP	k-ω SST	k-ε Realizable
21000	5.36	5.60	5.88
44000	4.70	4.42	4.41

where k is the turbulent kinetic energy, t is time, $\overline{u}_j$ is the average value of turbulence velocity, x_j is the coordinate component, ω is the special dissipation of turbulence, Γ_k, Γ_ω are the effective diffusion coefficient, F_k, F_ω are the turbulence production term, Y_k, Y_ω are the dissipation term of k and ω, D_ω is the diffusion term.

The corrected turbulent dynamic viscosity coefficient μ_t is:

$$\mu_t = \frac{\rho k}{\omega} \cdot \frac{1}{\max\left[\frac{1}{\alpha^*}, \frac{S_t F_2}{a_1 \omega}\right]} \tag{13}$$

where α^* is the low Reynolds number correction factor, a_1 is the empirical constant term, S_t is the shear force constant term, F_2 is the blending function.

4 Results Analysis and Discussion

In order to facilitate the clear description and analysis of the flow distribution characteristics in the latticework subchannels, Fig. 7 shows the latticework partition name and subchannel naming method completely. The latticework flow area is mainly divided into the inlet section (before the first turning channel), the longitudinal vortex section (both ends of the subchannel have turning channels), and the outlet section (after the last turning channel). Due to the inlet effect of the subchannel, the entrance of the inlet section is added to the subchannel inlet section.

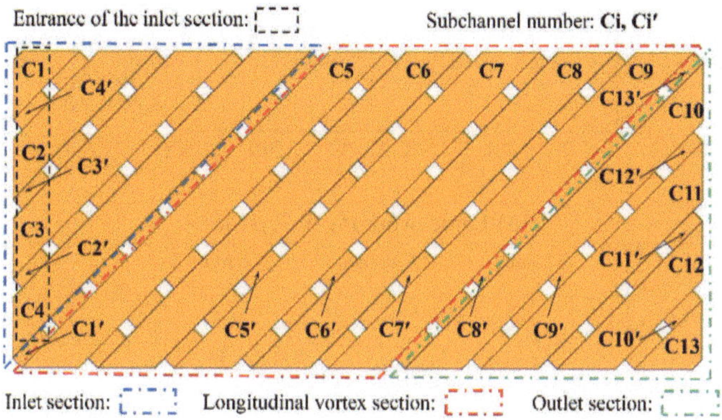

Fig. 7. Diagram of latticework partition and subchannel naming.

To facilitate a clear description of the relationship between the subchannels, each subchannel is marked in Fig. 7. The top C3 subchannel is taken as an example. When the cooling air flows to the C3 subchannel turning channel, it enters the bottom subchannel C7′ and flows along the C7′ subchannel. After the turning channel enters the C11 subchannel, it finally flows out along the C11 subchannel. The channel along the C3 subchannel is expressed as C3-C7′-C11, and the other subchannels are named in the same way as the C3 subchannel.

4.1 Flow Distribution Characteristics Along the Subchannel

Based on the average parameters along the cross-section of the latticework subchannel, the flow and heat transfer characteristics in the subchannel are studied. In order to facilitate the analysis, the average mass flow rate q_{ref} at the inlet of each subchannel is used as the reference value to convert the flow rate along the latticework subchannel. As shown in Fig. 8a, the flow distribution of each subchannel of latticework is not equal in the inlet section and the outlet section. As the length of the inlet section of the subchannel increases, the flow rate in the subchannel increases. However, in the longitudinal vortex section of the latticework, the flow rate between the subchannels is basically the same, indicating that the latticework turning channel has the effect of homogenizing the flow rate so that the flow rate of the longitudinal vortex section is equal. Which proves the uniform distribution of the heat transfer of the latticework. There is an obvious flow exchange process near the entrance and exit of the turning channel.

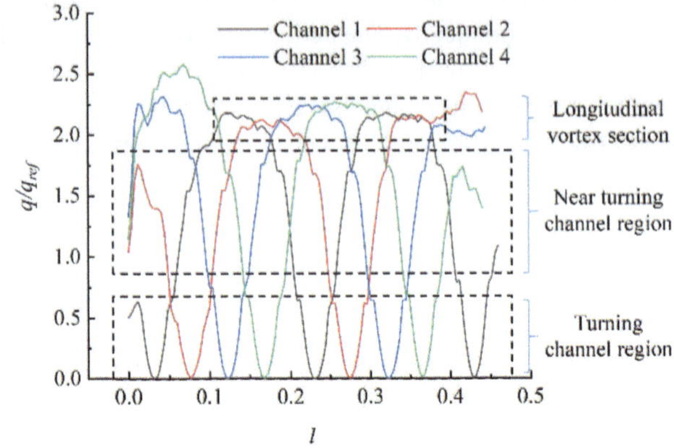

(a) Latticework t/w=0.3, β=45°

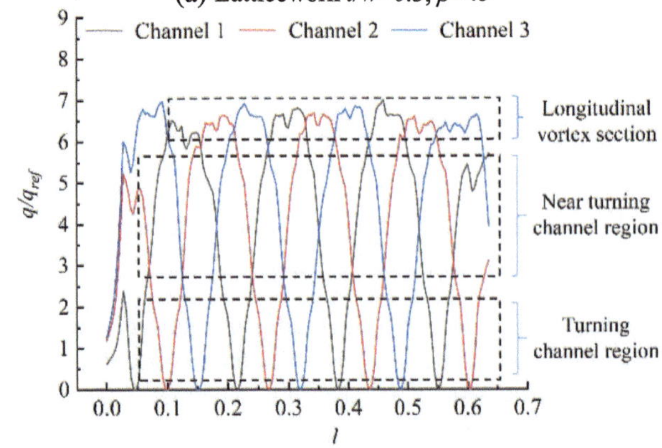

(b) Latticework t/w=0.3, β=60°

Fig. 8. Distribution of mass flow rate along the subchannel.

Because the number of turning channels of latticeworks with $\beta = 45°$ is small, the phenomenon of uniform flow is not obvious. In this regard, a schematic diagram of the flow distribution along the latticework subchannel with $\beta = 60°$ is added (see Fig. 8b). In the longitudinal vortex section of the latticework, the mass flow rate of each subchannel is compared with the average flow rate. The maximum fluctuation of the mass flow rate is about 5% of the average value, and the maximum fluctuation of the flow rate of the remaining sections is less than 3%. Therefore, it can be considered that the flow rate between the subchannels is equal in the longitudinal vortex section of the latticework.

Through the flow distribution characteristics of the latticework subchannels, it can be found that the areas with large differences in the flow distribution between the subchannels are: the entrance of the subchannel inlet section and the area near the turning channel. In this regard, the following will focus on the analysis of the flow exchange process at the entrance of the subchannel inlet section and near the turning channel.

4.2 Flow Exchange at the Entrance of the Subchannel Inlet Section

To study the flow distribution characteristics at the entrance of the subchannel inlet section, the cross-sections are made along the x direction. Through the comprehensive analysis of the velocity vector and pressure on the cross-section, the main reason for the flow exchange at the inlet of the subchannel is obtained.

Flow Exchange Process and Heat Transfer Characteristics. As shown in Fig. 9, there is an obvious flow exchange phenomenon at the inlet of the top and bottom layers of the latticework subchannel. Due to the separation of fluid flow at the inlet of the subchannel, a regularly distributed transverse vortex structure is formed at the inlet. It can be seen from the cross-section pressure distribution that under the blocking effect of the transverse vortex, a local high-pressure zone is formed at the inlet of the subchannel. When the length-diameter ratio of the subchannel increases, the range of the transverse vortex decreases and the pressure gradually decreases. Because the top and bottom subchannels of the latticework are symmetrically distributed on the interface, a pressure difference is formed between the top and bottom subchannels at the inlet. So a large amount of cooling air in the high-pressure area leaks to the low-pressure area, which eventually leads to uneven flow distribution between the subchannels. When x gradually increases, the interface between the top and bottom subchannels decreases, and the flow exchange between the top and bottom subchannels gradually decreases.

Figure 10 shows the flow distribution characteristics at the inlet of the latticework subchannel, where the flow ratio at the inlet of the subchannel is $q_{\text{pct}} =$ the flow at the inlet of the subchannel/the total inlet flow, and the flow ratio is used to measure the flow distribution characteristics of the subchannel. With the increase of the length-diameter ratio of the subchannel, the flow rate at the inlet of the subchannel gradually increases, and when it approaches the side wall, the flow rate gradually decreases.

In the rectangular cooling channel, due to the stagnation effect of the small length-diameter ratio subchannel, the pressure near the side wall is greater than the pressure in the middle of the channel, resulting in the flow rate near the side wall of the rectangular channel is less than the flow rate in the middle of the channel. Due to the flow exchange between the top and bottom subchannels at the inlet, the flow of the subchannel with a large length-diameter ratio near the side wall of the rectangular channel is compensated. With the increase of rib angle, the difference in flow distribution increases gradually, which indicates that the stagnation effect of subchannels with a small length-diameter ratio also increases gradually.

By calculating the relative Nusselt number and flow resistance coefficient at the inlet of the subchannel, the influence of the flow exchange process at the inlet on the flow and heat transfer characteristics at the inlet is analyzed. As shown in Fig. 11, when the rib angle increases, the heat transfer capacity and flow resistance of the cooling air at the inlet of the subchannel increase, and the difference in heat transfer and flow distribution between the subchannels also increases. With the increase of the length-diameter ratio of the subchannel, the heat transfer capacity and flow resistance of the cooling air at the inlet of the subchannel increase first and then decrease.

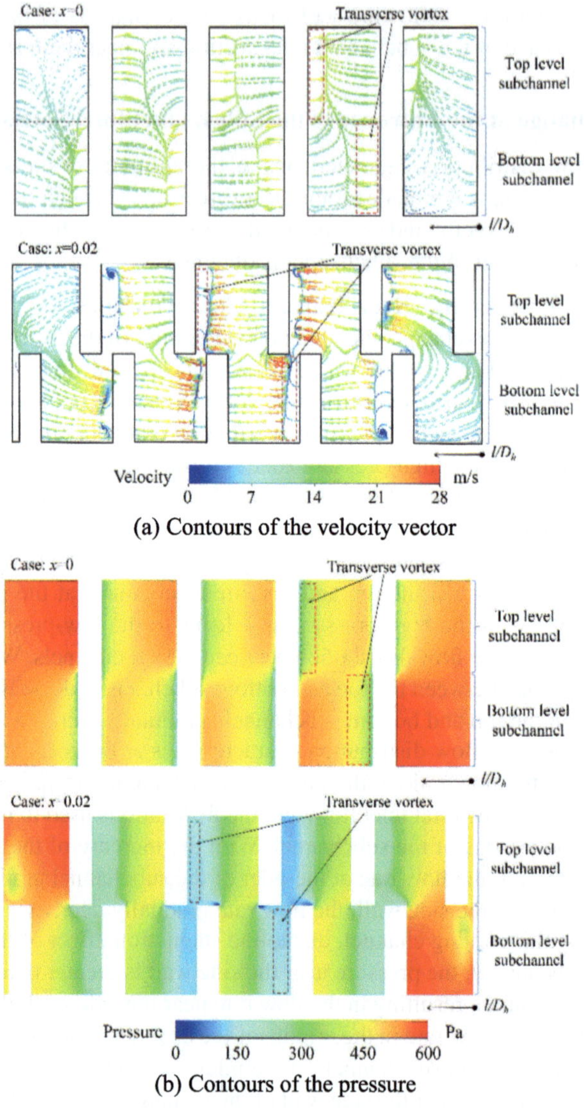

Fig. 9. Parameter distribution contours at the entrance of the subchannel inlet section.

Combined with Figs. 9 and 11, when the length-diameter ratio of the subchannel is small, the stagnation effect at the inlet of the subchannel is more obvious, and the pressure at the inlet of the subchannel is higher, resulting in the leakage of the flow at the inlet of the subchannel to the bottom subchannel, which reduces the heat transfer capacity in the subchannel. Meanwhile, because of the stagnation effect at the inlet of the subchannel, the cooling air velocity at the inlet of the subchannel is decreased, and the pressure loss at the inlet of the subchannel is smaller, resulting in a smaller flow resistance coefficient, resulting in a small flow resistance coefficient of the subchannel

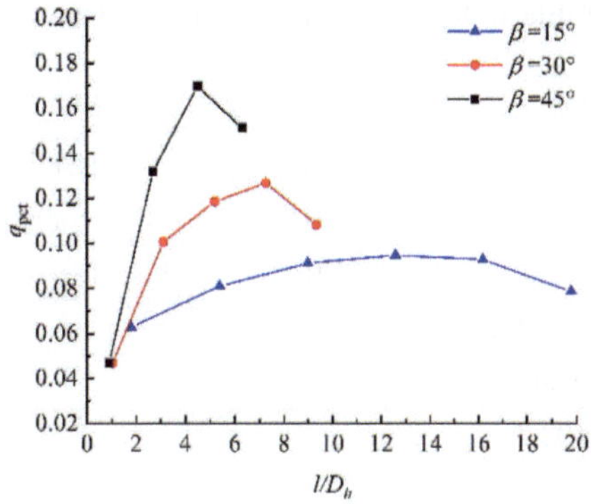

Fig. 10. Flow distribution at the inlet of subchannels (t/w = 0.3, h/S = 1.1).

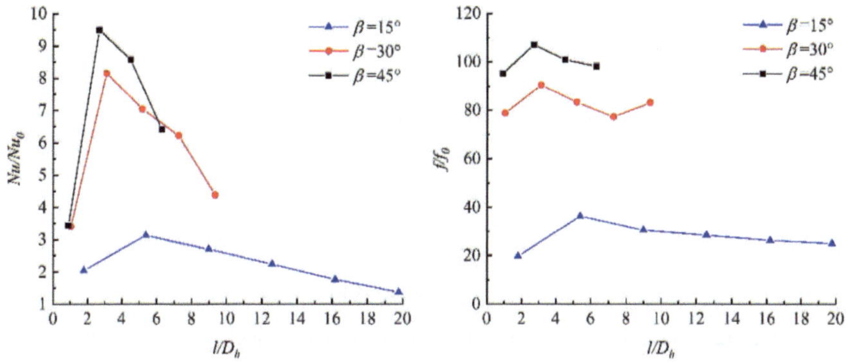

(a) Relative Nusselt number distribution (b) Flow resistance coefficient distribution

Fig. 11. Flow and heat transfer distribution at the inlet of subchannels (t/w = 0.3, h/S = 1.1).

and poor flow of cooling air. When the subchannel with a large length-diameter ratio corresponds to the subchannel with a small length-diameter ratio in the bottom layer, due to the large amount of cooling air flowing into the subchannel from the bottom layer, the flow mixing at the inlet of the subchannel is enhanced, the flow resistance coefficient is increased, and the heat transfer capacity is reduced.

By sorting out the mass flow values at the inlet of the latticework, a corresponding regression analysis model was established. The main influencing factors are rib angle β, rib width to rib spacing ratio t/w, subchannel height ratio h/S, and inlet Reynolds number Re. As shown in Table 3, except for the inlet Reynolds number, the significant value p of each influencing factor is less than 0.0001, the determination coefficient R^2 of each type of regression model is greater than 99%, and the difference is less than 20%, and the adequate precision is greater than 30. The validity of the regression model can be guaranteed.

From Fig. 12, it can be seen that the inlet Reynolds number has no effect on the mass flow distribution at the inlet of the latticework subchannel. The rib angle, rib width to rib spacing ratio, and channel height ratio are positively correlated with the subchannel

Table 3. Regression model variance analysis.

Index	t/w	β	h/S	Re
F	57.85	6714	164.5	0.003
p	<0.0001	<0.0001	<0.0001	0.96
VIF	1.0	1.0	1.0	1.0
Adjusted R^2	99.63%			
Predicted R^2	99.48%			
Adeq Precision	102.2			

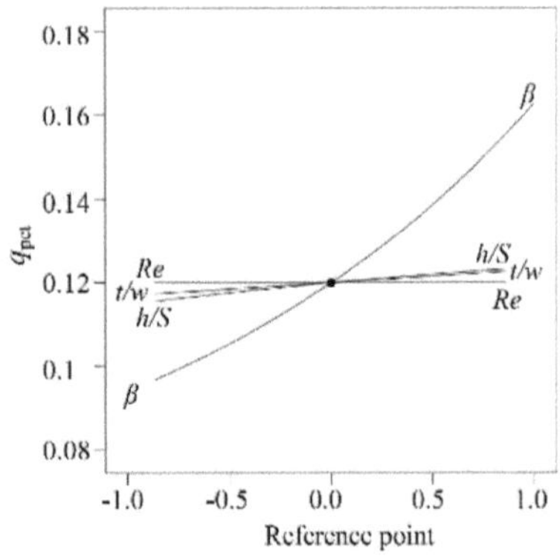

Fig. 12. Correlation between influencing factors and subchannel inlet flow.

flow ratio. The influence of rib angle on the inlet flow distribution of the subchannel is very obvious, and the influence of the subchannel height ratio is greater than that of rib width to rib spacing ratio.

Subchannel Flow Exchange Prediction Model. When the one-dimensional fluid network method is used to design the latticework cooling structure, the mass flow rate is usually used as the inlet boundary condition. Therefore, establishing the prediction model of mass flow rate at the inlet of the latticework can greatly reduce the design difficulty and cost of the latticework cooling structure. In order to accurately predict the flow rate of the inlet section of the latticework, based on the prediction model of the latticework feature subchannel (the length-diameter ratio of the top and bottom subchannels is equal), the mathematical model of the interaction of each subchannel is modified, and the flow prediction model at the inlet of the subchannel is finally obtained. Combined with Fig. 9, it can be seen that there is no obvious mass exchange between the top and bottom subchannels at the inlet of the feature subchannel, which can improve the accuracy of the subchannel prediction value.

Based on the above regression analysis model, the rib angle β, the ratio of rib width to rib spacing t/w, and the ratio of subchannel height h/S are used as independent variables, and the flow ratio at the inlet of the feature subchannel is used as the dependent variable

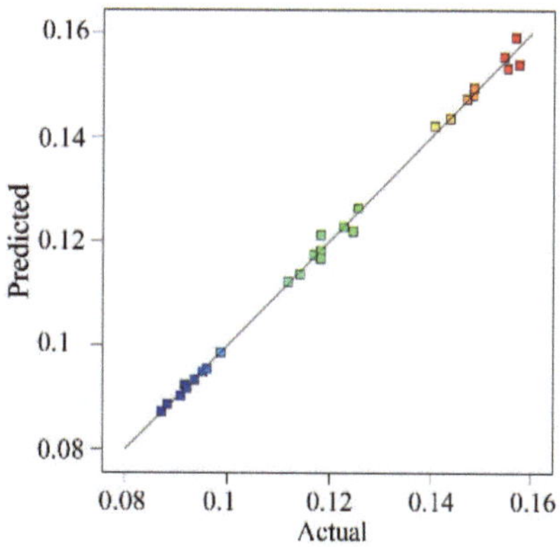

Fig. 13. Diagram of experimental value and predicted value distribution.

to establish a prediction model (see Eq. (14)). As shown in Fig. 13, the maximum error of the feature subchannel prediction model is 2.23%, which proves that the prediction model shown in Eq. 14 has better prediction accuracy.

$$q_{pct,0} = 0.06204 + 0.007427 \cdot \frac{t}{w} + 0.058323 \cdot \beta + 0.005781 \cdot \frac{h}{S} + 0.017707 \cdot \beta \cdot \frac{h}{S} + 0.036002 \cdot \beta^2 \tag{14}$$

Based on the prediction model of feature subchannels, the flow distribution at the inlet of each subchannel is obtained by the modifying model of the interaction of each subchannel. For latticeworks, if it is unreasonable to only consider the length-diameter ratio of the subchannels, the influence of the subchannels in the width direction (The *y* direction.) of the latticeworks must be considered at the same time. In this way, the latticework model can better satisfy the geometric similarity conditions and make the prediction model more general. In latticeworks, the length-diameter ratio of the feature subchannel is $l_0 = W/sin(\beta)$. It can be found that the feature subchannel length-diameter ratio has a good universality, so it can be used as a reference value to convert the length-diameter ratio of other subchannels to improve the similarity of the latticework subchannels.

Finally, the applicability of similar parameters, namely, the length-diameter ratio and the relative length-diameter ratio of subchannels is compared. As shown in Fig. 14, when the relative subchannel length-diameter ratio is used, the prediction ability of the modified model is better. Combined with the geometric characteristics in the inlet section of the latticework subchannels, the relative length-diameter ratio of the subchannels and the length-diameter ratio of the top and bottom subchannels are selected as independent variables to establish a modified model (see Eq. (15)).

$$N = 1.42421 - 0.201627 \cdot \frac{t}{w} - 1.16635 \cdot \beta - 0.401813 \cdot \frac{h}{S} - 0.421505 \cdot \frac{l}{l_0} + 0.048884 \cdot \frac{l^+}{l^-} + 0.209012 \cdot \frac{t}{w} \cdot \frac{l}{l_0} - 0.025091 \cdot \frac{t}{w} \cdot \frac{l^+}{l^-} + 1.03676 \cdot \beta \cdot \frac{l}{l_0} - 0.125623 \cdot \beta \cdot \frac{l^+}{l^-} + 0.41993 \cdot \frac{h}{S} \cdot \frac{l}{l_0} - 0.048121 \cdot \frac{h}{S} \cdot \frac{l^+}{l^-} + 0.025331 \cdot \beta^2 \tag{15}$$

In the actual latticework cooling structure, there are some latticeworks at the inlet of the top and bottom subchannels, as shown in Fig. 15. Due to the dislocation distribution,

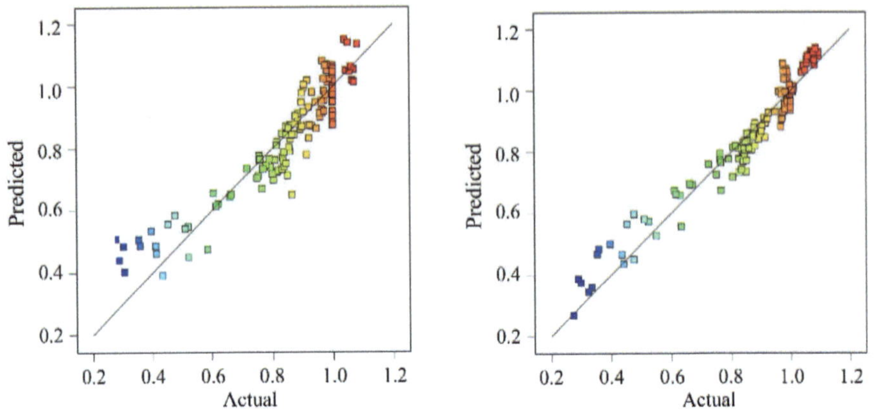

(a) Length-diameter ratio of subchannel (b) Relative length-diameter ratio of subchannel

Fig. 14. Diagram of the experimental value and the predicted value distribution.

the area of the inlet contact surface of the top and bottom subchannels is reduced, which leads to the change of the flow distribution characteristics. A modified model of the relative displacement δ is established (see Eq. (16)).

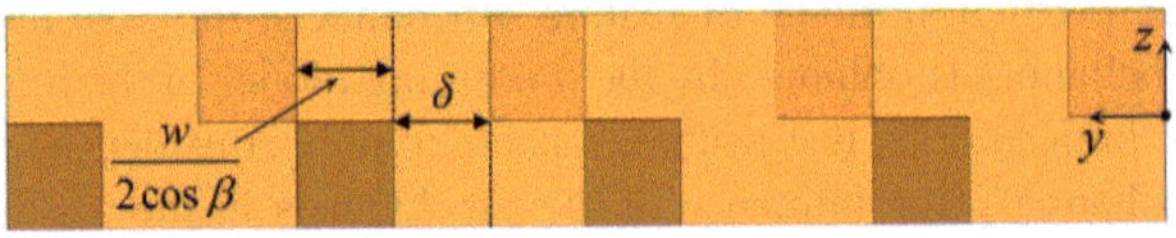

Fig. 15. Subchannel inlet dislocation distribution diagram.

$$n = 1.00921 - 0.007926 \cdot \frac{l}{l_0} + 0.000226 \cdot \frac{l^+}{l^-} + 0.128553 \cdot \frac{\delta}{w}\cos\beta - 0.06029 \cdot \delta \cdot \frac{l^+}{l^-} \tag{16}$$

$$q_{\text{pct}} = q_{\text{pct},0}\left(\beta, \frac{t}{w}, \frac{h}{S}\right) \cdot N\left(\beta, \frac{t}{w}, \frac{h}{S}, \frac{l}{l_0}, \frac{l^+}{l^-}\right) \cdot n\left(\delta, \frac{l}{l_0}, \frac{l^+}{l^-}\right) \tag{17}$$

Based on the above prediction model, the flow distribution prediction model at the inlet of the latticework channel is finally established (see Eq. (17)). Finally, the latticework models with different structures and experimental conditions are used to verify the accuracy of the prediction model. In the process of numerical simulation verification, the first type of heat transfer boundary condition is: adopted. The ideal gas inlet temperature is 293.15 K and the outlet pressure is 101315 Pa. Under the condition of given inlet Reynolds number, the flow rate at the inlet of the subchannel of the verification model is obtained. Table 4 is to verify the geometric parameters and experimental boundary conditions of the model.

According to Fig. 16, the average error of the verification model 1 is 1.15%, the average error of the verification model 2 is 8.15%, and the average error of the verification

Table 4. Verifies the geometric parameters and boundary conditions.

Index	Parameters								
Validation model	W	H	L	S	t/w	β	Re	T_W	δ
1	113.1	20	258.9	26.58	0.5	20°	100000	330	0.0094
2	60.96	20.7	182.9	10.78	0.3334	45°	35000	323.5	0
3	20.73	7	88.8	6	0.7143	30°	50000	340	0.0029

model is 5.65%. It shows that the subchannel inlet flow prediction model has good prediction accuracy and can be used for the design of latticework cooling structures.

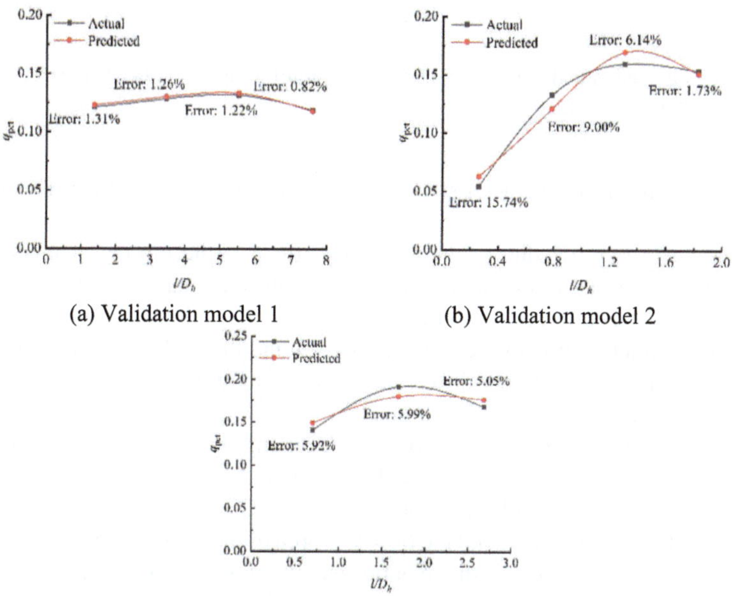

(a) Validation model 1

(b) Validation model 2

(c) Validation model 3

Fig. 16. Error analysis of prediction model.

4.3 Flow Exchange in Latticework Turning Channel

Flow Exchange Process and Heat Transfer Characteristics. In order to study the flow distribution characteristics of the turning channel section of the subchannel, the cross-section is made along the flow direction of the C3 subchannel in Fig. 7. Through the comprehensive analysis of the streamline and pressure distribution contours on the cross-section (see Fig. 17), the main reason for the flow dissection, the main reason

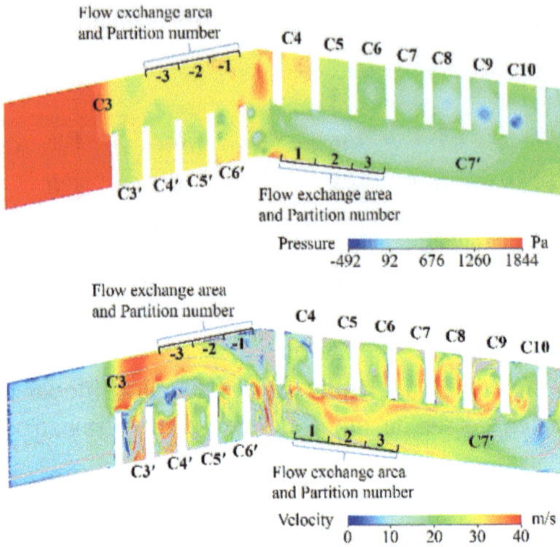

Fig. 17. Velocity vector and pressure distribution contours along the latticework subchannel ($t/w = 0.3, \beta = 45°, h/S = 1.1$).

for the flow distribution characteristics of the subchannel turning channel section was found.

As shown in Fig. 17, due to the throttling effect of the turning channel, a high-pressure area is formed in the inlet section of the turning channel, and a low-pressure area is formed in the outlet section of the turning channel, resulting in an obvious pressure gradient in the inlet section of the turning channel and the outlet section of the turning channel. Driven by the pressure difference, a large amount of cooling air leaks to the low-pressure area subchannel.

In order to facilitate the analysis, the flow exchange area before and after the sub-channel turning channel is partitioned and numbered. As shown in Fig. 17, in the inlet section of the turning channel, the cooling air in the C3 subchannel is leaked into the C4′, C5′, and C6′ subchannels, and the flow loss area of the C3 subchannel is partitioned and marked as −3, −2 and − 1. In the outlet section of the turning channel, the cooling air in channels C4, C5, and C6 is leaked into the C7′ channel, so that the C3-C7′-C11 channel obtains the flow, and the area of the flow inflow after the turning channel is marked as 1, 2 and 3. As the subchannel is close to the latticework turning channel, the throttling effect of the turning channel becomes more obvious, and the subchannel flow exchange value closer to the turning channel is larger.

In order to quantitatively analyze the subchannel, the flow change coefficient m_i is used to further analyze the flow exchange process along the subchannel. It is defined as follows:

$$m_i = q_{partition \pm i} / q_{average(c)} \tag{18}$$

where $q_{partition \pm i}$ is the flow variation at the corresponding position of the subchannel flow exchange area, and $q_{average(c)}$ is the average flow of each subchannel without flow exchange.

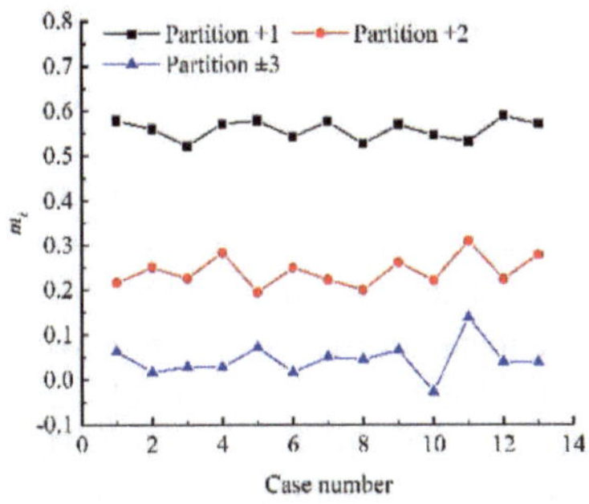

Fig. 18. Flow exchange distribution characteristics of latticework subchannels.

As shown in Fig. 18, through the statistics of the flow change value along the subchannel, except for the flow exchange area before and after the turning channel (partition±1, partition±2 and partition±3), the flow change coefficient m_i of the remaining positions in the subchannel is less than 0.01. It can be considered that in the subchannel, only the three subchannel areas adjacent to the turning channel are exchanged for flow, and there is no flow exchange between the top and bottom subchannels at the remaining positions. At each position corresponding to the distribution before and after the subchannel turning channel, the flow loss before the turning channel is approximately the same as the flow obtained after the turning channel. When the subchannel is closer to the turning channel, the flow exchange process is more obvious.

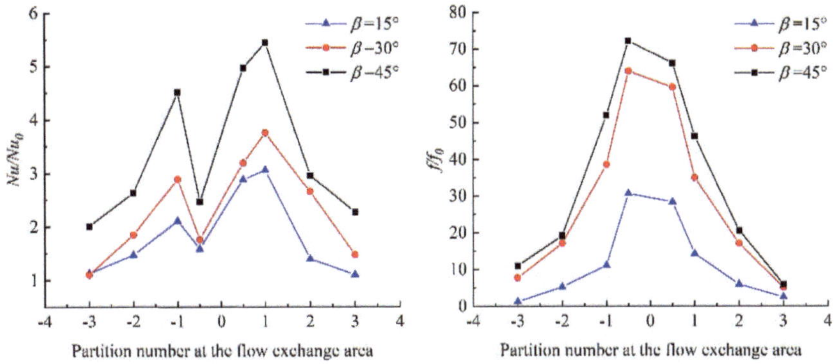

(a) Relative Nusselt number distribution (b) Flow resistance coefficient distribution

Fig. 19. Flow and heat transfer distribution at the latticework turning channel section ($t/w = 0.3$, $h/S = 1.1$).

Figure 19 shows the flow resistance and heat transfer distribution near the turning channel of the latticework subchannel. When the rib angle increases, the heat transfer capacity and flow loss in the subchannel increase. In the inlet section of the turning channel, as the cooling air in the subchannel gradually approaches the turning channel, the heat transfer intensity and flow resistance coefficient gradually increase. However, at the inlet of the turning channel, the heat transfer capacity of the cooling air decreases, because the speed of the cooling air at the inlet of the turning channel decreases, so the heat transfer capacity here decreases.

In the outlet section of the turning channel, as the cooling air in the subchannel gradually moves away from the turning channel, the heat transfer intensity and flow resistance coefficient gradually decrease. The heat transfer capacity of the outlet section of the turning channel is greater than that of the inlet section of the turning channel, indicating that the subchannel heat transfer capacity of the obtained flow is greater than the subchannel heat transfer capacity of the lost flow. The flow mixing phenomenon after the turning channel promotes the heat transfer in the channel. As the flow exchange process weakens, its heat transfer capacity also gradually decreases.

Figure 20 shows the influence of the subchannel height ratio on the distribution of flow and heat transfer characteristics in the turning channel section. With the increase of the height ratio of the subchannel, the heat transfer capacity of the inlet section of the turning channel decreases, the heat transfer capacity of the outlet section of the turning channel increases, and the overall heat transfer distribution of the turning channel section is more uniform. The subchannel height ratio mainly affects the distribution difference of flow and heat transfer characteristics in the inlet section of the turning channel, and has no obvious influence on the flow and heat transfer characteristics in the outlet section of the turning channel. When the subchannel height ratio increases, the flow resistance coefficient in the subchannel increases, and the sensitivity of the heat transfer characteristics in the subchannel to the throttling effect of the turning channel decreases.

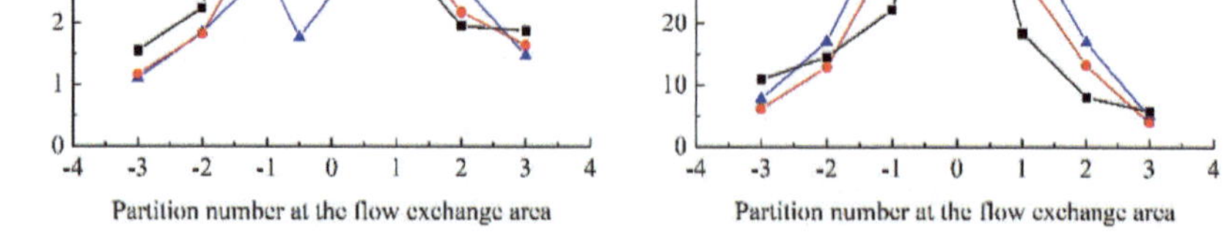

(a) Relative Nusselt number distribution (b) Flow resistance coefficient distribution

Fig. 20. Flow and heat transfer distribution in the turning channel section under the influence of subchannel height ratio ($t/w = 0.3$, $h/S = 1.1$).

Subchannel Flow Exchange Prediction Model. In order to deeply analyze the influence of the flow exchange process on the flow and heat transfer characteristics in the turning channel section, a regression analysis model of the flow variation coefficient m_i in the turning channel section is established. The main influencing factors are rib angle β, rib width to rib spacing ratio t/w, subchannel height ratio h/S, and inlet Reynolds number Re. As shown in Table 5, except for the inlet Reynolds number, the p values

Table 5. Regression model variance analysis.

Index	*t/w*	*β*	*h/S*	*Re*
F	11.35	535.82	80.78	0.012
p	0.0011	<0.0001	<0.0001	0.915
VIF	1.0	1.0	1.0	1.0
Adjusted R^2	88.61%			
Predicted R^2	86.29%			
Adeq Precision	31.44			

of all influencing factors are less than 0.0001, the coefficients of determination R^2 of the regression model are all greater than 86%, and the difference is less than 20%. The adequate precision is greater than 30, which can ensure the validity of the regression model. It can be seen from Table 5 that the significance value of the inlet Reynolds number is $p = 0.915$, indicating that the inlet Reynolds number has no obvious effect on the mass flow exchange of the subchannel turning channel.

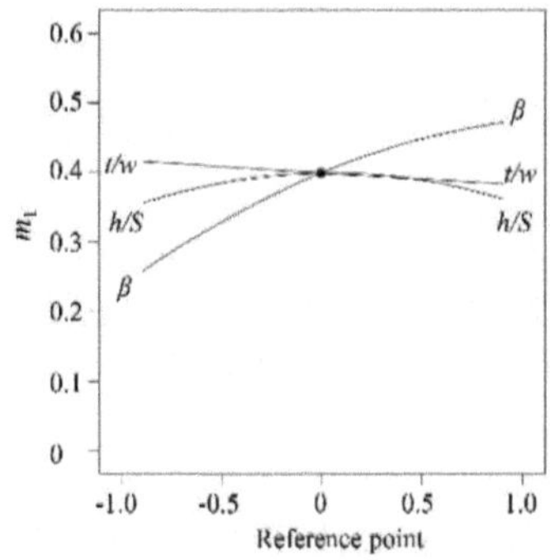

(a) Correlation between factors and m_1

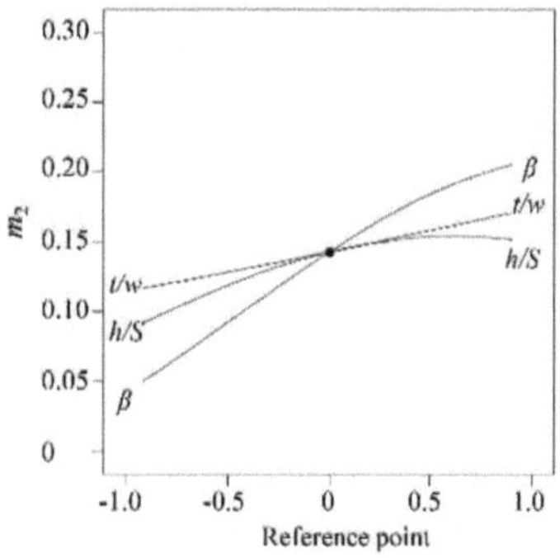

(b) Correlation between factors and m_2

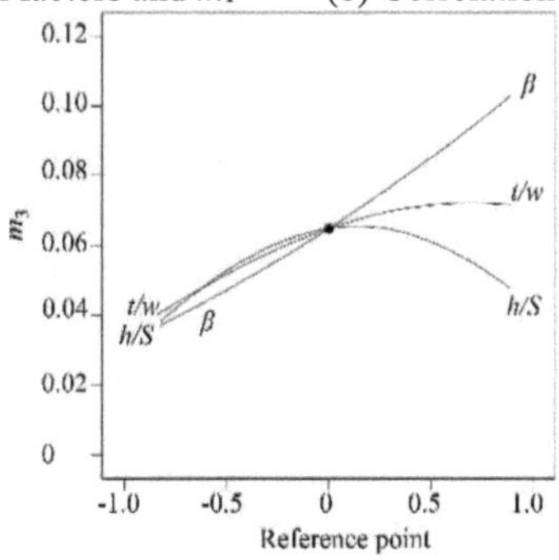

(c) Correlation between fctors and m_3

Fig. 21. Correlation between influencing factors and flow variation coefficient.

Figure 21 shows the flow variation coefficient of the subchannel near the subchannel turning channel. It can be found that if the subchannel is closer to the turning channel, the flow exchange process in the subchannel is more obvious. The subchannel flow variation coefficient m_1 is 20% ~ 50%, followed by m_2 is 6% ~ 24%, and the farthest subchannel flow variation coefficient m_3 is 3% ~ 10%. On the whole, the flow variation coefficient in the subchannels near the turning channel is 30% ~ 80%, indicating that there is a large amount of flow exchange between the top and bottom subchannels near the turning channel. Therefore, in the modeling process of the latticework one-dimensional fluid network, the flow exchange phenomenon in the turning channel section cannot be ignored.

Combined with Figs. 19 and 21, when the rib angle of the subchannel increases, the flow variation coefficient near the turning channel increases. With the increase of the ratio of rib width to rib spacing, the flow variation coefficient of the subchannel closest to the turning channel decreases slightly, while the flow variation coefficient of the subchannel far away from the turning channel increases. With the increase of the height ratio of the subchannels, the flow variation coefficient near the turning channel increases first and then decreases. Finally, combined with a large number of flow distribution parameters, the flow variation of the latticework turning channel section is summarized, and the flow variation coefficient prediction model of the turning channel section is obtained (see Eqs. (19), (20) and (21)).

$$\begin{aligned} m_1 &= -1.48046 - 0.069929 \cdot \tfrac{t}{w} + 4.58069 \cdot \beta + 2.51324 \cdot \tfrac{h}{S} + 0.231913 \cdot \\ &\tfrac{t}{w}\beta + 0.190341 \cdot \tfrac{t}{w} \cdot \tfrac{h}{S} - 4.68162 \cdot \tfrac{h}{S}\beta - 3.36128 \cdot \beta^2 - 0.934815 \cdot \left(\tfrac{h}{S}\right)^2 - \\ &0.61024 \cdot \tfrac{t}{w} \cdot \tfrac{h}{S}\beta + 3.4 \cdot \tfrac{h}{S}\beta^2 + 0.71483 \cdot \left(\tfrac{h}{S}\right)^2 \beta \end{aligned} \tag{19}$$

$$\begin{aligned} m_2^{0.54} &= -1.08465 + 0.817912 \cdot \tfrac{t}{w} + 1.95542 \cdot \beta + 1.37712 \cdot \tfrac{h}{S} - 1.00554 \cdot \\ &\tfrac{t}{w}\beta - 0.707617 \cdot \tfrac{t}{w} \cdot \tfrac{h}{S} - 0.733886 \cdot \tfrac{h}{S}\beta - 0.681063 \cdot \beta^2 - 0.431234 \cdot \\ &\left(\tfrac{h}{S}\right)^2 - 0.928319 \cdot \tfrac{t}{w} \cdot \tfrac{h}{S}\beta \end{aligned} \tag{20}$$

$$\begin{aligned} m_3^{0.54} &= -0.594947 + 0.411562 \cdot \tfrac{t}{w} + 0.280516 \cdot \beta + 1.19902 \cdot \tfrac{h}{S} - 0.230042 \cdot \\ &\left(\tfrac{t}{w}\right)^2 - 0.7137 \cdot \left(\tfrac{h}{S}\right)^2 \end{aligned} \tag{21}$$

Through verification model 1 and verification model 2 in Table 4, the flow change coefficient m_i prediction model near the turning channel is verified. As shown in Fig. 22a, the prediction error of the adjacent channel 1 of the verification model 1 is 5.67%, and the predicted values of the flow variation coefficients of the adjacent channels 2 and 3 are less than 0.01. It is considered that there is no flow exchange process, which meets the requirements of the prediction model. As shown in Fig. 22b, the prediction error of the adjacent channel 1 of the verification model 2 is 7.04%, the prediction error of the adjacent channel 2 is 5.96%, and the prediction error of the adjacent channel 3 is 18.34%. The distribution trend of the predicted value is basically consistent with the experimental value, and the average error is 10.54%. The error analysis shows that the prediction model has good prediction ability.

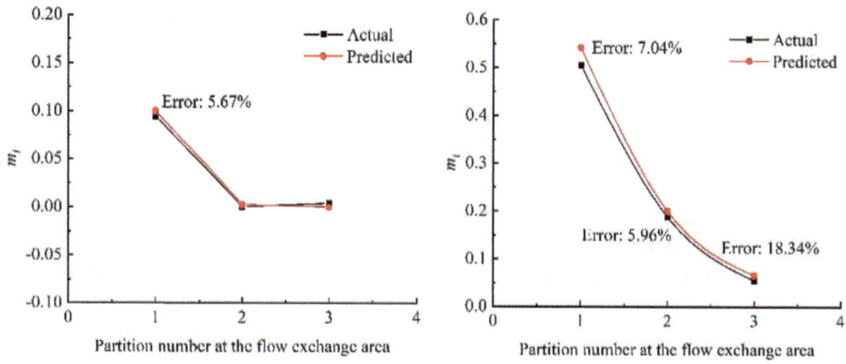

(a) Validation model 1
(b) Validation model 2

Fig. 22. Error analysis of prediction model.

5 Conclusion

Based on numerical simulation, this paper studies the flow distribution characteristics along the latticework subchannel, analyzes the influence of the subchannel flow exchange process on flow and heat transfer characteristics, and summarizes the influence of rib angle, rib width and rib spacing ratio, subchannel height ratio, and inlet Reynolds number on subchannel flow exchange. The main conclusions are as follows:

Under the blocking effect of the transverse vortex at the inlet of the subchannel and the stagnation effect of the subchannel with a small length-diameter ratio, there is an obvious flow exchange phenomenon at the inlet of the top and bottom subchannels. With the increase of the length-diameter ratio of the subchannel, the mass flow rate at the inlet of the subchannel increases first and then decreases. When the rib angle of the subchannel increases, the difference in flow distribution at the inlet of the subchannel increases. The inlet Reynolds number does not affect the flow distribution characteristics at the inlet of the subchannel.

When the length-diameter ratio of the subchannel increases, the heat transfer capacity and flow resistance of the cooling air at the inlet of the subchannel increase first and then decrease. Increasing the rib angle of the subchannel will promote the heat transfer of the cooling air at the inlet of the subchannel. The stagnation effect of subchannels with a small length-diameter ratio leads to a decrease in cooling air velocity and a slight decrease in pressure. Finally, the phenomenon of a small heat transfer capacity and small flow resistance coefficient at the inlet of subchannels with small length-diameter ratio appears.

In addition to the flow exchange with the three adjacent interfaces before and after the turning channel, there is no obvious flow exchange at other positions of the subchannel. With the increase of the distance from the turning channel, the flow variation coefficients of the subchannels are 20% ~ 50%, 6% ~ 24%, and 2% ~ 10%, respectively. At the corresponding positions before and after the turning channel, the flow loss before the turning channel is approximately the same as the flow obtained after the turning channel. When the rib angle increases, the flow exchange process near the turning channel gradually becomes obvious.

In the vicinity of the turning channel, the heat transfer capacity and flow resistance coefficient in the subchannel are positively correlated with the flow change in the subchannel. The heat transfer capacity of the outlet section of the turning channel is greater than that of the inlet section of the turning channel. The subchannel height ratio mainly affects the flow and heat transfer distribution difference in the inlet section of the turning channel, and has no obvious effect on the flow and heat transfer characteristics in the outlet section of the turning channel.

Acknowledgment. Thanks to Jingze Tong senior engineer and Yangen Wang senior engineer for the careful guidance of the verification method of this topic.

References

1. Han, J.C.: Turbine blade cooling studies at Texas A & M University: 1980–2004. J. Thermophys. Heat Transf. **20**(2), 161–187 (2006)
2. Chyu, M.K., Natarajan, V.: Heat transfer on the base surface of three dimensional protruding elements. Int. J. Heat Mass Transf. **39**(14), 2925–2935 (1996)
3. Terekhov, V.I., Kalinina, S.V., Mshvidobadze, Y.M.: Heat transfer coefficient and aerodynamic resistance on a surface with a single dimple. J. Enhance. Heat Transf. **4**(2), 131–145 (1997)
4. Skamniotis, C., Cocks, A.C.: 2D and 3D thermoelastic phenomena in double wall transpiration cooling systems for gas turbine blades and hypersonic flight. Aerosp. Sci. Technol. **113**(10), 106610 (2021)
5. Bunker, R.S.: Latticework (vortex) cooling effectiveness-part1: stationary channel experiments. In: Turbo Expo 2004: Power for Land, Sea, and Air on Proceedings, pp. 909–918. ASME, Austria (2004)
6. Won, S.Y., Ligrani, P.M.: Comparisons of flow structure and local Nusselt numbers in channels with parallel and crossed-rib turbulators. Int. J. Heat Mass Transf. **47**(8–9), 1573–1586 (2004)
7. Du, W., Luo, L., Jiao, Y., Wang, S., Li, X., et al.: Heat transfer in the trailing region of gas turbines–a state-of-the-art review. Appl. Therm. Eng. **199**(11), 117614 (2021)
8. Yeranee, K., Rao, Y.: A review of recent studies on rotating internal cooling for gas turbine blades. Chin. J. Aeronaut. **34**(7), 85–113 (2021)
9. Chen, W.: Experimental and Numerical Study of the Mechanism of Internal Cooling in Gas Turbine Blade. Tsinghua University, Beijing (2011)
10. Khalatov, A.A.: Vortex technologies in aerospace engineering. In: US–Ukrainian Workshop on Innovative Combustion and Aerothermal Technologies in Energy and Power Systems on Proceedings, pp. 20–25, Ukraine (2001)
11. Acharya, S., Zhou, F., Lagrone, J., Mahmood, G.: Latticework (vortex) cooling effectiveness: rotating channel experiments. J. Turbomach. **127**(3), 471–478 (2005)
12. Gillespie, D., Ireland, P.T., Dailey, G.: M.: detailed flow and heat transfer coefficient measurements in a model of an internal cooling geometry employing orthogonal intersectiong channels. In: Turbo Expo 2000: Power for Land, Sea, and Air on Proceedings, p. V003T01A104. ASME, Germany (2000)
13. Prajapati, A.N., Tariq, A.: Experimental investigation on heat transfer and friction factor in open matrix subchannels. J. Heat Transf. Trans. ASME. **143**(5), 052101 (2021)

14. Tsuru, T., Ishida, K., Fujita, J., Takeishi, K.: Three-dimensional visualization of flow characteristics using a magnetic resonance imaging in a lattice cooling channel. J. Turbomach. **141**(6), 061003 (2019)
15. Hagari, T., Ishida, K.: Numerical investigation on flow and heat transfer in a lattice (matrix) cooling channel. In: ASME Turbo Expo 2013: Turbine Technical Conference and Exposition on Proceedings, p. V03AT12A040. ASME, USA (2013)
16. Sun, H., Sun, T., Yang, L., Bu, S., Luan, Y.: Effect of bleed hole on internal flow and heat transfer in matrix cooling channel. Appl. Therm. Eng. **136**, 419–430 (2018)
17. Du, W., Luo, L., Wang, S., Liu, J., Sunden, B.: Heat transfer and flow structure in a detached latticework duct. Appl. Therm. Eng. **155**, 24–39 (2019)
18. Bu, S., Yang, L., Qiu, H., Luan, Y., Sun, H.: Effect of sidewall slots and pin fins on the performance of latticework cooling channel for turbine blades. Appl. Therm. Eng. **117**, 275–288 (2017)
19. Du, W., Luo, L., Wang, S., Liu, J., Sunden, B.: Numerical investigation of flow field and heat transfer characteristics in a latticework duct with jet cooling structures. Int. J. Therm. Sci. **158**(10), 106553 (2020)
20. Khalatov, A.A., Nam, C.D.: Aerothermal vortex technologies in aerospace engineering. J. Korean Soc. Mar. Eng. **28**(2), 163–184 (2004)
21. Luo, L., Du, W., Liu, J., Sun, P., Wang, S., et al.: Thermal performance in latticework ducts with various endwall shapes for aero-craft turbine cooling. Aerosp. Sci. Technol. **126**(10), 107558 (2022)
22. Oh, I.T., Kim, K.M., Lee, D.H., Park, J.S., Cho, H.H.: Local heat/mass transfer and friction loss measurement in a rotating matrix cooling channel. J. Heat Transf. **134**(1), 011901 (2012)
23. Shen, Z., Qu, H., Zhang, D., Xie, Y.: Effect of bleed hole on flow and heat transfer performance of U-shaped channel with dimple structure. Int. J. Heat Mass Transf. **66**, 10–22 (2013)

Research on Riveted Bolt Anti-loosening Structural in the Optimization Scheme for Wear of Heat Transfer Tubes in Steam Generators

Zhenhua Zhang[1(✉)], Lichen Tang[2], and Xiling Xie[1]

[1] State Key Laboratory of Mechanical System and Vibration, Shanghai JiaoTong University, Shanghai, China
zhenhuazhang@sjtu.edu.cn

[2] Shanghai Nuclear Engineering Research and Design Institute, Shanghai, China

Abstract. In the core industrial equipment of nuclear power plant steam generators, heat transfer tubes play a crucial role. They are not only critical components for heat transfer, but also have a significant impact on the safety and efficiency of the entire steam generation system. The steam generator operates in high temperature and irradiation environments, withstanding dynamic loads caused by water flow, earthquakes, etc. The Flow Induced Vibration often leads to wear and looseness of the connecting components, thereby affecting the integrity of the Pressurized Water Reactor (PWR) internal components. Therefore, the optimization and modification of heat transfer tube wear in the steam generator is a very noteworthy issue. Aiming at the optimization and renovation plan for the wear of heat transfer tubes in a certain third-generation nuclear power plant steam generator, the riveted bolt anti-loosening structural components is proposed. The method of determining equivalent vibration assessment test parameters based on the flow induced vibration test results of the internal component model is explained. High temperature rise and drop assessment tests, accelerated vibration tests, and bolt damage detection are carried out for the riveted bolt anti-loosening structural components, and conclusions that have reference significance for practical engineering are obtained.

Keywords: Nuclear power plant steam generators · Flow Induced Vibration · the Pressurized Water Reactor (PWR) internal components · riveted bolts · accelerated vibration test

1 Introduction

Eight steam generators of the four units of a certain third-generation nuclear power plant were found to have varying degrees of wear after two cycles of operation. The wear is mainly concentrated in the pipe gallery area (i.e. the small bend radius area), and most of it occurs at the contact between the upper 7th to 9th support plates and the heat transfer tube. This phenomenon has certain commonalities.

In response to this issue, the project team has designed a plan for optimizing the wear of heat transfer tubes in steam generators. The core concept is to seal the flow

Y. Zhao (Ed.): ICAMM 2025, SPM 96, pp. 334–343, 2025.
https://doi.org/10.1007/978-3-032-09862-7_30

channel and flow hole on the 10th support plate, and transform all the flow channels and flow holes with a diameter of 17.48 mm into flow holes with a diameter of 10 mm. The core of this renovation plan is to use rivet bolts to install the blocking plate on the 10th support plate. The core component that needs to be verified is the rivet bolt. From the perspective of failure mechanism, there may be three possible failure factors for rivet bolts: 1) loosening of fasteners caused by temperature rise and fall process; 2) The overall flow field causes vibration of the support plate and puts the rivet bolts under long-term fatigue loads; 3) The protruding part of the rivet bolt from the support plate vibrates under local flow field, causing fatigue damage to the rivet bolt [1].

Y. Weida [2] has conducted extensive experimental research and analysis on the water flow vibration of the reactor basket, accumulated a lot of experimental data, and obtained many valuable conclusions. The equivalent technique in vibration testing is mainly based on the cumulative damage theory, providing a determined damage criterion and making the total damage generated by different vibration experiences equal to establish equivalence. L. Xingyun [3] applied equivalent technology and based on the principle of consistent cumulative damage of anti-loosening structures in flow induced vibration environments during the lifespan of nuclear power plants, determined the parameters for accelerated vibration testing, and conducted vibration assessment tests on the anti-loosening structures of the power plant's connection components.

2 Test

2.1 Experimental Plan

Carry out high-temperature rise and fall temperature tests, accelerated vibration tests, and bolt damage detection for riveted bolt anti-loosening structural components:

1) The high-temperature temperature rise and fall test is aimed at reproducing the possible relaxation of vibration resistant structural components for riveted bolts during the temperature rise and fall process.
2) The accelerated vibration test is to examine the overall vibration effect of secondary fluid load on the riveted bolt anti-loosening structural components. The test meets the requirements of the "Fastener Test Methods - Vibration" (GJB 715.3A-2002) standard.
3) After the accelerated vibration test is completed, a penetration test is carried out on the riveted bolt anti-loosening structural components to detect any cracks or other issues (Fig. 1).

There are three specifications of riveted bolt anti-loosening structural components, named as anti-loosening structural component I, anti-loosening structural component II, and anti-loosening structural component III. Each specification of anti-loosening structural component has three components of the same size.

2.2 The High-Temperature Rise and Fall Test

Thermal transfer simulation analysis was conducted using ABAQUS finite element software (K-temperature is used in the simulation) (Fig. 2).

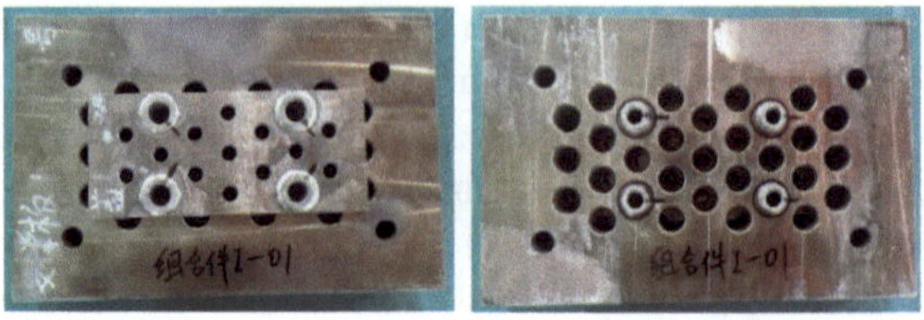

Fig. 1. Front and back of riveted bolt anti-loosening structural component I-01

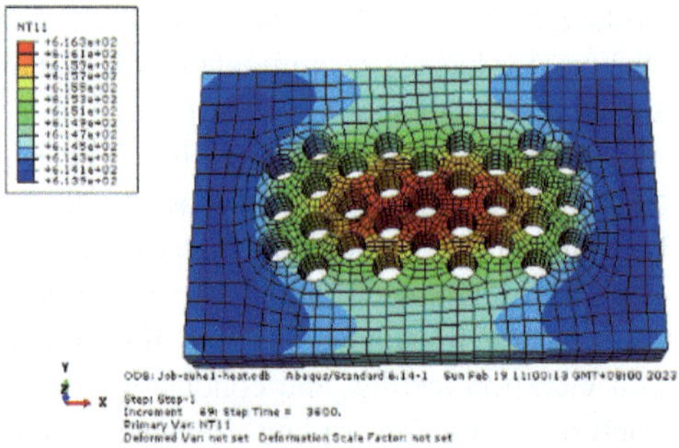

Fig. 2. The temperature field cloud map of support plate 1 after 1 h

The temperature field cloud map of support plate 1 after 1 h is shown above. The highest temperature is 616.3 K, which is 343.3 °C, located near the central hole; The minimum temperature is 613.9 K, which is 340.9 °C, located near the short edge of support plate 1.

Select 4 nodes along the height direction of the center hole of the support plate 1 model, and extract their time-temperature data. The specific data curve is shown in the following figure (Fig. 3).

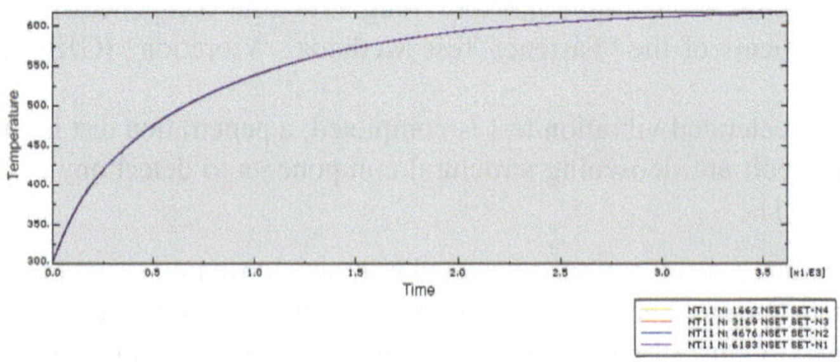

Fig. 3. The time-temperature data of the center hole of the support plate 1 model

According to the heat transfer simulation, it can be concluded that after entering the high-temperature sintering test chamber for 1 h, the test pieces can meet the requirement of greater than 320 °C.

Place the riveted bolt anti-loosening structural components in a high-temperature sintering test chamber and conduct a temperature rise and fall test from room temperature to an operating temperature not lower than 320 °C on the secondary side. The number of temperature rise and fall cycles is 200 (Figs. 4, 5 and 6).

Fig. 4. Riveted bolt anti-loosening structural components for the temperature rise and fall test

Fig. 5. Test piece equipped with K-type thermocouple

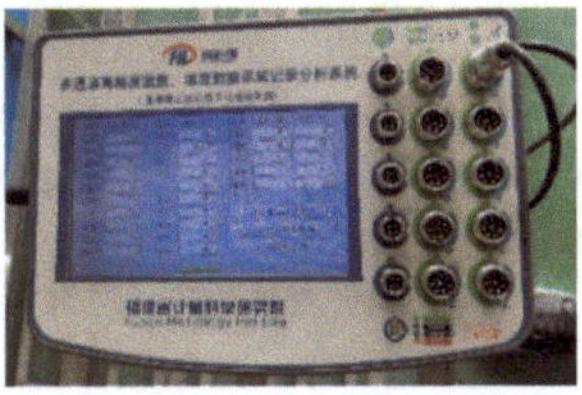

Fig. 6. The real-time temperature monitoring system

During the experiment, real-time temperature monitoring was conducted on the test piece, The thermocouple model is K-type, and the temperature and humidity calibration system is type Vtest-1101X. Through real-time temperature monitoring data, it is confirmed that the test piece meets the index requirement of over 320 °C after entering the high-temperature sintering test chamber for 1 h.

After temperature rise and fall, the riveted bolt anti-loosening structural components are used for vibration anti-loosening tests.

2.3 The Accelerated Vibration Test

According to the hydraulic vibration test of the 1:5 hanging basket model in the power plant, the relationship curve between the acceleration a_m within 1 min and the number of vibration cycles N_1 was obtained According to literature [2–4], the similarity between the water flow vibration test model of the reactor basket and the actual object is as follows:

$$\begin{cases} a_s = \frac{1}{5} a_m \\ N_s = \frac{1}{5} N_1 \end{cases} \tag{1}$$

In the formula: a_s is the acceleration of the physical object; a_m is the acceleration of the model; N_s is the number of vibration cycles of the physical object, N_1 is the number of vibration cycles in the model.

Conduct an analysis of the vibration response of the support plate based on the Flow Induced Vibration (FIV) load (Fig. 7).

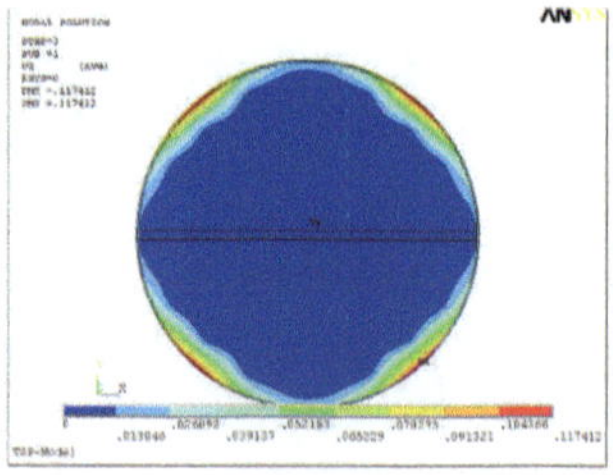

Fig. 7. The vibration response of the support plate based on the FIV load

Obtain the Power spectrum density (PSD) acceleration spectrum near the point of maximum displacement in the bolt installation area. Convert the PSD acceleration spectrum into a time history curve, and use the average cross peak counting method to obtain the statistical value of vibration cycles based on the response time history data of the support plate under FIV load (Fig. 8).

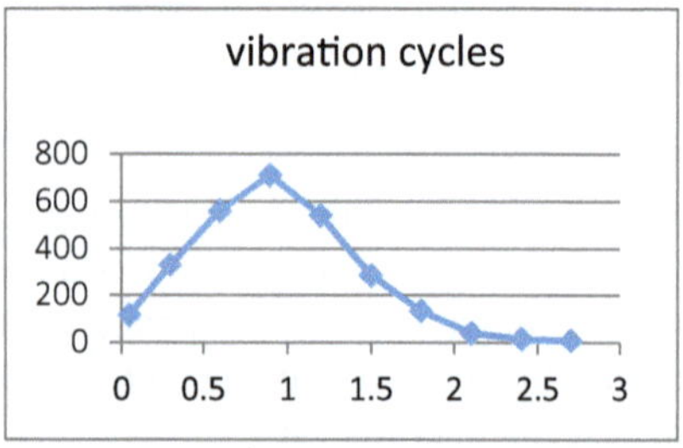

Fig. 8. The statistical value of vibration cycles

According to the Miner linear cumulative damage theory in reference [5], the number of equivalent accelerations per minute is:

$$N_Q = \frac{1}{5^{1+b}} \frac{\sum_{i=1}^{511} \left(N_1 a_m^b\right)_i}{a_Q^b} \tag{2}$$

In the formula: $\left(N_1 a_m^b\right)_i$ is the product of the i-th acceleration obtained from the statistical analysis of the model experiment and its corresponding degree; $1/b$ is the slope of the material design fatigue curve on logarithmic coordinates; a_Q is the acceleration of the equivalent acceleration test.

The design lifespan of this nuclear power plant is 40 years, and the number of vibration cycles that occur in the hanging basket during the 60 year lifespan, N_P, can be calculated.

When setting the equivalent acceleration test frequency f_Q and the acceleration is a_Q, the total assessment time t_Q (in hours) required for equivalent physical vibration is:

$$t_Q = \frac{N_P}{3600 \times f_Q} \tag{3}$$

Considering the riveted bolt anti-loosening structural components, based on the design fatigue curve values in ASME BPVC-1 Appendix and the design fatigue curve values in ASME Code Case N-208-1 (both considering the influence of average stress), calculate the value of b and substitute it into formulas (2) and (3), obtaining a vibration assessment test time of 156 h (Fig. 9).

Fig. 9. The Electric vibration testing system

The accelerated vibration test is carried out after the high-temperature rise and fall temperature test (Fig. 10).

Set the vibration table to input sine wave vibration, with a vibration frequency of 30 Hz and an acceleration of 20g [6]. Conduct experiments to verify the anti-loosening performance of the riveted bolt anti-loosening structural components. The experimental duration is 156 h. Before the experiment, mark the riveted bolts and plates (cover plates, support plates) of each batch of structural components, and conduct a pretest status inspection and record.

Fig. 10. The accelerated vibration test

2.4 The Penetration Testing

Conduct initial state Penetration Testing (hereinafter referred to as PT testing) on the riveted bolt anti-loosening structural components; After each accelerated vibration test is completed, PT testing is carried out on them. Using SKL-SP2 penetrant, mainly targeting the weld ends on the front and back of the structural components, to detect surface opening defects and cracks.

2.5 The Complete Test Process

1) Conduct initial state penetration testing on the structural components;
2) The first stage of temperature rise and fall test, with 60 cycles of temperature rise and fall;
3) The first stage of the accelerated vibration test, with a duration of 46 h;
4) The first stage of the PT testing;
5) The second stage of temperature rise and fall test, with 70 temperature rise and fall cycles (total to 130 times);
6) The second stage of the accelerated vibration test, with a duration of 55 h (total to 101 h);
7) The second stage of the PT testing;
8) The third stage of temperature rise and fall test, with 70 temperature rise and fall cycles (total to 200 times);
9) The third stage of the accelerated vibration test, with a duration of 55 h (total to 156 h);
10) The third stage of the PT testing.

3 Test Results

After the completion of temperature rise and fall tests and the accelerated vibration tests in each stage, on-site inspections were conducted on the structural components, including:

1) Whether the riveted bolts have broken or fallen off.

2) Whether the riveted bolt undergo significant plastic deformation.
3) Check if the riveted bolts are displaced from the marking on the plate (cover plate, support plate).
4) Use a magnifying glass to check for cracks at the solder joints.

After the first, second, and third stages of temperature rise and fall tests, the accelerated vibration tests, no abnormalities were found during on-site inspections (Fig. 11).

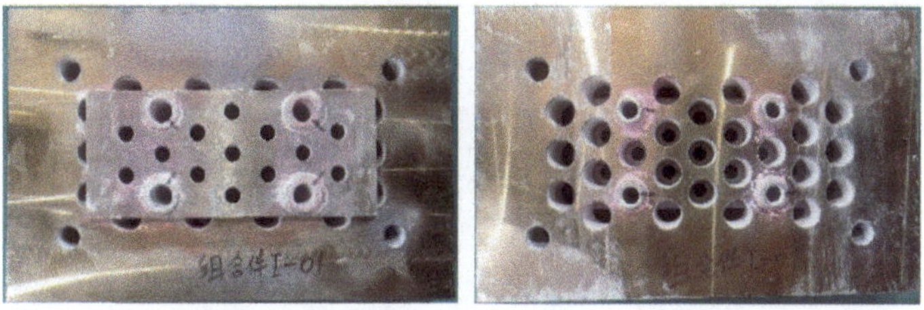

Fig. 11. Front and back of riveted bolt anti-loosening structural component I-01after all tests

The overall thickness (from the upper surface of the cover plate to the lower surface of the support plate) of the test piece was measured, including the initial state of the test piece and the size measurement after every test. Specifically by measuring each test piece at three different positions (Table 1).

Table 1. The overall thickness of the test piece after all tests (unit: mm)

Piece Number	Thickness 1	Thickness 2	Thickness 3
I-01	34.20	34.18	34.12
I-02	33.76	33.80	33.90
I-03	33.56	33.60	33.64
II-01	33.70	33.64	33.78
II-02	33.92	**33.98**	**33.80**
II-03	33.86	33.94	33.78
III-01	33.80	33.82	33.80
III-02	33.82	33.80	33.78
III-03	33.92	33.90	33.90

The thickness dimensions of each test piece were measured at three different positions, and the combination test piece II-02 had the largest difference of 0.18 mm. After completing the temperature rise and fall assessment test and the three-stage accelerated vibration test, the test piece did not show significant deformation in the thickness direction.

An initial 100% PT test was conducted on the test piece. It was found that there is a gas hole and an arc crater crack on the front surface of test piece I-01. The remaining 8 test pieces showed no surface opening defects or cracks.

After the completion of each batch of the accelerated vibration test, 100% PT testing was carried out on them. The weld ends on the front and back of the structural components were tested according to the acceptance standard NB/T47013.1/5-2015. Except for one air hole and one arc crater crack on the front of test piece I-01, no new surface opening defects or cracks were found. The test conclusion is qualified, and the qualification level is Grade I (Fig. 12).

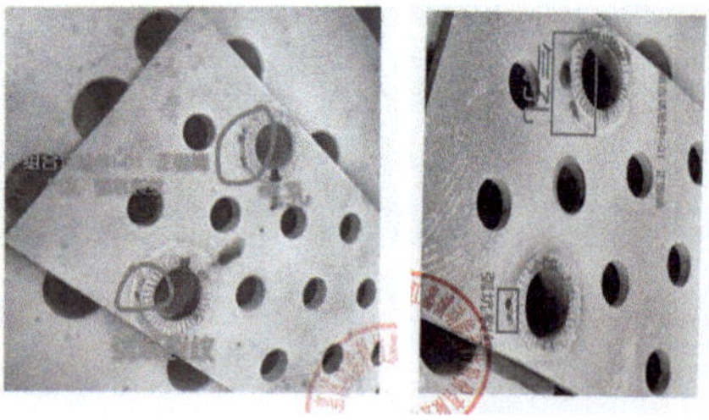

Fig. 12. The initial state PT testing report (Left);Third stage PT testing report (Right)

4 Conclusion

After each stage of temperature rise and fall assessment test, on-site inspection is carried out on the vibration proof structural components of the riveted bolts; After the completion of the accelerated vibration test, on-site inspection and PT testing were carried out on the structural components.

The on-site inspection and PT testing results show that if the connection quality between the rivet bolts and the plate can be ensured according to the process requirements, Riveted bolt anti-loosening structural components mentioned above can pass the temperature rise and fall assessment test and the accelerated vibration test, and their connection performance is reliable. The optimized modification structure for the wear of the steam generator heat transfer tubes can meet the corresponding design requirements.

Acknowledgment. This work was supported by the Shanghai Nuclear Engineering Research and Design Institute.

References

1. Youguang, H., Weida, Y.: Design and development of reactor internal components. China Nucl. Sci. Technol. Report, J. **3**, 72–75 (1997)
2. Weida, Y., Guolin, S., Nanyan, J., Yongyong, P., Feijuan, Z., Yufeng, W., et al.: Research and analysis of water flow vibration in reactor baskets. Nucl. Sci. Eng. J. **9**(01), 8–17 (1989)

3. Xingyun, L., Chihu, W., Lei, G., Yongcheng, X., Renan, Y.: Vibration assessment test for anti loosening performance of internal components in pressurized water reactor nuclear power plants. Chin. J. Mech. Eng. J. **7**(3), 326–329 (2009)
4. U.S Nuclear Regulatory Commission (NRC) Regulatory Guide 1.20, Rev.2, Comprehensive Vibration Assessment Program for Reactor Internals during the Preoperational and Initial Startup Testing (1976)
5. F. Warren C.: Equivalence technique for vibration testing. Y. Qixing, Translation. National Defense Industry Press, Beijing (1979)
6. GJB 715.3A-2002 Fastener Test Methods Vibration. National Defense Science Technology and Industry Committee (1976)

Author Index

Y. Zhao (Ed.): ICAMM 2025, SPM 96, pp. 345–346, 2025.
https://doi.org/10.1007/978-3-032-09862-7

MIX
Papier aus verantwortungsvollen Quellen
Paper from responsible sources
FSC® C105338

If you have any concerns about our products,
you can contact us on
ProductSafety@springernature.com

In case Publisher is established outside the EU,
the EU authorized representative is:
Springer Nature Customer Service Center GmbH
Europaplatz 3, 69115 Heidelberg, Germany

Printed by Libri Plureos GmbH
in Hamburg, Germany